Arduino™ Projects for Amateur Radio

Arduino™ Projects for Amateur Radio

Dr. Jack Purdum, W8TEE
Dennis Kidder, W6DQ

New York   Chicago   San Francisco
Athens   London   Madrid
Mexico City   Milan   New Delhi
Singapore   Sydney   Toronto

McGraw-Hill Education books are available at special quantity discounts to use as premiums and sales promotions or for use in corporate training programs. To contact a representative, please visit the Contact Us page at www.mhprofessional.com.

Arduino™ Projects for Amateur Radio

Copyright © 2015 by McGraw-Hill Education. All rights reserved. Printed in the United States of America. Except as permitted under the United States Copyright Act of 1976, no part of this publication may be reproduced or distributed in any form or by any means, or stored in a database or retrieval system, without the prior written permission of the publisher.

3 4 5 6 7 8 9 10 LCR 20 19 18 17

ISBN   978-0-07-183405-6
MHID   0-07-183405-2

This book is printed on acid-free paper.

Sponsoring Editor Roger Stewart	**Project Manager** Asheesh Ratra, MPS Limited	**Indexer** Francine Cronshaw
Editing Supervisor Stephen M. Smith	**Copy Editor** Rashi Sinha, MPS Limited	**Art Director, Cover** Jeff Weeks
Production Supervisor Lynn M. Messina	**Proofreader** Susan Higgins	**Composition** MPS Limited
Acquisitions Coordinator Amy Stonebraker		

The Arduino name and logo and the graphics design of its boards are a protected trademark of Arduino and its partners.

McGraw-Hill Education, the McGraw-Hill Education logo, TAB, and related trade dress are trademarks or registered trademarks of McGraw-Hill Education and/or its affiliates in the United States and other countries and may not be used without written permission. All other trademarks are the property of their respective owners. McGraw-Hill Education is not associated with any product or vendor mentioned in this book.

Information contained in this work has been obtained by McGraw-Hill Education from sources believed to be reliable. However, neither McGraw-Hill Education nor its authors guarantee the accuracy or completeness of any information published herein, and neither McGraw-Hill Education nor its authors shall be responsible for any errors, omissions, or damages arising out of use of this information. This work is published with the understanding that McGraw-Hill Education and its authors are supplying information but are not attempting to render engineering or other professional services. If such services are required, the assistance of an appropriate professional should be sought.

Jack Purdum: To Hailey, Spencer, and Liam

Dennis Kidder: To Helen and Bud

About the Authors
Dr. Jack Purdum, W8TEE, has been a licensed ham since 1954 and is the author of 17 programming books. He retired from Purdue University's College of Technology where he taught various programming languages.

Dennis Kidder, W6DQ, has been a licensed ham since 1969. He is also an electrical engineer with a distinguished career in major engineering projects throughout the world, working for companies such as Raytheon and Hughes.

Contents

Preface		xvii
Acknowledgments		xix

1 Introduction ... 1
Which Microcontroller to Use? ... 1
    We Chose Arduino, So Now What? ... 3
    Interpreting Table 1-1 ... 5
    Making the Choice ... 5
What Else Do You Need? ... 6
Software ... 7
    Downloading and Installing the Arduino Integrated
        Development Environment ... 8
    Installing the Software ... 9
    Running Your First Program ... 11

2 I Don't Know How to Program ... 17
I Don't Need No Stinkin' CW! ... 17
    Like CW, Like Programming ... 18
The Five Program Steps ... 18
    Step 1. Initialization ... 18
    Step 2. Input ... 19
    Step 3. Processing ... 19
    Step 4. Output ... 19
    Step 5. Termination ... 19
Arduino Programming Essentials ... 20
    The Blink Program ... 20
    Data Definitions ... 22
    Where's the *main()* Function? ... 22
    The *setup()* Function ... 24
    The *loop()* Function ... 25
    I Thought There Were Five Program Steps? ... 26
    Modifying the Blink Sketch ... 26
Saving Memory ... 32
    Remove Unused Variables ... 32
    Use a Different Data Type ... 32
    Avoid Using the *String* Class ... 33
    The *F()* Macro ... 33

        The *freeRam()* Function 34
    Conclusion 34

3 The LCD Shield Project **35**
    Libraries: Lessening the Software Burden 36
        Not All LCDs Are the Same 36
    LCD Shield Parts List 37
    Assembling the LCD Shield 39
        Breakaway Header Pins 40
        Soldering Components to the Shield 42
    Adding Components Using a Schematic 46
    An Alternative Design 51
    Loading the Example Software and Testing 52
    A "Code Walk-Through" of the "HelloWorld" Sketch 56
        Explore the Other Examples 59
    Using Your LCD Display with the TEN-TEC Rebel 59
        Under the Rebel Hood 60
        Software Modifications 61
    Conclusion 65

4 Station Timer **67**
    Software Version of ID Timer 68
    Magic Numbers 70
    Preprocessor Directives 71
        Fixing Bad Magic Numbers: *#define* 71
        A Second Way to Remove Magic Numbers: *const* 73
    Fixing Flat Forehead Mistakes 73
    Encapsulation and Scope 74
        Fixing Our Program Bug 75
        The *static* Data Type Specifier 76
    Using a Real Time Clock (RTC) Instead of a Software Clock 78
        The Inter-Integrated Circuit (I^2C or I2C) Interface 78
    The I2C and the DS1307 RTC Chip 79
        BCD and the DS1307 Registers 80
    Constructing the RTC/Timer Shield 81
        The Adafruit RTClib Library 85
    Initializing the RTC 89
        Running the Program 97
    The RTC Timer Program 98
        The *loop()* Function 98
    A Software Hiccup 99
    Conclusion 100

5 A General Purpose Panel Meter **101**
    Circuit Description 102
    Construction 104
        An Alternate Design Layout 106
    Loading the Example Software and Testing 110

Code Walk-Through	113
Instantiating the *lcd* and *lbg* Objects	113
The *loop()* Code	114
Testing and Calibration of the Meter	115
Changing the Meter Range and Scale	116
Voltmeter	116
Ammeter	117
Changing the Scale	117
Conclusion	117

6 Dummy Load 119
Mechanical Construction	120
Resistor Pack Spacing	121
Fabricating the Lid Connections	122
Attaching the Lid to the Resistor Pack	123
Electronic Construction	124
Doing the Math	124
Software	126
Conclusion	130

7 A CW Automatic Keyer 131
Required Software to Program an ATtiny85	133
Connecting the ATtiny85 to Your Arduino	134
The Proper Programming Sequence	137
Some Things to Check If Things Go South	137
Using the Digispark	138
Compiling and Uploading Programs with Digispark	140
The CW Keyer	143
Adjusting Code Speed	144
Capacitance Sensors	144
The *volatile* Keyword	150
Construction	151
Conclusion	153

8 A Morse Code Decoder 155
Hardware Design Considerations	155
Signal Preprocessing Circuit Description	157
Notes When Using the Decoder with the TEN-TEC Rebel	159
Decoder Software	160
Search a Binary Tree of ASCII Characters	160
Morse Decode Program	162
Farnsworth Timing	169
Conclusion	171

9 A PS2 Keyboard CW Encoder 173
The PS2 Keyboard	173
Testing the PS2 Connector	175
The PS2 Keyboard Encoder Software	176
Adding the PS2 Library Code to Your IDE	176

    Code Walk-Through on Listing 9-1 189
    Overloaded Methods 190
    The *sendcode()* Method 190
    Some Bit-Fiddling 192
  Isolating the Arduino from the Transmitter 194
  Testing 196
  Other Features 197
    Change Code Speed 197
    Sidetone 198
    Long Messages 198
  Conclusion 198

10 Project Integration 199
  Integration Issues 200
  The Real Time Clock (RTC) Shield 201
  CW Decoder Shield 202
  PS2 Keyboard Keyer 202
  The Expansion Board 203
  Software Project Preparation 205
  C++, OOP, and Some Software Conventions 206
    C++ Header Files 207
  Class Declaration 209
    public and *private* Members of a Class 209
    Function Prototypes 209
  cpp Files 210
    Class Constructor Method 211
  IntegrationCode.ino 211
    Header Files 212
    Constructors 214
    How the Terms *Class, Instantiation*, and
     Object Relate to One Another 214
    The Dot Operator (.) 215
  The *loop()* Function 217
  Conclusion 218

11 Universal Relay Shield 219
  Construction 221
    Circuit Description 221
    Construction of the Relay Shield 222
  Testing the Relay Shield 224
    Test Sketch "Walk-Through" 225
  Conclusion 226

12 A Flexible Sequencer 227
  Just What Is a Sequencer? 228
  The Sequencer Design 228
    Timing 228

Constructing the Sequencer 229
    A Purpose-Built Sequencer 230
Programming and Testing the Sequencer 234
    Initial Testing of the Sequencer 234
Loading the Sequencer Program and Testing 235
    Sequencer Code "Walk-Through" 238
    Modifying the Sequence Order and Delay Time 239
    Configuring the Jumpers for Different Situations 239
Modifying the Relay Shield from Chapter 11 240
    Alternate Listing for the Relay Shield Sequencer 241
Conclusion 244

13 Rotator Controller 245
The Arduino Antenna Rotator Controller 246
    Supported Rotators 246
    Relay Shield 247
    Panel Meter Shield 248
    The Control Panel 253
    Adding the I2C Interface to the Relay
      Shield from Chapter 11 256
Connecting the Rotator Controller 256
    Early Cornell-Dublier Electronics (CDE) Models 257
    Later Models from HyGain, Telex, and MFJ 258
    Yaesu Models G-800SDX/DXA, G-1000SDX/DXA,
      and G-2800DXA 259
Software 260
    Arduino Beam Heading Software 260
    Moving the Beam 277
    Setting a New Heading 279
    Storing a New Heading in EEPROM 279
World Beam Headings 279
    Finding the Coordinates for a QTH 279
    Finding a Beam Heading 280
Conclusion 282

14 A Directional Watt and SWR Meter 283
SWR and How It Is Measured 284
Obtaining the Antenna System SWR 284
    Detectors 286
Constructing the Directional Watt/SWR Meter 286
    Design and Construction of the Directional
      Coupler/Remote Sensor 288
    The Sensor Board 292
    Final Assembly of the Coupler/Sensor 296
    Interface Shield Construction 298
    LCD Shield Options 299
    Final Assembly 301

Testing the Directional Wattmeter/SWR Indicator 304
    Calibrating the Directional Wattmeter 304
Software Walk-Through 307
    Definitions and Variables 324
    setup() 325
    loop() 326
Further Enhancements to the Directional
   Wattmeter/SWR Indicator 329
Conclusion 329

15 A Simple Frequency Counter 331
Circuit Description 333
Constructing the Shield 334
    An Alternate Design for Higher Frequencies 337
Code Walk-Through for Frequency Counter 338
Displaying the Tuned Frequency of Your Display-less QRP Rig 342
    Double Conversion Applications 342
    Adding a Frequency Display to the MFJ Cub
      QRP Transceiver 343
    Adding a Frequency Display to a NorCal 40 345
    Direct Conversion Applications 346
    Other Radio Applications 347
Conclusion 347

16 A DDS VFO 349
Direct Digital Synthesis 350
The DDS VFO Project 350
DDS VFO Circuit Description 352
    The Analog Devices AD9850 Breakout Module 352
    Constructing the DDS VFO Shield 353
    Adding an Output Buffer Amplifier for the DDS VFO 353
    The Front Panel and Interconnection 356
DDS VFO Functional Description 357
    Overview 357
    EEPROM Memory Map 357
    SW1, the User Frequency Selection Switch (UFSS) 358
    SW2, the Band-Up Switch (BUS) 360
    SW3, the Band-Down Switch (BDS) 360
    SW4, Plus Step Switch (PSS) 361
    SW5, Minus Step Switch (MSS) 361
    SW6, the Encoder Control 361
The DDS VFO Software 361
EEPROM Initialization Program 362
The KP VFO Software (VFOControlProgram.ino) 366
    setup() 367
    loop() 368
Testing the DDS VFO 369

	Calibrating the DDS VFO	370
	Using the DDS VFO with Your Radio	371
	The Pixie QRP Radio	372
	Blekok Micro 40SC	374
	CRKits CRK 10A 40 meter QRP Transceiver	374
	Other Applications of the DDS VFO and Additional Enhancements	376
	Conclusion	377
17	**A Portable Solar Power Source**	**379**
	The Solar Sensor	381
	Solar Charger Controller	384
	Panel Positioning and Stepper Motor	385
	Stepper Wiring	385
	Stepper Motor Driver	386
	Control Inputs	388
	Solar Panel Support Structure	389
	Stepper Motor Details	390
	Mounting the Stepper Motor	391
	Solar Panel Connections	395
	Placing the Quick Connectors	396
	The Motor Controller Shield	396
	Routing Power Cables	397
	Motor Controller Shield Wiring	397
	Altitude Positioning	398
	The Software	399
	Final Assembly	403
	Assembly and Disassembly	403
	Conclusion	404
A	**Suppliers and Sources**	**405**
B	**Substituting Parts**	**419**
C	**Arduino Pin Mapping**	**423**
	Index	**429**

Preface

Microcontrollers are cropping up everywhere, from the car you drive to the washing machine that makes you look good for work. More importantly, they are showing up in our transceivers, keyers, antenna analyzers, and other devices we use as ham radio operators. This book has two primary objectives: 1) to present some microcontroller-based projects that we hope you will find both interesting and useful, and 2) to show you just how easy it is to use these devices in projects of your own design. As you will soon discover, microcontrollers are pretty easy to use and bring a whole lot to the feature table at an extremely attractive price point.

Why Should I Buy This Book?

First, we think there is a sufficient variety of projects in this book that at least several of them should appeal to you. The projects result in pieces of equipment that are both useful around the shack and inexpensive to build when compared with their commercial counterparts. Not only that, but we are pretty sure that many of you will have an "ah-ha" moment where you can think of extensions of, or perhaps even new, projects. If so, we hope you will share your ideas on our web site.

Finally, when you finish this book, we feel confident that you will have a better understanding of what microcontrollers are all about and how easy it is to write the software that augments their power.

For all these reasons, we hope you will read the book from start to finish. In that same vein, we assume there is no urgency on your part in reading this book. Take your time and enjoy the trip.

Errata and Help

Dennis, Jack, Beta testers, and scores of editorial people at McGraw-Hill have scoured this book from cover to cover in every attempt to make this book perfect. Alas, despite the best efforts by all of those people, there are bound to be some hiccups along the way. Also, Jack does not profess to be the world's authority on software development nor does Dennis presume he has cornered the market on brilliant hardware design. As hiccups show up, we will post the required solutions on the Web. McGraw-Hill maintains a web site (www.mhprofessional.com/arduinohamradio) where you can download the code in this book and read about any errors that may crop up. Rather than type in the code from the book, you should download it from the McGraw-Hill web site. That way, you know you have the latest version of the software. Likewise, if you think you have found an error, please visit the web site and post your discovery. We will maintain our own web site too. This web site, www.arduinoforhamradio.com, will serve as a clearing house for project hardware and software enhancements, new ideas and projects, and questions.

Acknowledgments

Any book is a collaborative work involving dozens of people. However, we would especially like to single out a number of people who helped us in many different ways with this book. First, we would like to thank Roger Stewart, our editor at McGraw-Hill, whose leap of faith made this book possible. The editorial staff at McGraw-Hill also did yeoman's work to polish our drafts into a final work. We would also like to thank John Wasser for helpful guidance on some interrupt issues. A special thanks to Leonard Wong, who served as a special Beta reader for the entire text. His keen eye caught a number of hiccups in both the narrative and schematics.

We also appreciate the efforts of Jack Burchfield (K4JU and President of TEN-TEC), who mentored Bill Curb (WA4CDM and lead TEN-TEC project engineer) on the Rebel transceiver, and Jim Wharton (NO4A and Vice President at TEN-TEC), whose vision helped make the Rebel an Open Source project. Thanks, too, to John Henry (KI4JPL and part of the TEN-TEC engineering team) for his help. Their early commitment to our book made it possible for us to have an advanced Beta of the Rebel long before the general public had access to it. That access affected the way in which we developed this book and, we hope, the way other manufacturers work with Open Source projects.

Michele LaBreque and Doug Morgan of Agilent Technologies were able to provide us with the long-term loan of one of their recent MSO 4000 series oscilloscopes. The MSO-X 4154A is an incredibly versatile tool that made much of the hardware testing a breeze. In the time that it would take to set up a test using conventional instruments, Dennis could set up and take multiple measurements, with variations, greatly reducing the time required to complete testing.

We also owe special thanks to all the companies mentioned in Appendix A. In many cases, their efforts made it possible for us to test our work on a variety of equipment that otherwise would not have been possible.

Each of us would also like to single out the following people for their thoughts, ideas, and encouragement during the development of this book.

> Jack Purdum: Special thanks and appreciation to Katie Mohr, John Purdum, Joe and Bev Kack, John Strack, and Jerry and Barb Forro. A special note of thanks to Jane Holcer, who let me hole up in my basement office while there were a bazillion tasks around the house that needed attention, but she handled on her own.
>
> Dennis Kidder: A personal thanks goes to Janet Margelli, KL7MF, Manager of the Anaheim HRO store, for her support during development of the rotator controller. Also, a lot of thanks to my friends who have seen very little of me for the past 10 months but nonetheless have provided a great deal of encouragement and support.

To everyone, our sincere thanks and appreciation for your efforts.

CHAPTER 1

Introduction

Many, many, years ago, Jack was a member of the local Boy Scouts group in his home town. Jack's scout leader had arranged for the troop to spend some time at the home of a local merchant named Chuck Ziegler who was a ham radio operator. As Jack recalls, Chuck had a schedule with his son every Sunday afternoon. What really impressed Jack was that Chuck was in Ohio and his son was in South Africa! In the weeks and months that followed, Jack spent many hours watching Chuck twiddle the dials on his all-Collins S-Line equipment feeding a 50-ft-high tri-band beam. It wasn't too long after that initial meeting that Chuck administered Jack's Novice license exam. Jack has been licensed ever since ... almost 60 years now.

Our guess is that each ham has their own set of reasons about what attracted them to amateur radio in the first place. In our case, we both really enjoy the potential experimentation in electronics as well as the communications elements. Lately, we have also become more intrigued by emergency communication and *QRP* (i.e., low-power communications using less than 5 W of power). In essence, that's the core of this book: making QRP communications even more enjoyable via microcontroller enhancements. While many of the projects are not actually "QRP only," it is just that a lot of them are features we wish inexpensive transceivers had but usually don't. Many other projects presented in this book are just plain useful around the shack.

Microcontrollers have been around since the early 1970s, but they have been slow to penetrate the amateur radio arena. However, a number of things are beginning to change all of that. First, the unit cost of many popular microcontroller chips is less than $10, putting them within the price range of experimenters. Second, several microcontrollers are *Open Source*, which means there is a large body of existing technical information and software available for them at little or no charge. Finally, despite their small size, today's microcontrollers are extremely powerful and capable of a wide variety of tasks. Most development boards are not much bigger than a deck of cards.

Which Microcontroller to Use?

There is no "right" microcontroller for every potential use. Indeed, showing preference of one over another is sort of like telling new parents that their child has warts. Each family of microcontrollers (we'll use µC as an abbreviation for "microcontroller" from now on) has a knot of followers who are more than willing to tell you all of the advantages their favorite µC has over all the rest. And, for the most part, they are telling you the truth. So, how do you select one over all the others?

FIGURE 1-1 Arduino-compatible microcontrollers.

The *Arduino* μC board began in 2005 by a group of students in Italy using an 8-bit Atmel AVR μC. The students' goal was to develop a low-cost development board that they could afford. The original hardware was produced in Italy by Smart Projects. Subsequently, SparkFun Electronics, an American company, designed numerous Arduino-compatible boards. Atmel is an American-based company, founded in 1984, that designs and produces μCs which form the nucleus of the Arduino boards.

Figure 1-1 shows several Arduino-compatible μC boards. The chipKIT Uno32 shown in the upper right of the picture is actually not part of the Arduino family of boards, but it can run all of the programs presented in this book. It costs a little more, but has some impressive performance characteristics. It is also at the heart of a new Open Source *Rebel* transceiver from *TEN-TEC*.

Well, perhaps the more important question is why we bothered to pick one μC over another in the first place. Since many of them have similar price/performance characteristics, why make a choice at all? As it turns out, there may be some pretty good reasons to select one over another.

Part of the reason probably has to do with the Jack-of-All-Trades-Master-of-None thingie. While the size, cost, and performance characteristics of many μCs are similar, there are nuances of differences that only get resolved by gaining experience with one μC. Also, the entry price point is only the tip of the development cost iceberg. For example, how robust are the support libraries? Is there an active support group behind the μC? Is the μC second-sourced? How easy is it to get third-party support? Are there add-on boards, often called *shields*, available at reasonable cost? What's the development language? No doubt we've left out a host of other important considerations you must make when selecting a μC for your next project.

Clearly, we ended up selecting the Arduino family of μCs. We did, however, consider several others before deciding on the Arduino family. Specifically, we looked long and hard at the *Netduino*, *PIC*, *Raspberry Pi*, and *pcDuino* μCs. The PIC family is actually similar to the Arduino on most comparisons, including cost, language used for development, libraries, books, etc. However, when looking for add-ins, like sensor shields, motors, and other external sensing devices, there seem to be fewer available and those that are available are more expensive than the Arduino alternatives.

The Netduino was especially tempting because its price point (about $35) is lower than the Raspberry Pi and pcDuino but has a much higher clock rate (120 MHz versus the Arduino's 16 MHz) and memory size (60 kb SRAM versus 8 kb) than the Arduino family. An even bigger draw from Jack's perspective is the fact that the Netduino uses Microsoft's Visual Studio Express (VSE) with the C# programming language. (Jack used VSE and C# when teaching the introductory programming courses at Purdue, and has written several Object-Oriented Programming texts centered on C#.) The debugging facilities of VSE are really missed when using the Arduino programming environment. Still, the availability of low-cost development boards and supporting shields for the Arduino family pushed the decisions toward the Arduino boards.

At the other extreme, both the newer Raspberry Pi and pcDuino are often an H-Bomb-to-kill-an-ant for the projects we have in mind. In a very real sense, both are a full-blown Linux computer on a single board. They have a relatively large amount of program memory (e.g., 512 Mb to 2 Gb) and are clocked at higher speeds than most Arduino µCs. While the support for Raspberry Pi is widespread, it's a fairly new µC having been introduced in 2011, even though its development began as early as 2006. Its price varies between $25 and $45 depending on configuration. The more powerful pcDuino is newer and has a $60 price point. Because of its newness, however, the number of add-on boards is a little thin, although this may change quickly as it continues to gain followers.

We Chose Arduino, So Now What?

We ultimately ended up selecting the Arduino family of µCs for use in this book. Why? Well, first, the ATmega328 µC is extremely popular and, as a result, has a large following that portends a large number of benefits to you:

1. They are cheap. You can buy a "true" 328 (from Italy) for about $30, but you can also buy knockoffs on eBay for less than $15. All of the projects in this book can also be run on most of the Arduino family of µCs, including the Duemilanove, Uno, ATmega1280, and ATmega2560. Their prices vary, but all can be found for less than $25.
2. There are lots of resources for the Arduino family, from books to magazine articles. Search Arduino books on Amazon and over 400 entries pop up. Google the word Arduino and you'll get over 21 million hits.
3. A rich online resource body. Arduino supports numerous forums (http://forum.arduino.cc/) covering a wide variety of topic areas. These forums are a great place to discover the answers to hundreds of questions you may have.
4. Free software development environment. In "the Old Days," you used to write the program source code with a text editor, run another program called a compiler to generate assembler code, run an assembler program to generate the object code, and then run a linker to tie everything together into an executable program. Today, all of these separate programs are rolled into a single application called the Integrated Development Environment, or *IDE*. In other words, all of the steps are controlled from within a single program and Arduino makes the Arduino IDE program available to you free of charge.
5. Open Source with a large and growing community of active participants. Open Source is actually a movement where programmers give their time and talent to help others develop quality software.
6. Uses the C language for development.

Item	Flash	SRAM	EEPROM	I/O	Price
ATmega328P, Duemilanove	32K	2K	1K	14 (6 provide PWM)	$16.00
UNO, R3	32K	2K	1K	14 of which 6 are analog	$17.00
ATmega1280	128K	8K	4K	54 (14 provide PWM and 16 analog)	$19.00
ATmega2560	256K	8K	4K	Same as 1280	$18.00
ChipKIT Uno32	128K	16K		42 (Note: System is clocked at 80 MHz instead of Atmel 16 MHz)	$28.00

TABLE 1-1   Table of Arduino Microcontrollers

Arduino gives you some choices within its family of μCs. In the beginning, the price points for the different boards were more dramatic. Now, however, clones have blurred the distinctions considerably. Table 1-1 presents some of the major choices of Arduino μCs that you might want to consider. (There is also an ATmega168, but has about half the memory of the ATmega328 yet costs about the same. Although most projects in this book can run on the 168, the difference in price is under a dollar, which seems to be penny-wise-pound-foolish.)

We should point out that the chipKIT Uno32 (pictured in Figure 1-1) is not part of the Arduino family. It is produced by Diligent but is Arduino-compatible in virtually all cases. One reason we include it here is that it is used in the new Rebel transceiver produced by TEN-TEC. To its credit, TEN-TEC has made the Rebel an Open Source project and actively encourages you to experiment with its hardware and software. TEN-TEC even includes header pins for the chip and a USB connector that makes it easy to modify the software that controls the Rebel, which is also Open Source. The Uno32 also has a fairly large amount of SRAM memory and is clocked at 80 MHz versus 16 MHz for the Atmel chips. We have more to say about the chipKIT Uno32 later in the book.

By design, the list presented in Table 1-1 is not exhaustive of the Arduino family. For example, the Arduino Pro Mini is essentially an ATmega328, but it leaves a few features off the board to make it smaller and less expensive. Most notably, the Mini does not have the USB connector on the board. While you can easily work around this, we have enough on our plate that we don't need to address this issue, too. The absence of a USB port on the board is an important omission because you will transfer the programs you write (called *sketches*) from your development PC to the Arduino over the USB connection. Further, by default, the Arduino boards draw their working voltages from the USB connector, too. If more power is needed than can be supplied by the USB specs, most Arduino boards have a connector for an external power source. (In Figure 1-1, the "silver box" in the upper left of most boards is the USB connector and the black "barrel shaped" object in the lower left corner is the external power connector.) Therefore, we encourage you to purchase a board from the list in Table 1-1 if for no other reason than to get the onboard USB connector.

As this book is being written, Arduino has announced the Arduino Due board. The Due is the Ferrari of the Arduino boards. It supports 54 I/O ports (12 of which can be used as PWM outputs), 12 analog inputs, 4 UARTs, an 84-MHz clock, a mega-munch of memory plus a host of other improvements. Given all of these cool features, why not opt for the Due? The reason is because the Due is so new, the number of shields and support features just aren't quite in place yet. Also, it is at least three times as expensive and many of the new features and the additional horsepower will just be idle for the purpose of our projects. Finally, the Due has a maximum pin voltage of 3.3 V, where the rest of the family cruises along at 5 V, making many existing shields unusable on the Due without modification. While we really like the Due, for the reasons detailed here, it is not a good choice for our projects.

Interpreting Table 1-1

So, how do you decide which µC to purchase? Let's give a quick explanation of what some of the information in Table 1-1 means. First, Flash is the number of kilobytes of Flash memory you have for your program. While 32K of memory doesn't sound like much, it's actually quite a bit since you don't have the bulk of a heavy-duty operating system taking up space. Keep in mind that Flash memory is *nonvolatile*, which means it retains its state even if power is removed. Therefore, any program code you load into Flash memory stays there until you replace it or there is some kind of board malfunction.

SRAM is the static random access memory available to the system. You can think of it as memory that normally stores variables and other forms of temporary data used as the program executes. It's a fairly small amount of memory, but since a well-designed program has data that ebbs and flows as it goes into and out of scope, a little thought about your data and what seems like a small amount is usually more than adequate.

EEPROM is the electrical erasable programmable read-only memory. Data stored in EEPROM is also nonvolatile. As stated earlier, most of your program data resides in SRAM. The bank of EEPROM memory is often used to store data that doesn't get changed very often but is needed for the program to function properly. For example, if your application has several sensors that have to be initialized with specific values on start-up, EEPROM may be a good place to put those start-up data values. On the downside, EEPROM memory can only be rewritten reliably a finite number of times before it starts to get a little flaky. We'll have more to say about each of these memory types as we progress through the book.

The 328 and Uno µCs have a fairly small number of input/output (I/O) lines available to you. Most are digital lines, but analog lines are also provided. Both of these boards are going to cost a little north of $15. However, if you're willing to work with a clone from China, these are available for around $10 each. The ATmega1280 and 2560 are similar boards, except for a larger amount of Flash memory and a greater number of I/O pins that are provided. The Diligent chipKIT Uno32 is like the ATmega1280 and 2560 except for a slightly smaller I/O line count and a much higher clock speed. Given that it is the clock speed that plays such an important part in the throughput of the system, the Uno32 is going to perform a set task faster than an Arduino board in most cases.

Making the Choice

Now that you have a basic understanding of some of the features of the various boards available, you should be totally confused and no closer to knowing which µC choice to make. Our Rule of Thumb: The more memory and I/O lines you have, the better. Given that, simply select one that best fits your pocketbook. Most of the projects don't come close to using all available memory or I/O lines, so any of those in Table 1-1 will work. If you have a particular project in mind, skip to that chapter and see if there are any special board requirements for that project. Otherwise, pick the best one you can afford. (In later chapters, we will show you how to "roll your own" board using a bare chip. This approach is useful when the chip demands are low and the circuitry is simple.)

Having said all that, we *really* hope you will pick the ATmega1280 or "higher" board, at least for your experimental board while reading this book—the reason being the increased memory and I/O pins. If you develop a circuit that's pretty simple and a bare-bones 328 would do, you can always buy the chip, a crystal, and a few other components and roll your own 328 board for under $10. (A new µC called the Digispark from Digistump has a one-square-inch footprint yet has 6 I/O lines, 8K of Flash, a clever USB interface yet sells for $9!) However, some of the advanced projects in this book make use of the additional I/O lines, which simplifies things considerably. Therefore, we are going to assume you're willing to sell pencils on the street for a few days until you get the additional couple of dollars to spring for the 1280 or 2560. You won't regret it.

By the way, there are a ton of knockoff Arduino's available on the Internet, mainly from China and Thailand, and we have purchased a good number of them. We have yet to have a bad experience with any foreign supplier. (However, some of them appear to have *bootloader* software [e.g., the software responsible for moving your program from the host PC to the Arduino] that only work on pre-1.0 Arduino IDEs. Check before you buy.) On the other hand, many times we need a part quickly and domestic suppliers provide very good service for those needs. Appendix A lists some of the suppliers we have used in the past.

What Else Do You Need?

There are a number of other things you need to complete the various projects in this book. One item is a good breadboard for prototyping circuits (see Figure 1-2). A breadboard allows you to insert various components (e.g., resistors, capacitors, etc.) onto the board to create a circuit without actually having to solder the component in place. This makes it much easier to build and test a circuit. The cost of a breadboard is determined in large part by the number of "holes," or tie points, on the board. The cost of a reasonably sized breadboard is around $20. Most of the breadboards we use have over 1500 tie points on them, although we don't think we have ever used even 5% of them at once. The board pictured in Figure 1-2 is from Jameco Electronics, has over 2300 tie points, and sells for around $30. Notice the binding posts at the top for voltage and ground

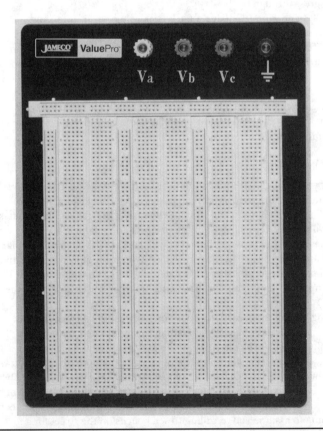

FIGURE 1-2   An inexpensive breadboard. (*Breadboard courtesy of Jameco Electronics*)

connections. You can buy smaller boards with about half the tie points for about half the price. A good quality board should last for years while a really cheap one will wear out and provide weak connection points over time.

The next thing you must have is a soldering iron for when you wish to finalize a circuit. Select an iron with a temperature control and a fairly small tip (see Figure 1-3). You can see a small light area beneath the iron, which is actually a sponge that is soaked with water and then the tip can be scraped on it to keep the tip clean. The small dial allows you to adjust the temperature of the iron. Such an iron can be purchased for around $25 or less.

You will also want a bunch of jumper wires that you will use to tie various components on the board together. Usually, you want jumper wires that run from one breadboard hole to another. These wires have a pin attached to both ends and are called male-to-male jumpers. In other cases, you will want to attach the lead of a device (perhaps a sensor) to the breadboard. In this case, you'll want one end of the jumper with a small socket-like hole where the sensor lead can be inserted while having a male pin at the other end. These are female-to-male jumpers. Finally, you may have instances where you want both ends to be sockets, or female-to-female jumpers. Jumpers come in various colors and lengths. Personally, it seems we run out of the longer (10 in.) male-to-male jumpers most often. We like the quality of Dupont jumpers (see Figure 1-4) the best.

You will also need a variety of resistors, capacitors, wire, solder, cable ties, and a host of other things, depending on your area of interest. Again, your local electronic components store will have most of the components you need. If you can't find what you need at your local supply store, check Appendix A for a list of suppliers we have found useful.

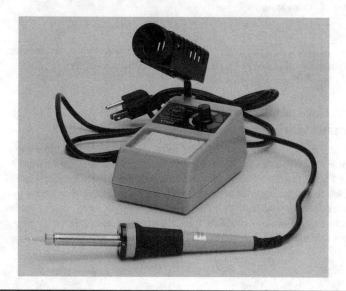

FIGURE 1-3   Adjustable temperature soldering iron.

Software

For most µC projects, software is the glue that holds the project together. In simple terms, there is software that was written by others that you use (e.g., editors, compilers, debuggers, linkers, an IDE, libraries) and there is the software that you write to tell the hardware and other pieces of software what to do. Together, you form a team that is geared toward the solution of some specific

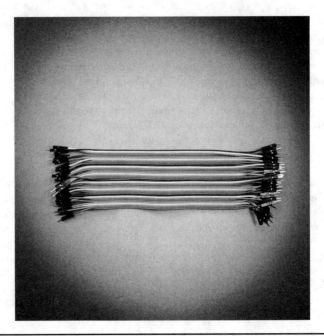

FIGURE 1-4   Dupont jumpers.

problem. In a nutshell, that's what software development is all about: solving problems. We'll have a lot more to say about software design and solutions in Chapter 2. For now, we want to get you up and running with a new set of tools.

Downloading and Installing the Arduino Integrated Development Environment

Downloading and installing the Arduino Integrated Development Environment, or IDE, is the first piece of software you need to be able to write your own programs. As this book was being written, Version 1.0.5 is the release number for the current IDE. Now, Version 1.5.6 of the IDE is currently being beta tested, but has not been labeled "stable" yet. However, we have made the switch to 1.5.6 and have not had a problem.

Personally, before downloading the IDE, we prefer to create a matching root directory named in such a way that it is easily identified, such as:

```
Arduino156
```

We use this directory as the place to install the software rather than the default directory (which usually ends up under the program directory). The reason for doing this is that it makes it easier to locate certain program, library, and header files should we need to look at them at some time in the future. Note this is just our preference. You are free to place the files wherever you want them.

You can access the Arduino IDE download site by loading your Internet browser (e.g., Internet Explorer, Chrome, Firefox) and then typing the following URL into the address bar:

http://arduino.cc/en/Main/Software

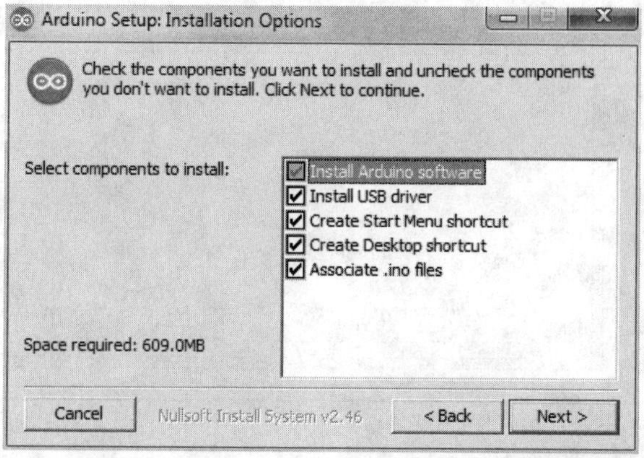

FIGURE 1-5 Installation options.

When you get to the download page, you are given download choices for Windows, Mac OS X, or Linux. Select whichever one applies to you. Once started, it may take a while to download the software depending on your Internet download speed. The IDE file, even in its compressed format, is over 56 Mb and even with a fairly fast connection, it still took us almost 10 minutes to download the software.

Installing the Software

When the software is downloaded, move the file (e.g., arduino-1.5.6-windows.exe) to your new working directory (Arduino156 or whatever directory name you have chosen). Now double-click on the file to begin its execution. You will be asked if you want to grant permission to run the software. Press the "Continue" button.

The installer will then ask if you agree to the terms of the license. This is a standard GNU Open Source contract, so you should click the "I agree" button. You should soon see the dialog presented in Figure 1-5.

You should accept all of the programs to be installed by default. Click "Next."

The next dialog asks where you wish to install the software, and presents the default folder for installation, as seen in Figure 1-6.

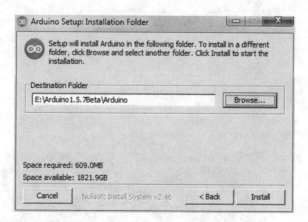

FIGURE 1-6 Default installation folder.

10　Arduino Projects for Amateur Radio

FIGURE 1-7　Install USB device driver.

However, since we do not want to use the default installation folder, press the "Browse" button and navigate to the folder you created earlier (e.g., Arduino156). Now click the "Install" button.

It takes a few minutes for the installer to unpack and install all of the associated files. When finished, our installation presented another dialog, as shown in Figure 1-7.

We checked the "Always trust software from 'Arduino LLC'" box and then clicked the "Install" button. Within a minute or so, the dialog said "Completed" and we clicked the "Close" button and the installation was complete.

Or was it?

Upon looking in our Arduino156 directory, we found a subdirectory named Arduino that the installer had created automatically. Looking inside of that directory, we found the IDE application file named arduino.exe. We double-clicked the file and were quickly rewarded with the dialog shown in Figure 1-8.

FIGURE 1-8　The Arduino IDE.

Once you see something similar to Figure 1-8, you can be pretty sure you have installed the Arduino IDE successfully. (You may wish to create a desktop shortcut for running the IDE, since you will be using it a lot.)

Running Your First Program

While seeing the image depicted in Figure 1-8 is pretty good *a priori* evidence that you have installed the IDE correctly, there's no better proof of the puddin' than to run a small test program. Fortunately, the IDE comes with a bunch of sample programs you can compile and run.

Connecting Your Arduino to Your PC

The first step in getting your system ready for use is to connect the Arduino board you purchased to your PC. Most Arduino vendors include a USB cable that connects the Arduino board to the PC via the supplied USB cable. This cable supplies power to the Arduino board, plus allows you to transfer your compiled program code from the PC into the Arduino's Flash memory. Connect the standard USB connector to your PC and the mini connector on the other end to the USB connector on the Arduino board.

Because the Arduino IDE is capable of generating code for different Arduino boards, you need to tell the IDE which board you are using. Assuming you have the IDE up and running, select the Tools → Board → Arduino Mega 2560, as shown in Figure 1-9.

If you purchased a different board, select the appropriate board from the list. If you ever switch boards at a later date, don't forget to change the board setting to match the new board.

The IDE senses the serial port at this time, too. While our sample Blink program does not take advantage of the serial port, be aware that programs that do have serial communications between the Arduino and your PC must be in sync across the serial port. Although we repeat it later, the default serial baud rate is 9600. Anytime you seem to be getting no serial output from a

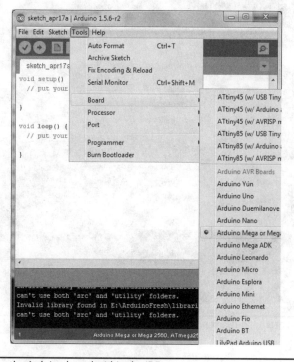

FIGURE 1-9 Selecting the Arduino board within the IDE.

program, first check to make sure you have the right COM port selected. On the other hand, if the serial communications produce output, but it looks like it came from someone whose native language is Mandarin, chances are the baud rate in your program doesn't match the serial monitor's baud rate. You can change the baud rate specified in your program or you can change the baud rate using the serial monitor dialog box. Our preference is to change the baud rate in the program.

Having set up the hardware connection between the IDE and your board, you can now load in the program you wish to compile. Like a bazillion other articles and books, select the sample Blink program that is distributed with the IDE. To load this program, use the File → Examples → 01.Basics → Blink menu sequence as shown in Figure 1-10.

As soon as you select the Blink program, your IDE should look like that shown in Figure 1-11. While Figure 1-11 shows the Blink program in the IDE, you will also notice an "empty" second IDE in the background. This is normal, as the IDE makes a fresh copy of the IDE when you actually add program source code to it.

What you see in the IDE is the program's source code. *Source code* refers to the human-readable form of the program as it is written in its underlying source code language. For almost all of the programs in this book, we use the programming language named C to write our programs. While you do not have to know how to program in C to use the projects in this book, knowing C makes it easier to edit, modify, enhance, debug, and understand μC programs. In a totally

FIGURE 1-10   Loading the Blink program into the IDE.

unabashed plug, Dr. Purdum's *Beginning C for Arduino* from Apress Publishing is an introductory C programming language book that assumes no prior programming experience. If you want a deeper understanding of what the code does in a given project or simply have a better grasp of how to write your own programs, *Beginning C for Arduino* is a good starting point. Simon Monk's *Programming Arduino* (McGraw-Hill) is also a popular introductory programming book.

Once the program's source code is visible in the IDE, as in Figure 1-11, you can compile the program. The term *compile* refers to the process by which the source code language program seen in Figure 1-11 is converted into the machine code instructions that the Atmel processors can understand and execute. You compile the source code by clicking the check mark icon that appears just below the File menu option. Alternatively, you can also use the Sketch → Verify/Compile menu sequence or the Ctrl-R shortcut. If there are no program errors, the IDE should look like that shown in Figure 1-12. Notice the message at the bottom of the IDE. It said it has done compiling and that the Blink program uses 1116 bytes of program space out of a maximum of 30,720 bytes.

Wait a minute!

If we selected an ATmega2560 with 256K of memory as my board choice, why is there only about 30K of memory left after the program only uses a little over 1K? The reason is: we lied. As we write this, our 2560 boards are "tied up" in other projects, so we're really using a smaller ATmega328 (often simply called a "328") we had lying around. Because we really hope you are

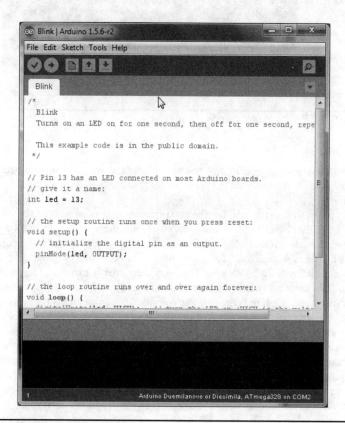

FIGURE 1-11 The Blink program in the IDE.

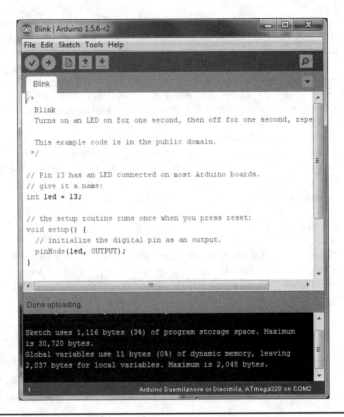

FIGURE 1-12   The IDE after a successful compile.

using an ATmega2560, which was the example board we used in the instructions above. The compiler actually did generate the correct code, but for the board we actually do have attached to the system.

You probably also noticed that the maximum program size is 30,720 bytes. However, if the 328 has 32K (32,768) bytes of Flash memory, where did the approximately 2K bytes of memory go? Earlier we pointed out that µCs don't have a large operating system gobbling up memory. However, there is a little nibbling away of available memory to the tune of about 2K used up by a program stored in the µC called a bootloader. The *bootloader* provides the basic functionality of the µC, like loading and executing your program and handling some of the I/O responsibilities.

At this stage, the IDE is holding the compiled program and is ready to send it over the USB cable to the Arduino board. To do this, click the Upload icon, which looks like a circle with a right-pointing arrow in it, just below the Edit menu option. You can also use the File → Upload menu sequence or the Ctrl+U shortcut. Any one of these options moves the compiled program code from the IDE on your PC into the Flash memory on the Arduino board. If you look closely at the board during the upload process, you can see the transmit/receive LEDs flash as the code is being sent to the board. If you're in a hurry, you can also simply click the right-pointing arrow and skip the (compile-only) check mark icon. If the program source code has changed, the IDE is smart enough to recompile the program and upload the code even though you didn't click the compile icon.

Once all of the code is sent to the board, the code immediately begins execution. For the Blink program, this means that an LED on the Arduino board begins to blink about once every second. Congratulations! You have just downloaded, installed, and run your first Arduino program.

Now the real fun begins ...

CHAPTER 2

I Don't Know How to Program

Do you have to know how to program to make use of the projects presented in this book? No. Will you enjoy the projects more if you understand a little bit about programming? Probably, yes. The purpose of this chapter is not to teach you how to program. There are plenty of books available to do that. (As mentioned in Chapter 1, Dr. Purdum's *Beginning C for Arduino* is written for those who have no prior programming experience. Simon Monk's *Programming Arduino* is also a popular introductory programming text.) Rather, the goal is to get you to think a little bit like a programmer so you can at least follow what the code is doing. That appreciation alone may be enough for you to start tinkering with µCs and their associated code yourself.

When we hear hams express their concerns about programming, we think about our own experiences when we first started to learn Morse code.

I Don't Need No Stinkin' CW!

Jack remembers well getting ready for his General Class license exam. He had been a Novice for almost a full year and was sick and tired of Morse code. Like most Novice class operators, he couldn't wait to get on AM or maybe even that new SSB stuff ... someday. Still, way back in 1954, you had to pass a code exam of 13 words per minute ... a blistering pace of 65 characters a minute! Jack's friend, Charlie McEwen, and he studied for the exam together and were pretty confident about the theory and regulations part of the exam. The code part ... not so much.

They took the bus from their hometown of Medina into Cleveland about 30 miles away. They left the bus, climbed the stairs into the federal building, and found the exam room. It was probably 50-ft square with six or seven rows of dark wood tables and chairs to match. One wall held tall windows that reached for very high ceilings while the other walls were painted government green and decorated with some bad photos of public figures of the day. At the front of the room was a blackboard and just in front of that a small table with a square box sitting on it with an AC line cord running from it into the wall. It had to have been the room used to design the Bates Motel and felt just as friendly.

In a few moments, a man came in and told us we would be taking the receiving part of the code exam first. It required that we copy at least 1 minute of perfect code out of the 5 minutes that were to be sent at 13 words per minute. If we passed the receiving test, then we would take the sending part and, passing that, the written part of the exam. Jack was sweating bullets and Charlie didn't look much better. The examiner then asked if there were any questions. There were none.

Then, out of nowhere, Morse code erupted from the little square box, passing Jack's ears with unrecognized speed before crashing into the back walls only to ricochet off the side walls just in time to collide with the next letter being sent from the little square box. The room had the acoustics of a waste can. Jack spent the first 30 seconds of that exam trying to hear something other than his own heart pounding in his ears. At the end of 5 minutes, he handed his paper in and took a seat, still trying to get his pulse rate below 1000. In a relatively short period of time, Jack was told he had passed the receiving exam and could take the sending part next. The sending part was a snap and, upon hearing he passed it, Jack told himself: "I will never use Morse code again!"

Really?

Here we are five decades later actually enjoying CW again. Indeed, QRP and CW have given amateur radio a new breath of life for many of us. But why?

We think the hang-up with Morse code was based on fear of the unknown. As a Novice, Jack never pushed himself hard enough to get comfortable with code at 20 wpm, which is what everyone was telling us we needed to be able to copy to have enough margin for error to pass the code exam. Jack simply lacked the confidence to try.

Like CW, Like Programming

After talking with a lot of hams about using µC, we hear the same story: "I don't know how to program µCs so I don't use them." When you first thought about getting your ticket, did you know Morse code? Could you draw a schematic of a Colpitts Oscillator? Many of us couldn't. But we learned enough to get our tickets, and that was the goal. No, we were not EEs at the time, but we learned enough electronics to reach the goal we sought ... our General Class ticket.

Just like we said we never wanted to hear another dit or a dah all those years ago, many of you say you don't want to learn how to program, and that's fine. However, even though you may not want to command a cruise ship by yourself, the trip might still be more enjoyable if you know what you're looking at during the journey. And that's what you'll learn in this chapter ... an appreciation for what the program code is doing. And trust us ... programming and the power it conveys can be very addictive!

The Five Program Steps

No matter how complex a program might be, it can be distilled to essentially five steps. Knowing the Five Program Steps serves multiple purposes. First, you've probably been in a situation where you've been assigned a project, but you just don't know where to start. When it comes to software, thinking about the Five Program Steps is perhaps the first step you should take toward a solution. Also, once you understand the Five Program Steps, it gives you an organized way to examine someone else's code ... like ours. It also gives you a broad perspective on how to start designing your own programs, should you decide to give programming a try.

Step 1. Initialization

The purpose of the Initialization Step is to establish the environment in which the program is to run. If you've ever run a Windows program like Word or Excel, you know that the File menu keeps track of the last four or five files you've used. Finding those files and adding them to the File menu before you see the first screen displayed would be part of the Initialization Step. Other programs might be concerned with opening database connections, activating printer ports, turning on your rig so it can warm up, plus a host of other "pre-operating" tasks. Generally, then, the Initialization Step creates the operating environment for the program and it all takes place before the user sees anything change due to the program running.

Step 2. Input

If you think about it, every computer program takes data in one form, crunches on it in some way, and outputs a new set of data. The Input Step concerns itself with getting the data into the program. The source of the data could be a keyboard or keypad, a fiber optic connection, a database, a voltmeter, a CW keyer, an antenna signal ... whatever. The point is that data in some form is introduced into the program.

Also note that getting data into a program is an imperfect process. People do hit the wrong keys, connections can break down, there can be voltage spikes, plus a host of other things that can possibly go wrong. For that reason, the data presented in the Input Step is often validated for potential errors before it gets passed along to the next step.

Step 3. Processing

As we mentioned earlier, data are presented to the program in one form, reworked, or "crunched," in some manner, and then presented in a different form. The "crunch" step is the Processing Step that is used to transform the data from its input form to its final form. Often, the Processing Step involves applying some formula, or algorithm, to the data that manipulates the input data mathematically. However, that is not the only way processing can take place. For example, you may have two vats of chemicals that have sensors attached to them and when their contents reach certain temperatures, the two vats are combined together in a mixing process. The precise way in which the data inputs are combined is determined by the underlying methodology required to produce the desired result. You will also hear this methodology, or process, referred to as the algorithm. An *algorithm* is nothing more than a detailed, step-by-step, description of how the data are transformed from its initial state to its final state.

Generally, therefore, the Processing Step means taking the data from Step 2 and combining it to change the data into a desired new form.

Step 4. Output

Once the data have been processed into whatever new form is desired, that new data is output in some way. In many cases, this means displaying the output on a monitor, an LED or LCD display, perhaps on a printer, or even using a WiFi link to send the output somewhere else. The output can also be digital signals that are used to close switches, solenoids, relays, or other electronic devices. Outputting the results can also mean writing the data to a database or SD card for later use in some subsequent process, or as part of a script that will be run as a batch process at some predetermined time later in the day. The point is that the Output Step represents the desired state of the program's input data, assuming no errors occurred along the way. The Output Step represents a "solution" to your programming problem.

Step 5. Termination

Well-behaved software should "clean up after itself." What this means is that, if the Initialization Step opened a database connection, the Termination Step should make sure the connection is closed and its associated resources reclaimed. Memory leaks, dangling file handles, and the like are common examples of programs that didn't clean up after themselves.

If your program maintains a list of files that were last used, like Word or Excel does, the Termination Step should ensure that the file list is updated before the program ends. In a real sense, the Termination Step often "undoes" whatever was done in the Initialization Step. You will see that most µC programs are an exception to the rule in that they are really not written to terminate themselves. We'll have more to say about that later in this chapter.

Arduino Programming Essentials

All Arduino programs have certain elements present in them. First, almost all of them are written in the C programming language. C is a simple, yet robust, language that has been around since the 1970s. It has relatively few *keywords; words that have special meaning in the language*. C has no intrinsic I/O commands like many other languages have.

However, don't be misled. The genius behind C was that its inventors made the I/O external to the language itself, and instead placed the I/O functionality into what is called the C Standard Library. The great thing about C is that, if you don't like the way a certain piece of I/O functionality works, you are free to write your own. As you read more programs, you'll start to gain an appreciation for what this means, since it allows you to handle "those special cases" exactly the way you want to, not the way someone else forces you to.

Actually the Arduino IDE uses the C++ compiler from the Open Source group, which fully supports C++ and all that its Object Oriented Programming (OOP) paradigm brings to the table. Still, for the most part, the Arduino programs, or *sketches* as they are called, are usually written in plain old C. If you need help understanding a program, the first place to look would be a book on C, and probably not C++. C++ is a significantly more complex language than C. Indeed, someone once wrote that C gives you the power to shoot yourself in the foot, while C++ lets you blow your whole leg off.

Let's just stick with C for now.

The Blink Program

In Chapter 1, you loaded and ran a program that is distributed with the Arduino IDE called Blink. Listing 2-1 presents the source code for the Blink sketch. Let's examine this simple program line by line.

Comments in C

The first thing that you see in the sketch is a slash-asterisk (/*) pair of characters. Several lines farther down, you find an asterisk-slash (*/) pair of characters. These two character pairs mark the start (/*) and end (*/) of a multiline program comment. That is, the comment can span as many lines as it needs as long as those lines fall within the boundary formed by the comment pair. *Program comments are notes written by the programmer to anyone who may read the program's source code with the goal of helping the reader understand what the code is doing.* Everything between the two comment character pair is ignored by the compiler. In other words, long wordy comments have absolutely no impact on the execution or performance of the program. The programmer put them there for information only. In this example, the multiline comment simply explains what the Blink program does.

```
/*
  Blink
  Turns on an LED for one second, then off for one second,
  repeatedly. This example code is in the public domain.
*/
```

LISTING 2-1   The Blink sketch.

```
// Pin 13 has an LED connected on most Arduino boards.
// give it a name:
int led = 13;

// the setup routine runs once when you press reset:
void setup() {
  // initialize the digital pin as an output.
  pinMode(led, OUTPUT);
}

void loop() {
  digitalWrite(led, HIGH);   // turn LED on (HIGH is voltage level)
  delay(1000);               // wait for a second
  digitalWrite(led, LOW);    // turn LED off (voltage is LOW)
  delay(1000);               // wait for a second
}
```

LISTING 2-1   The Blink sketch. (*continued*)

After the multiline comment you see the following two lines:

```
// Pin 13 has an LED connected on most Arduino boards.
// give it a name:
```

These two lines are examples of single-line comments. *Single-line comments begin with two slash characters (//) at the start of the comment and end at the end of the same line.* No special closing comment characters are needed. These two single-line comments could just as well have been written using the multiline comment characters.

When to Comment

When Jack was teaching computer programming classes, he saw both extremes of commenting. Students submitted programs that had source code lines that looked like:

```
k = k + 1;      // Take the original value of k and add 1 to it
```

Really? If the reader can fog a mirror they should be able to figure what the code statement does without the comment. At the other extreme, we've also seen code like:

```
sol = math.sin(w) + 2.0 * ladder - myFunction(nodding == FULL?10:5);
```

and no comment in sight ... anywhere. If either programmer worked for us, they both would have been fired. Why?

First, commenting every line of code is almost never necessary. Indeed, such commenting actually makes it harder to read the code. Second, complex code like the second statement above should probably never be written that way in the first place. Because there is a fairly high chance that statement ultimately needs to be debugged, we would write it as:

```
double temp1 = math.sin(w);                // w is the window angle
int temp2 = (nodding == FULL) ? 10 : 5;    // If nodding is fully open
                                           // set to 10, 5 otherwise
```

```
double temp3 = myFunction(temp2);      // Check that temp3 is
                                       // within range
sol = temp1 + (2.0 * ladder) - temp3;  // Solve for sol
```

Writing the code as a series of smaller steps makes it easier to test and debug the complex expressions. If there is some serious performance hit by breaking the statement into smaller parts, perhaps because the expression is buried in a tight loop, keep the statements apart until the code is fully debugged and then combine it into its complex form ... and test again.

We usually place multiline comments at the start of a function to explain what the purpose of the function is, what data (if any) need to be passed with the call, and what value (if any) is returned from the function call.

So ... when to comment and when not to comment? It's a personal decision, but we think that, if it takes you more than 10 or 15 seconds to figure out what a statement is doing, it likely will benefit from a comment. Otherwise, the comment is probably a waste of time.

Data Definitions

Returning back to the Blink program, the next line in the program's source code is:

```
int led = 13;
```

This line is a C statement that *defines* an integer data type and assigns the variable name *led* to it. It also initializes the value of the variable to 13. Because each integer variable requires two bytes of memory for storage, this statement sets aside two bytes of memory and tags it with the name *led*. Having done that, the compiler then assigns the integer value 13 into those two bytes of memory. Part of the compiler's job is to keep track of where each variable is stored in memory so its value can be recalled when needed.

Do *NOT* fall into the trap of using the words *define* and *declare* to mean the same thing. They do not have the same meaning. Most programmers are extremely sloppy when using these words. In its simplest terms, a *data definition* creates an attribute list (e.g., a description of the variable, such as: "an integer variable named *led*") that describes the data item <u>but also allocates memory for that variable.</u> A *data declaration*, on the other hand, also creates an attribute list for a variable <u>but does not allocate storage for it</u>. Because no memory is allocated with a data declaration, there is no memory location set aside where the variable can store data. There are good reasons for this distinction, but we aren't ready to discuss them yet. (If you're interested, a more complete explanation can be found at: "Computer Programming and Precise Terminology," *Information Week*, July, 2008. Also reprinted in *Dr. Dobbs Journal*, July 2008; http://www.drdobbs.com/cpp/computer-programming-and-precise-termino/208808373.)

Where's the *main()* Function?

A lot of people use the Arduino µCs because they know that the Arduino IDE uses C as its basis. However, people familiar with C, C++, or Java are a little confused because an Arduino program has no *main()* function in it. If you have ever used C, C++, or Java you know that all programs must have a *main()* function, as that is the starting point for program execution. What happened to *main()* in Arduino sketches?

The *main()* function is actually still there, but it is hidden from direct view. If you look in your directory where you installed the Arduino IDE, you should be able to track a path similar to:

```
C:\Arduino1.0.5\hardware\arduino\cores\arduino
```

In that last arduino subdirectory you will find a number of header files (files ending in ".h") and source code files (ending in .cpp, for C-Plus-Plus, or C++ files or .c for ordinary C source code files). The header files contain a lot of information that the compiler uses to set the environment for compiling a program and the source code files are used in conjunction with your own Arduino "sketch" files. (As mentioned earlier, program source code files that you write within the Arduino IDE are called *sketches* and have a secondary file extension of *.ino.) The fact that the Arduino directory is filled with both *.c and *.cpp files tells us that we are free to mix C and C++ files in our programs.

If you look closely in the arduino directory, you can see a file named main.cpp. The source code for main.cpp is presented in Listing 2-2.

```
#include <Arduino.h>

int main(void)
{
    init();

#ifdef (USBCON)
    USBDevice.attach();
#endif

    setup();

    for (;;) {
        loop();
        if (serialEventRun) serialEventRun();
    }

    return 0;
}
```

Listing 2-2   The main.cpp file.

The source code in main.cpp is pretty simple. Within the main.cpp file is the definition for the *main()* function. The *main()* function begins with a call to the *init()* method, which is used to establish some compiler settings. (Hmmm ... sounds like the Step 1 Initialization Step from our Five Program Steps.) As you probably know, *init()* is called a function in C. *Functions in C are small pieces of code that are designed to perform one specific task.* Next is a *#ifdef* preprocessor directive that messes around with the USB connection device, if one is present. (Not all Arduino boards have USB devices.) Next a function named *setup()* is called one time, followed by a *for* loop that repeatedly calls a function named *loop()*. Under certain circumstances, the code may also call a function named *serialEventRun()*.

Normally, a programming *for* loop has three expressions:

```
for (expression1; expression2; expression3) {
    // statement(s) controlled by the for loop
}
```

An example might be:

```
for (i = 0; i < MAXVAL; i++) {
    // statements...
}   // end of for loop
```

In the *for* loop above, the first expression initializes a variable named *i* to 0, the second expression checks to see if the value of *i* is less than (the '<' character) whatever MAXVAL is. If *i* is less than MAXVAL, the statements between the opening and closing braces (the '{' and '}') are executed. Once those statements are executed, the *i* is incremented (expression3, or *i++*) and then expression2 (*i < MAXVAL*) is evaluated again. This *for* loop keeps executing until *i* has been incremented to a value that equals or exceeds MAXVAL, at which time the *for* loop ends.

In Listing 2-2, the *for* loop omits all three expressions. Because the three expressions are missing, there are no expressions present that can be tested to terminate the *for* loop. This creates an infinite loop: a loop that never ends. Stated differently, it means that the *loop()* function in Listing 2-2 is called forever or until power is removed from the μC or the system fails in some way. (It is also possible that the *serialEventRun()* function call could end the program.)

Distilled to its simplest form, therefore, all Arduino programs do have a *main()*, albeit it's tucked away from view a little bit. The primary purpose of *main()* is to: 1) establish the basic environment for compiling a program (i.e., the function call to *init()*), 2) process any special stuff that you want your program to do once via the single call to *setup()*, and 3) keep calling *loop()* until the cows come home. Therefore, it might be useful for us to look at the *setup()* and *loop()* functions in a little more detail.

The *setup()* Function

Referring back to our Blink program in Listing 2-1, after the definition of the variable named *led*, you find the following four lines:

```
                    // the setup routine runs once when you load a new
                    // sketch or press the reset button
void setup() {
                    // initialize the digital pin as an output.
  pinMode(led, OUTPUT);
}
```

The actual task a function performs is determined by the statements contained within the body of the function. A function body begins with the opening curly brace following its name ({) and ends with the closing curly brace at the bottom of the function (}). In our example, the *setup()* function body has only two lines in it: 1) a comment line, and 2) a call to another C function named *pinMode()*. It appears from the name of the function that *pinMode()* is designed to use the variable *led* in some way to affect the pin's mode and sets it to OUTPUT. If you look up the library documentation for *pinMode()* you would find that it is used to set a specific Arduino pin (*led*, or pin 13 in this case) to a specific mode, *OUTPUT* in this case. As you might guess, pin 13 is tied to the Arduino's onboard LED.

So what does the word *void* in front of *setup()* mean? In C, functions have the ability to send a value back to whatever section of code invoked the function. For example, you might write a function that passes in the number of yards of cotton cloth are available on one or more bolts of cloth and your function returns the number of shirts that many bolts can produce. In that example, you might invoke the function like:

```
shirts = BoltsToShirts(boltsAvailable);
```

The variable *shirts* is probably an integer data type that was defined earlier in the program. When the *BoltsToShirts()* function finishes its task, *shirts* holds the number of shirts that can be produced with the bolts of cloth available. The reason is because the programmer probably designed the *BoltsToShirts()* function to return an integer value that is the number of shirts that *boltsAvailable* yards of cloth can produce. In programming, integer values cannot have a decimal, or fractional, component. Integers are always whole numbers. In this example, most people have no use for a "fraction of a shirt," so the programmer made *shirts* an integer data type.

However, not all functions need to return a value. If no value is returned from the function, the word *void* appears before the function name, as it does for *setup()*. The word that appears before the function name is called a *function type specifier*, and it indicates the type of data being returned from the function. Since we see *void* in front of *setup()*, we know nothing useful is returned from the function call to *setup()*.

The *setup()* Function is Special

Every Arduino program has a *setup()* function that is called from the (hidden) *main()* function. Indeed, *setup() is the first function that every Arduino program sketch calls directly.* In terms of the Five Program Steps, *setup()* is usually Step 1, the Initialization Step. As we said earlier, Step 1 sets the environment in which your program runs. In our simple Blink program, the only change to the default environment we need to make is to set pin 13 to be an *OUTPUT* pin. Once we do that, Step 1 for this program is done.

In other programs where you want to monitor values being produced by your program, you will likely use the *Serial.print()* function to pass information back to your PC via the serial monitor. To do that, your program and the serial monitor must have the communication's rates set to the same baud rate. Therefore, it is common to see *Serial.begin(9600)* as a program statement in *setup()*. As our programs get more complex, most of the functions also become more complex. Still, the basic ideas remain the same.

The *loop()* Function

The remainder of the Blink program is fairly short:

```
// the loop routine runs over and over again forever:
void loop() {
  digitalWrite(led, HIGH);   // turn LED on (HIGH is voltage level)
  delay(1000);               // wait for a second
  digitalWrite(led, LOW);    // turn LED off (voltage is LOW)
  delay(1000);               // wait for a second
}
```

The function named *loop()* (also called from within the hidden *main()* function) is preceded with the word *void*. As mentioned earlier, a *void* function type specifier means that nothing useful is returned from the function call to *loop()*. Every program written using the Arduino IDE must have a *loop()* function. The statement body of the *loop()* function is executed "forever." That is, after the *setup()* function is executed once, the *loop()* function is endlessly called as the program executes. Normally, the *loop()* function continues to execute until: 1) power is removed from the board running the program, 2) you upload a new μC program to replace the current program, or 3) the board fails because of some malfunction. As we mentioned earlier, the *loop()* function normally executes forever, which creates an *infinite loop*; a loop that never ends.

Within the *loop()* function, the first statement is:

```
digitalWrite(led, HIGH);
```

The words *HIGH* and *LOW* are defined for the Arduino environment and you can think of them as representing turning the *led* pin on (*HIGH*) or off (*LOW*). In this program, the *digitalWrite()* function is serving as Step 2 of the program by providing the input values (e.g., *led* and HIGH) to be used by the *digitalWrite()* function. Once those inputs are given to *digitalWrite()*, the function can process those inputs (Step 3) and proceed to turn on the LED (Step 4) and keep it on long enough for us to observe by calling the *delay(1000)* function. Because the *delay()* function is passed a value representing milliseconds, the call to *delay(1000)* keeps the LED illuminated for 1 second. If the *delay()* call was not made, the LED wouldn't be on long enough to observe its change in state from LOW to HIGH.

After 1 second has passed, the call:

```
digitalWrite(led, LOW);
```

is made. This means the program is repeating Step 2 and supplying a new set of inputs to *digitalWrite()*, which turns the LED off (Step 3). Once again, the call to *delay(1000)* allows us to observe that the LED is now off and remains so for 1 second.

Because of the way the *loop()* function is implemented in the Arduino IDE, once the last statement in the *loop()* function (e.g., the second *delay(1000)* call) is executed, the program jumps back up to the first *digitalWrite(led, HIGH)* at the first statement in the *loop()* function and the entire process repeats itself. Indeed, this is why the function is called *loop()* ... the statements within the *loop()* function repeat themselves forever, or until power is lost or there is a malfunction.

I Thought There Were Five Program Steps?

Wait a minute? Didn't we tell you that there were Five Program Steps? What happened to Step 5, Termination? Well, that's kinda what makes µC programs different from most other applications. Unlike a word processing program where you write your letter, save it, and you're done and you shut the word processor down, most µC applications are designed to run "forever." You might, for example, write a µC program that monitors several hundred fire sensors in a building. Every few seconds each of the sensors is visited and checked to see if a fire is present or not. If not, the code proceeds to check the next sensor in the list. The code continues to do this sensor checking *ad infinitum* until there is a fire or there is some kind of power or system failure. After all, how useful would the fire sensor system be if we shut it down when everyone left at 5PM, only to start it back up when they started to return to work at 8AM?

Therefore, most µC programs concentrate on the first four program steps, perhaps even hoping never to reach the Termination Step.

Modifying the Blink Sketch

Let's make a few minor changes to the Blink program and observe the impact those changes have on the behavior of the program. Listing 2-3 shows the modified program.

```
/*
  Blink: Turns on an LED for one second, then off for one second,
  repeatedly.

  This example code is in the public domain.
*/

// Pin 13 has an LED connected on most Arduino boards.
```

LISTING 2-3   A modified Blink program.

```
// give it a name:
int led = 13;

// the setup routine runs once when you press reset:
void setup() {
  Serial.begin(9600);
  // initialize the digital pin as an output.
  pinMode(led, OUTPUT);
}

// the loop routine runs over and over again forever:
void loop() {
  Serial.println("Turning LED on...");
  digitalWrite(led, HIGH);   // turn LED on (voltage is HIGH)
  delay(1000);               // wait for a second
  Serial.println("Turning LED off...");
  digitalWrite(led, LOW);    // turn LED off (voltage is LOW)
  delay(1000);               // wait for a second
}
```

LISTING 2-3   A modified Blink program. (*continued*)

The program is exactly as it was before, except for the three highlighted lines. The first line appears in the *setup()* function, so it is part of the Initialization Step. Without getting too complicated, one of the Arduino Libraries available to you is called the Serial library and it has a number of functions associated with it that you can use in your programs. (You can find details about the library in the Arduino Reference section of the web site: http://arduino.cc/en/Reference/Serial.)

Actually, the Serial library is a C++ class that you have access to in your programs. One of the functions (also called methods in OOP jargon) that is part of the Serial class is *begin()*. The purpose of the Serial class's *begin()* function is to set the baud rate for communication between your Arduino program and your PC. Figure 2-1 shows how to load the Arduino Serial Monitor using the Tools → Serial Monitor menu sequence. Note that the serial monitor does not become available for use until after a sketch has been uploaded to your Arduino board. (You can see the monitor's menu option, but clicking the option does nothing until after the new program has been uploaded and starts to execute.)

When you select the Serial Monitor, you should see something similar to Figure 2-2 appear on your PC display. The Serial Monitor does allow you to both send and receive data. The top textbox seen in Figure 2-2 is used when you want to type in data to be sent to the Arduino sketch that is running at the time. You would type in the data and click the Send button. The Monitor program then uses the serial data link (your USB cable) to transfer the data in the serial monitor textbox from your PC to the Arduino program. Our program doesn't need to send any data, so we ignore sending data to the Arduino for the time being.

In the lower right corner of Figure 2-2 you can see a drop down list box with 9600 baud in it. This is the default baud rate for the Serial Monitor. Therefore, the monitor program is expecting the data from the Arduino program to be arriving over the serial link at a rate of 9600 baud. The statement in *setup()* says:

```
Serial.begin(9600);
```

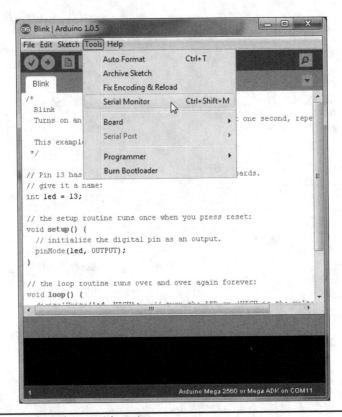

FIGURE 2-1 Loading the Arduino serial monitor.

which means our Arduino program is going to be pumping out the serial data at 9600 baud while the Serial Monitor running on your PC has its 9600 baud earphones on, so the Arduino and the PC can communicate with each other.

What if you don't match the two baud rates? If this happens, you will see small wisps of bluish-white smoke starting to drift up from ... Naw ... just kidding! What likely happens is that you will see a bunch of gibberish appear on the Serial Monitor. When that happens, it's almost always a sign that the baud rates for the Monitor and your program are not matched. The quickest fix is to change the value in the drop down list box value of the Serial Monitor program and restart the Arduino program. (Arguably, the proper fix is to pick the desired baud rate and make sure the program uses that rate. After all, some programs may be time-sensitive and you need the highest baud rate available. Regardless of the rate, if you change the rate in the program, you must recompile and upload the new program version. The Arduino board family has a small reset button on the board that you can press to restart the program.)

With the Serial Monitor's Autoscroll box checked as it is in Figure 2-2, the program displays the data without pause. The "Newline" dropdown box gives you several options for sensing when a complete line of data has been sent. Because we used the method named *Serial.println()* (note the "ln" at the end of the word *print*), the method is saying to print the information and then send a newline character so the next set of data appears on its own new line. If you had used *Serial.print()* instead, all of the output would appear on one, very, very, long line.

After making the baud rate change and setting the COM port correctly, the output from the new program should look similar to that shown in Figure 2-3.

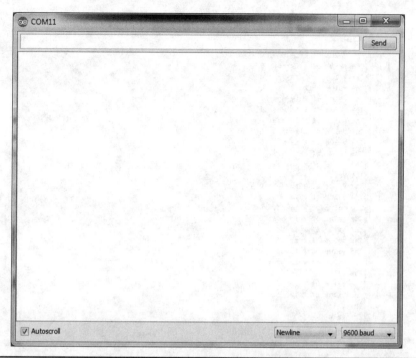

FIGURE 2-2   The Arduino serial monitor.

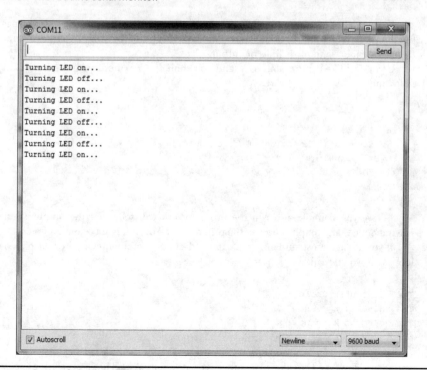

FIGURE 2-3   Output from the modified Blink sketch.

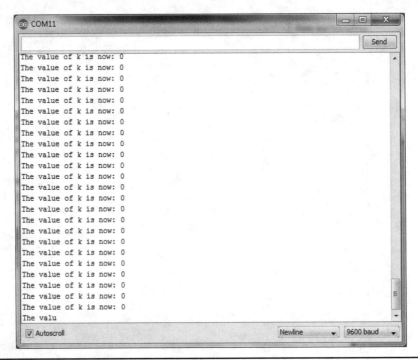

FIGURE 2-4   Displaying an integer value.

If you uncheck the Autoscroll feature on the Serial Monitor, it fills up with one page of output, but then vertical scroll bars would appear on the right side of the monitor as the program continued to generate more and more lines of output.

If you feel like experimenting ... and you should ... try modifying the *loop()* function to the following:

```
void loop() {
   int k = 0;
   Serial.print("The value of k is now: ");
   Serial.println(k);
   k = k + 1;
}
```

Before you compile and run the program, ask yourself what the output should look like. Having done that, compile and run the program. The output is shown in Figure 2-4.

If you are like most students, you expected *k* to be incremented by 1 on each pass through the loop, so it would read:

```
The value of k is now: 0
The value of k is now: 1
The value of k is now: 2
The value of k is now: 3
```

and so on. However, the value of *k* is always 0 (see Figure 2-4). The reason it does not work the way most people might expect is because the code redefines and initializes *k* to 0 on each pass through the loop. Therefore, *k* never has a chance to display a value other than 0.

Chapter 2: I Don't Know How to Program   31

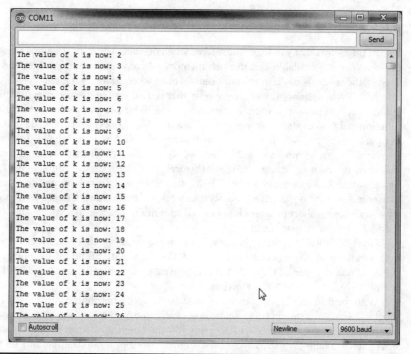

FIGURE 2-5   Moving the definition of *k outside* the *loop()* Function.

Let's make one small change to our program and see what happens. Move the definition of variable *k outside* the *loop()* function and just before it, as shown below:

```
int k = 0;
void loop() {
   Serial.print("The value of k is now:");
   Serial.println(k);
   k = k + 1;
}
```

Now compile and run the program. The output should look like that in Figure 2-5.

Now the output looks more like we expected it to look in the first place. So what's the moral of the story? Simply stated, it does make a difference where you place the definitions of your data. Because *loop()* is designed to repeat itself forever, placing the definition of *k* and initializing it to 0 within *loop()* means its value won't change in the print statements. Placing the definition of *k* outside the *loop()* function means it is not reinitialized on each pass. Like the *setup()* function, statements outside of *loop()* are only executed once.

CAUTION: *There may be times when you want to cut-and-paste code from an article you are reading in your word processor into the Arduino IDE. Usually, this works just fine. However, double quotation marks sometimes do not translate properly when moved into the Arduino IDE. Often this translation problem manifests itself as an error message stating: "Stray '(' in program." If you see this error message, try retyping the line with the double quotation marks in it from within the Arduino IDE. Chances are, the error message will disappear.*

Saving Memory

One of the factors that you likely considered when deciding which μC to purchase was the amount of Flash memory available. It is in Flash memory where your program resides and is nonvolatile. That is, the contents of Flash memory remain even when power is removed.

The SRAM memory is volatile memory that is used for temporary storage while the program is running. For example, if you call a function using an *int* as a parameter to the function call, each function call pushes a specific amount of data into SRAM over and above the memory taken for the parameter data. When the function finishes, the SRAM is reclaimed. However, if you call *function3()* from within *function2()* from within *function1()*, all that overhead for each function call starts to add up. Functions that call themselves (i.e., a recursive function call) can really play havoc with SRAM memory space. SRAM, therefore, is a scarce commodity that you want to conserve. If your program mysteriously dies while executing, check to see how much SRAM you have available. (A short program is presented later in this chapter that you can use to help monitor the SRAM memory space left.)

Finally, EEPROM memory is nonvolatile memory that retains whatever data you place there even when power is removed from the board. One issue with EEPROM memory, however, is that it has a finite number of times that it can be written to before it gets flaky. EEPROM is usually pretty stable for the first 100,000 writes.

With the different types of memory available to you, what should you do when you run out of memory? First, you should try to determine which type of memory is giving you issues. When you compile a program, the Arduino IDE gives you the amount of Flash memory you are using and what's still available for use as program memory. If your program is impinging on that limit, start looking for changes that reduce your program size. If you have lots of Flash memory available but still have unexpected program crashes or other issues, SRAM may be the problem. Quite often, running out of SRAM causes spectacular program failures while you can see that there is plenty of Flash memory left. If your code uses recursive function calls ... don't. Recursive function calls are very likely to fail. Finally, if you are doing reads and writes to EEPROM, try commenting out those statements and replace them with temporary SRAM definitions and see if the problem goes away. If so, you could be running out of EEPROM memory or it isn't taking writes reliably any longer.

What follows are a few ideas of how to save some memory if the need arises.

Remove Unused Variables

While this seems so obvious, it's easy to overlook a variable that remains unused in a program. While the compiler does a very good job of removing unused variables, it may not catch all of them depending upon the context in which they are used. If for no other reason, you should remove unused variables simply because they add clutter to your code.

Use a Different Data Type

If you are using an *int* data type to store a logic *true* or *false* condition, you're wasting memory. An *int* data type uses two bytes of memory while a *boolean* data type only uses one byte. Likewise, how many times have we seen something like this:

```
int bodyTemperatures[200];
```

It seems unlikely that someone's body temperature is going to vary between plus or minus 32,000 degrees. Using the *byte* data type:

```
byte bodyTemperatures[200];
```

immediately saves you 200 bytes of memory. Also, if your data require recording the temperature as 98.6 and you really need the decimal faction, the definition:

```
float bodyTemperatures[200];
```

is probably still wasteful simply because the array takes 800 bytes of memory. Instead, store each temperature as a three digit *int* data type and divide by 10 when you need to use it. Therefore, 98.6 gets stored in an *int* as 986, but used as 986/10. Saving just two bytes for a large array of data can add up quickly.

If you really get in a crunch storing *boolean* data, you could store each value as a bit in an 8-bit byte and use the bitwise operators to extract the data. This can get a little complex and may use more code to extract the data than is saved. You may have to experiment to see if it's worthwhile.

Avoid Using the *String* Class

Try this experiment: Start the Arduino IDE, load the Blink sketch, and add the following as the first line in the *loop()* function:

```
String message = "This is a message to display";
```

and compile the program. The Blink program size with this statement added is 2560 bytes on a Duemillanove. Now, change that line to:

```
char message[] = "This is a message to display";
```

and recompile the program. The program now occupies 1084 bytes, or a savings of 1476 bytes of Flash memory. Why such a large difference? The reason is because, any time you use the keyword *String* in your program, you cause the compiler to bring in the C++ *String* class. Although the *String* class has some nifty features, you may not need them and the resulting program bloat may be unnecessary.

The *F()* Macro

Suppose you have the following statement in your program code:

```
Serial.println("Don't forget to activate all of the external sensors.");
```

The message contained within the double quotation marks in the statement is called a string literal. A *string literal is a sequence of text characters that does not change as the program executes.* These string literals are imbedded in your program's memory space and, hence, do use up Flash memory. The problem is that the compiler sees these string literals *and copies them* into SRAM just before the program starts executing! In other words, the same string literal is duplicated in Flash and SRAM memory! With the message above, you just wasted 53 bytes of precious SRAM.

However, if you change the statement to:

```
Serial.println(F("Don't forget to activate all of the external sensors."));
```

note that the string literal is contained within the parentheses of the *F()* macro. Without going into the mechanics of how this works, the end result is that the *F()* macro prevents the compiler from copying the string literal to SRAM.

If you think you are running low on SRAM, look for string literals and see if using the *F()* macro saves the day.

There are other memory-saving techniques that you can use, but a good number of them are pretty advanced and may work for one program but not the next. For example, the HardwareSerial.ccp source file defines a 64-byte buffer that is used in serial communications. If your program is running on the ragged edge and doesn't need high-speed communications, you can reduce the size of the buffer. However, you may forget about this change a few months later when you are using high-speed communications and can't figure out why the code isn't performing as expected. (Also, while it may look like a single buffer is being changed, actually multiple buffer sizes are affected.)

The *freeRam()* Function

The *freeRam()* function is available on the Arduino web site, but it's so short, you can just copy it from here:

```
int freeRam()
{
  extern int __heap_start, *__brkval;
  int v;
  return (int) &v - (__brkval == 0 ? (int) &__heap_start : (int) __brkval);
}
```

Because the amount of SRAM available ebbs and flows and the program executes, you should inject the *freeRam()* function at whatever points you think are necessary. A typical use might be:

```
Serial.print(F("SRAM available = "));
Serial.println(freeRam());
```

While not a perfect tool, at least *freeRam()* can give you some idea of what's happening to your SRAM as your program executes.

If you need additional ways to save some memory and those presented here don't solve the problem, Google: "Saving memory in Arduino." We got over 2 million hits. Chances are at least one of those hits can save you a few more bytes of memory someplace.

Conclusion

As you read through the rest of this book, think about the Five Program Steps and try to figure out for yourself where those steps are in the code. It will help you understand what the program is doing. You might also load some of the other example programs that come with the Arduino IDE and try to figure out what they do. (You can find these programs using the File → Examples menu sequence.) Try to modify them, too. (You won't be able to save your changes under the original program name because they are marked "Read-Only.") This is a great way to get a feel for what programming is all about without actually having to "learn" programming in a formal manner. Still, our guess is that, once you get started with programming, you'll discover just how much enjoyment can be derived from it and you'll want to learn more about programming on your own. Be sure to spend some time with the Arduino library reference pages. You already have hundreds of prewritten functions waiting to do work for you. You should spend a little time looking at them so you don't inadvertently reinvent the wheel.

CHAPTER 3
The LCD Shield Project

An important aspect of any µC system is the ability to display data in either text or numeric form. As you learned in Chapter 2, Step 4 (Output) of the Five Program Steps is designed to display the results of a program. One of the most common methods to display µC results is to output these data to a Liquid Crystal Display or "LCD." LCDs are inexpensive and available in many configurations, from an array of dot-matrix characters to bit-mapped color displays with built-in touch screens. As our first construction project, we chose to assemble a 16 × 2 dot-matrix LCD display.

We chose the LCD display as a first project for several reasons. First, the construction and software for this project is simple and a good starting point for our discussion of amateur radio projects. Second, this display is used in several subsequent projects that only need one or two lines of text for display. Third, the overall size of the 16 × 2 LCD display is small, thus lending itself to small enclosures that are compatible with many QRP rigs. Finally, these displays are easy to find and fairly inexpensive, usually costing less than $5. You could modify the design presented in this chapter for a 20 × 4 display without too much difficulty, but the display would be somewhat larger and two or three times more expensive. The finished display shield is shown in Figure 3-1.

FIGURE 3-1   The completed LCD shield piggybacked onto an Arduino.

The Arduino development system is designed as a flexible prototyping environment. The Arduino board has a series of connectors around the edges that may be used to connect to *"shields,"* the term used in the Arduino community to describe "add-on" modules. If assembled properly, shields can be plugged into the Arduino and stacked one on top of the other, further enhancing the capabilities of the system.

Libraries: Lessening the Software Burden

The LCD shield is designed to use the standard Arduino software libraries that are distributed as part of the Arduino's IDE installation. In Chapter 2 you learned that functions are the basic building blocks from which applications are created. You can think of a software library as a collection of functions designed for some specific purpose. In many ways, an Arduino library is much like a single book. The book has a title that reflects its general area of interest (e.g., LCD). Opening the book and inspecting its table of contents (TOC), you discover more details about the book's general area of interest. Each chapter title in the book's TOC provides more information about the general area of interest (e.g., *setCursor()*, *write()*, *clear()*). By reading each chapter, you learn how to use the LCD feature that is the subject of that particular chapter. Each chapter provides you with the details you need to know on how the functionality of that particular chapter is used within your own programs.

The Arduino environment provides you with dozens of libraries. Some are shipped with the IDE, others are available as free downloads from various web sites. Later chapters use some of these additional libraries. Whenever we use a "non-IDE" library, we tell you where you can find the library for download. It's worth your time to read about the libraries that already are provided with the Arduino IDE (see http://arduino.cc/en/Reference/Libraries). Knowing what the libraries contain helps prevent you from reinventing the wheel. Also, any time you think you need to write your own library because you don't think a library exists for that area of interest, do a Google search first. You'll be surprised how often you'll be able to stand on the shoulders of someone else who has already done most, if not all, of the work for you. That's what Open Source is all about ... sharing your work.

In this chapter, you use the LCD software library, named LiquidCrystal, that is distributed with the Arduino IDE. At the completion of this project, you will have learned how to assemble the shield, use the correct LCD library functions, and execute the "HelloWorld" example program sketch that is included with the LCD library. You will also learn how to use various library functions to display different data types. Virtually all of the software used in this project is already written for you.

Not All LCDs Are the Same

An important thing to note is that the LCD display libraries are written to use a specific LCD controller chip, the HD44780. Developed by Hitachi, the HD44780 controller greatly simplifies the hardware and software needed to use the LCD display. Make sure that the LCD you use employs the HD44780 controller or its equivalent. If you must use a different LCD controller, make sure the vendor supplies Arduino-compatible libraries with it. Most online vendors make a statement if their LCD is compatible with the HD44780 controller. If you don't see such a statement in their ad, you should write and ask if their LCD is compatible with the HD44780 and, if not, do they supply Arduino-compatible libraries. While there is nothing to prevent you from writing your own library for a noncompatible display, clearly more time and effort on your part is required to do so. However, if you do write a new library, we encourage you to make it available to Open Source.

As pointed out earlier, we chose to use an LCD that has 2 rows of 16 characters (commonly referenced as a 2 × 16 LCD). The displays are available in other form factors; more rows, or more characters. It is perfectly fine to use one of these other displays but you will have to modify our program code to accommodate the different form factor. There are articles in the Arduino Forums that should help you to use these other LCD formats (see http://forum.arduino.cc/index.php?board=7.0).

Also be aware that some LCD displays do not have backlighting. *Backlighting* refers to the illumination that appears as a background for the LCD characters. Without backlighting, the display depends upon ambient lighting to provide enough contrast to read the display. While backlighting of the LCD display is not required, we think it is definitely worth having. Nonbacklit displays are a little cheaper, but not enough to make it worthwhile. We suggest you always purchase backlit displays.

LCD Shield Parts List

All of the parts needed to assemble the LCD shield are presented in Table 3-1.

Clockwise from the upper left in Figure 3-2, the parts are: R1, a 10 kΩ potentiometer, a prototyping shield to hold the parts, header pins for connecting the shield to the Arduino, the LCD display module, and the 16-pin header for connecting the LCD to the shield. Note that in the construction projects presented in this book, a potentiometer (or "pot") is a small, printed circuit board (PCB) type you adjust with a screwdriver, not the larger type you find on audio equipment that you adjust with a knob. Also, there is considerable variance in the cost of parts because some suppliers, especially foreign suppliers like those often found on eBay, are extremely inexpensive. The downside is that you may have to wait a while longer to get your parts since they often come from China or Thailand. Because you probably will not want to stack another shield on top of the LCD display, we recommend using the "breakaway" headers to interconnect this shield to any shield that may be under it. You can see these headers near the top of Figure 3-2. However, you can optionally make the LCD removable using stackable headers.

Prototyping shields come in many forms as can be seen in Figure 3-3 and are available preassembled or as a kit. The kit gives you some flexibility in how the connectors are installed, either as a stacked shield (i.e., one built to connect directly to the Arduino board) or not. A prototype shield, or *proto shield*, has a number of isolated holes into which you can insert components. The holes are "plated-thru" and allow you to solder your components to the board. *Plated-thru holes* is a term that means that the holes that accept the components on the board are all lined with conducting material so the connections extend from the top of the board through to the other side of the board. Some boards may have holes that are interconnected with other holes, which can help simplify wiring. Some have pads for installing Dual In-line Package (DIP) or Surface Mount Devices (SMD) or other parts, while others duplicate the layout of a solderless prototyping board as described in Chapter 1. These solderless prototyping boards often have two "buses"

Ref	Description	Source (See Appendix A)	Cost
	Arduino prototype shield (includes headers, reset switch)	eBay, Maker Shed, Adafruit, SparkFun, ArduinoMall	$3–$15
DISPLAY	2 × 16 HD44780 LCD	eBay, Maker Shed, Adafruit, SparkFun, Jameco, Seeed Studio	$3–$20
	16-pin "breakaway" header, .1 in. centers	eBay	
R1	10 kΩ potentiometer, PC mount	eBay, Radio Shack, others	< $2
R2	220 Ω, ¼ W resistor	eBay, Radio Shack, others	< $0.25
	Hookup wire, 26-28 AWG solid	Radio Shack, others	

TABLE 3-1   LCD Shield List of Parts

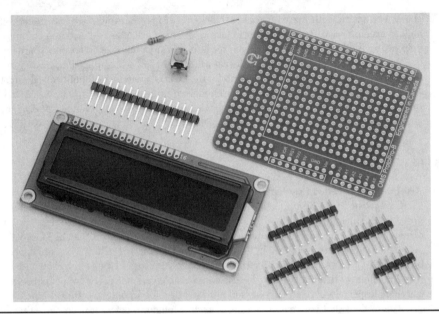

FIGURE 3-2   LCD shield parts.

(i.e., heavy lead traces) running down the center (perfect for V+ and ground) and columns of interconnected pins. If you look closely at Figure 3-3, you can see buses on the boards at the top-left and lower-right side of the figure.

Selection of a prototyping shield depends on the nature of the project you want to build. For instance, how many parts are needed for the project? Will all of the parts fit on a single board? Do

FIGURE 3-3   Examples of prototyping shields.

you need to mount any SMD parts? And let's not forget personal preferences. We have an affinity toward using the shields that are similar to the solderless prototyping boards so our choice would be this style of proto shield in kit form. The shield we are using for this and many other projects in this book come from Omega MCU Systems in Canada (see http://oms.onebytecpu.com). We obtained the proto shield for this project from eBay as a bare board.

Assembling the LCD Shield

The LCD shield is shown as a schematic diagram in Figure 3-4. The schematic shows how the LCD shield parts are connected together.

One thing we find very helpful in assembling a shield is a layout drawing. If you have access to some type of drawing package or even if you use a sheet of quad-ruled paper, laying the parts out ahead of time will save you time in the long run. Make sure that there are enough solder points for the LCD header and that the 10K pot can be installed so it doesn't interfere with the display or is hidden behind it. There is nothing so painful as assembling a number of components on a prototyping shield and then realizing the last part won't fit. Reworking the parts placement is time consuming, you risk damaging the shield and the components, and is generally zero fun. It is much better to take your time, sketch out a layout, and test-fit the components prior to ever applying the soldering iron. We have used Microsoft Visio to create the layout examples for this book. You can also do an Internet search for free schematic drawing downloads and find a number of free drawing tools, such as TinyCAD (http://sourceforge.net/projects/tinycad/).

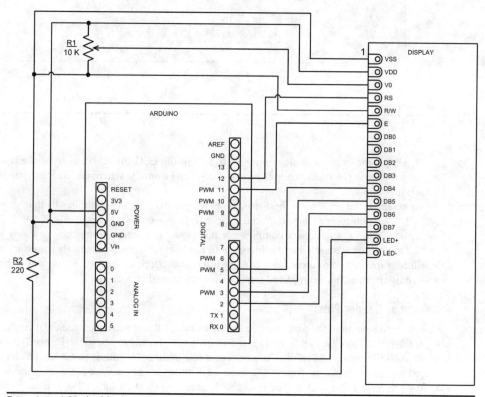

FIGURE 3-4 LCD shield schematic.

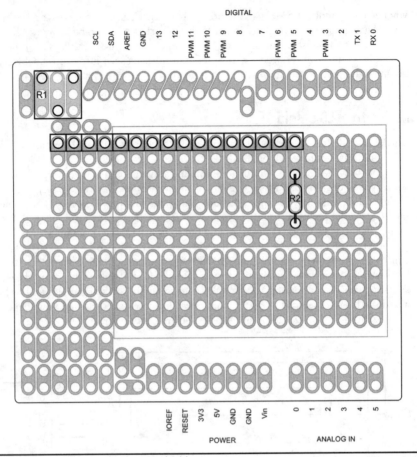

FIGURE 3-5   A layout drawing for the LCD shield.

Figure 3-5 shows an example of one way to lay out the LCD shield. We also use the layout to help with the wiring by transferring the interconnections from the schematic diagram as is shown in the layout drawing.

To assemble the shield you first need a "bare" prototyping shield similar to that shown in Figure 3-2. An example of a bus mentioned earlier can be seen near the center of Figure 3-5. The buses make it easier to connect components that share a common connection (e.g., +5 V and GND). Once a part is inserted and soldered onto the board, do not immediately trim the leads as these leads provide a convenient place to make interconnections with other components. Trim the excess only after you have completed the last connection to that part.

Breakaway Header Pins

Now that you have all of the parts needed for the LCD shield and you have a rough layout, you can begin assembly. The first step is to prepare and install the breakaway headers. Figure 3-6 shows a standard 40-pin breakaway header. You can use a pair of diagonal cutters to cut the header to the desired length as shown in Figure 3-7. Figure 3-8 shows a comparison between stackable and standard headers and a socket. The stackable header (top right in Figure 3-8) not only provides connection to the Arduino or shield underneath, but also allows another shield to be inserted from

Chapter 3: The LCD Shield Project   41

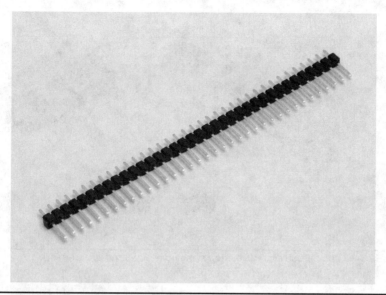

FIGURE 3-6   Standard 40-pin breakaway header.

the top. We make use of this feature in future projects, but for this project you might consider using the socket header (top left in Figure 3-8) for mounting the LCD. This allows easy access to the area under the LCD should you want to add some circuitry at a future time. It really beats trying to unsolder the LCD, which we find similar to putting toothpaste back into the tube.

The first step in assembling the LCD shield is to solder the headers that connect the shield to the Arduino. Note that in Figure 3-6 the header pins have one side with a fairly long lead, while the other side of the plastic has a short, stubby lead. The best approach for soldering the headers to the shield is to insert the longer side of the header pins into your Arduino header sockets as shown in Figure 3-9.

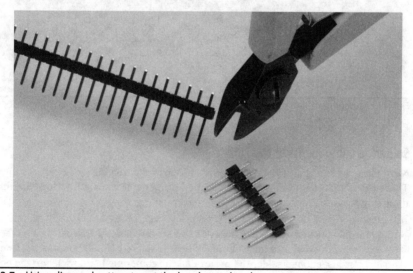

FIGURE 3-7   Using diagonal cutters to cut the breakaway header.

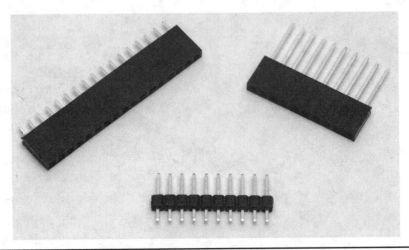

FIGURE 3-8 Comparison of stackable and standard headers and a header socket.

FIGURE 3-9 Using Arduino to hold proto shield header pins.

Soldering Components to the Shield

Next, place the shield over the stubby pins and start soldering the stubby pins to the proto shield. This ensures that the header pins area aligned properly. If you are new to soldering, just follow these simple rules:

1. Apply the heat to the component being soldered, not the solder.
2. Apply the solder to the component being soldered, NOT the soldering iron.
3. Use just enough solder to flow around the component and wick into the hole.
4. Allow the parts to cool before moving them and especially before touching them.

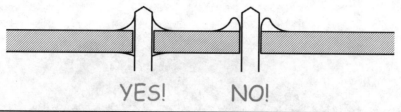

FIGURE 3-10   Comparison of good (left) and bad (right) solder joints.

Figure 3-10 shows what good and bad solder joints look like. A good solder joint is bright and shiny and fully "wets" the surfaces being soldered together, say, a pad on a prototype board and a component lead. A bad (cold) solder joint appears to have a grainy surface and looks more like the joint on the right in Figure 3-10. A bad solder joint doesn't adhere to the component lead and the lead may in fact be loose.

Also, beware of solder "bridges." As the name implies, a *solder bridge* is where two adjacent pins are connected by extra solder causing a short circuit, hence the term "bridge." The bridges are created because of the surface tension of the liquid solder. The liquid solder tends to want to naturally create bridges! Solder bridges are sometimes hard to control but are easy to avoid and easy to clean up. Avoid bridges by only using enough solder to make the connection. Any extra invariably leads to trouble, like a solder bridge. If you should happen to create a solder bridge, clean it up using solder wick, a "solder sucker," or even get the part hot with the iron and give the board a sharp "rap" on the work surface to remove the excess solder. Be careful with the latter technique as it doesn't work well with delicate components and the excess solder always seems to seek bare skin. The latter has the same effect as catching a small meteorite with your bare hands ... not good.

Figure 3-11 shows the proto board in place on top of the header pins, with the "stubby" ends of the pins protruding through the plated-thru holes. When you have soldered all of the stubby pins to the board, gently rock the shield slightly as you pull the proto board from the Arduino

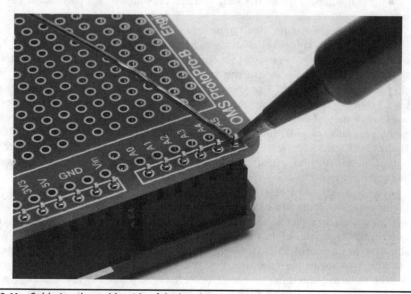

FIGURE 3-11   Soldering the stubby side of the header pins to the shield.

FIGURE 3-12 LCD header, potentiometer, and resistor placement.

board below. When you have separated the two boards, you will see the longer ends of the header pins sticking out from the bottom of the proto shield.

Figure 3-12 shows our parts placement for the LCD project. Compare Figure 3-12 with Figure 3-5 to see how the components are placed on the shield. It is best to add one part at a time to the shield, soldering as you go along.

Also note that in Figure 3-12 you can see the long pins "below" the board and the stubby ends of the pins are on "top" of the board. For this reason, we refer to the side with the stubby pins showing as the "top" of the proto shield, and the side with the long pins showing as the "bottom" side of the proto shield.

Insert the resistor in the appropriate location and solder it in place as shown in Figure 3-12 (and Figure 3-5), making sure that the body of the resistor is flush against the shield. You can use a pair of needle nose pliers to pull the component lead ever so slightly on the bottom side of the shield and gently bend it to one side (see Figure 3-13). This holds the resistor in place long enough to solder the leads. Once a part is inserted and soldered to the shield, do not trim the leads as we may use them in making interconnections to other parts. Trim the excess only after you have completed the last connection to that part. (If you are using a kit that came with a reset pushbutton this is a good time to insert and solder the pushbutton in place.) You can see in Figures 3-11 and 3-12 how the holes along the top edge where our header pins are installed are interconnected on this particular shield. We will make use of this to reduce the number of wires and complexity.

The next step is to solder the 16-pin header you just cut from the 40-pin strip to the shield. (You can see the placement of the 16-pin header near the top of Figure 3-12.) There is no "best way" to solder this header in place, but the method we use is to insert the stubby pins of the 16-pin header into the top side of the shield, flip the shield over to the bottom of the shield, and rest the shield on the workspace, letting the weight of the shield hold the 16-pin header in place. As shown in Figure 3-14, solder one stubby lead near the center of the strip (e.g., stubby lead number 7) and then lift the board to examine the header to see if it is perpendicular to the shield. If it is, solder the remaining stubby leads. If the shield is not perpendicular, then reheat the soldered pin while using

Chapter 3: The LCD Shield Project   45

FIGURE 3-13   Soldering the resistor to the shield.

a free finger to adjust the header by moving it with one of the end pins. This approach keeps you from burning your finger! Apply a little pressure on the end pin and heat the soldered pin while pushing the header into place. With the header properly aligned, you can solder the rest of the stubby leads. Eventually the LCD is installed on these pins, but don't do that just yet. We'll discuss mounting the LCD a little later on. We have some more work to do first.

Continue by inserting and soldering the potentiometer in place. Once again, you can use the technique of soldering one lead and pressing the part into place to get a flush fit. Once you are happy with how the potentiometer looks, solder the remaining two leads.

FIGURE 3-14   Soldering the 16-pin header for the LCD.

Adding Components Using a Schematic

Review the schematic in Figure 3-4 and familiarize yourself with the Arduino IO pins and the LCD. If you are new to wiring from a schematic, one helpful trick is to use a colored highlighter pen to mark the wires you have installed on the schematic. (Yellow is a good color choice for this. Black ... not so much.) This helps to eliminate wiring errors. As mentioned earlier, showing the interconnections from the schematic on the layout drawing you have made is also very helpful. This prevents you from making other mistakes. Figure 3-15 shows the interconnecting wires to be added to the shield. Note that the wiring diagram is reversed from the parts placement diagram, Figure 3-5.

Begin adding the connecting wires. Speaking of wires, the gauge of the wire you select can make things more or less difficult to fabricate. Clearly, the shield works with low voltage and amperage, so there's no need for Romex cable. We prefer 24 AWG wire because it is easily bent for routing and quick to solder. Specifically, many of the shields shown in this book are wired with 24 AWG solid, bare wire and then covered with Teflon tubing of the appropriate size. With one end of the wire soldered in place, we cut the Teflon tubing to the needed length and slip it over the wire and then solder the remaining connection. This wiring technique makes for a clean, finished

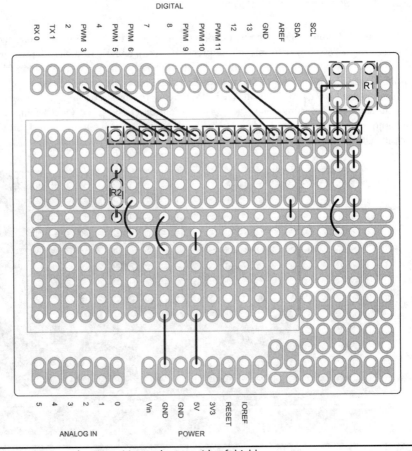

FIGURE 3-15 Diagram showing wiring on bottom side of shield.

FIGURE 3-16   Detail of wrapping a wire around a header pin prior to soldering.

product as the Teflon tubing will not burn or melt when soldering the wire. Heavier gauge wire (e.g., 20 AWG) works fine, but is considerably thicker and more difficult to route. Another option is to use colored wires to denote different signals, for instance, red wire for the positive voltage lead and black wire for ground. Obviously, this is not a requirement, but it can make it easier to visually trace a circuit if something goes wrong.

There is no right way to wire a prototype board. Each builder adds their own individual touches. The sequence of photographs in Figures 3-16 through 3-19 shows one method for connecting a wire to an LCD header pin. Figure 3-16 shows a connection being made to the Arduino header for digital IO pin 4. The prototyping shield we are using here conveniently has holes connected to the adjacent header pins connecting to the Arduino. On the side of the shield away from the header we first solder a short piece of wire to the hole adjacent to Arduino IO pin 4.

FIGURE 3-17   Soldering the wire ...

FIGURE 3-18   ... Soldered ...

We then cut a piece of Teflon tubing to the proper length and slip it over the wire. The loose end of the wire is then wrapped around the LCD header pin using a pair of needle nose pliers and then clenched to hold it in place. Figure 3-17 shows solder being applied to the connection with Figure 3-18 showing the pin after soldering. Because no additional components are wired to this component, the excess wire is trimmed away as in Figure 3-19 and the connection is completed.

The process is repeated for the remaining connections. Each subsequent wire is cut to the proper length with Teflon tubing for insulation. We do this so as not to cause a short to the holes

FIGURE 3-19   ... And trimmed!

Chapter 3: The LCD Shield Project 49

FIGURE 3-20 Completed wiring for the potentiometer, jumpers, and LCD header.

the wire may cross over. Once the wire is in place it can be soldered and trimmed flush on the other side of the shield.

Add the jumper wires to the bottom of the board as shown in Figure 3-15. Note that two adjacent holes connect with a bare jumper but for any other connection we use an insulated wire. Many times we can use the component leads (resistors, capacitors, etc.) to make the interconnections. Note how the jumpers take advantage of the existing circuits etched on the shield. When completed, the shield should look something like what is shown in Figure 3-20.

Once you have completed the wiring and before installing the LCD it is a good idea to check your work. A continuity tester is useful in this task. You can purchase a cheap multimeter for less than $10 and use it to check the resistance measurements for shorts. But lacking that, a visual inspection is sufficient. Since both of us are two years younger than dirt, we find a magnifying glass useful when checking for solder bridges. Check against the schematic diagram to see that each wire is connected to the correct points on the shield.

When you are confident that the shield is correctly wired, install the LCD. You should have decided by this point whether to permanently install the LCD or use header socket to make the display removable. If you are making it permanent, then set the LCD over the header pins and support it so that it doesn't touch the shield itself. You can use a spacer between the LCD and the shield such as a piece of cardboard. Go back to Figure 3-1 to get an idea of what this should look like. Hold the display in place on the pins and solder just one of the pins somewhere near the middle. Now you can adjust the display to make it level and even on all of the pins, just as you did when first installing the headers on the shield. Once you are satisfied with the display's position, solder the remaining pins. Again, watch for solder bridges shorting adjacent pins. Figure 3-21 shows the header pins being soldered to the LCD. Figure 3-22 shows how the LCD should look after being soldered in place.

If you decided to make the LCD display removable, then you can use the same technique we used for installing the LCD header to the shield. Rather than a header, we use a header socket, as shown in Figure 3-8, to slip the LCD over the header installed on the shield. Insert the header socket pins from the back side of the LCD and solder one pin toward the center, align the LCD, and

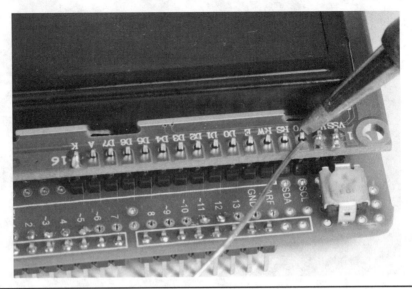

FIGURE 3-21   Soldering the LCD display into place.

then solder the remaining pins, as shown in Figure 3-23. The completed shield with removable LCD is shown in Figure 3-24.

As a final touch after assembly, it is a good idea to remove the solder flux from your project. The flux residue is mildly sticky and will attract dirt and contaminants. It also detracts from the appearance of the finished project and that may be the greater reason. The easiest way to remove the flux residue is using laboratory-grade Isopropanol, or anhydrous isopropyl alcohol. This is not the stuff you get at the local pharmacy, "rubbing alcohol," which is mostly water. Anhydrous isopropyl alcohol has very little water content and is ideal for removing flux. Apply the alcohol with a stiff brush, such as an "acid brush," wiping the flux downward into a rag. Be generous with

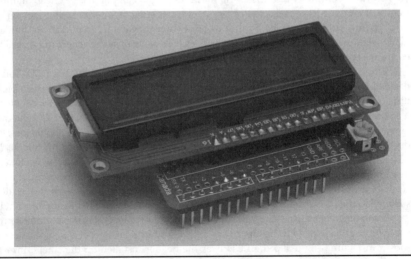

FIGURE 3-22   LCD mounting completed.

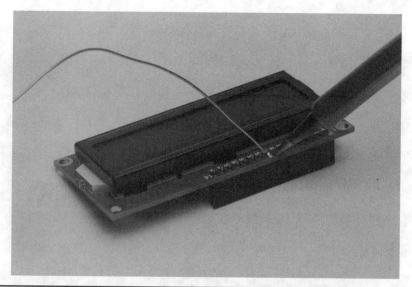

FIGURE 3-23   Soldering a header socket to the LCD.

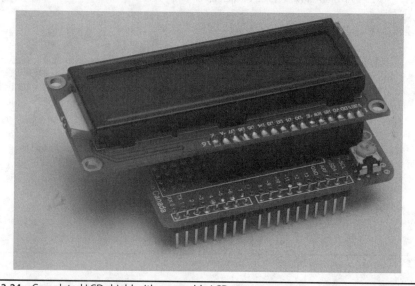

FIGURE 3-24   Completed LCD shield with removable LCD.

the alcohol. This type of alcohol, as well as the brushes, can generally be found at a good paint or hardware store. You will be pleased with the results.

An Alternative Design

As we mentioned earlier, there is no one "right way" to assemble the projects in this book. We provide examples of ways that they can be assembled. Here we show an alternative method using a different prototyping shield. We won't go into the details of the construction, but the photographs

FIGURE 3-25 An alternative version of the LCD shield.

FIGURE 3-26 Side view of the alternative design.

in Figures 3-25–3-27 show different views of the alternative construction. In this example, the shield is constructed in such a way that it provides very few "built-in" wiring traces. As a result, all of the circuitry is added with individual wires. You can see them on the side view as they connect the LCD header to the Arduino digital IO header.

Loading the Example Software and Testing

All that remains is to load a test program and see that your work has been successful. You may need to adjust the 10K pot to get the correct contrast on the screen. The actual setting varies depending on the individual LCD, ambient lighting, and other factors.

FIGURE 3-27   Bottom view of the alternative design.

Make sure your PC is connected to the Arduino board using a USB cable as discussed in Chapter 1. Now plug the LCD shield into the sockets on the Arduino board. You should see a series of "blocks" across the top row of the display. If not, adjust the LCD potentiometer until you see the blocks. The actual setting will vary depending on the individual LCD, ambient lighting, and other factors. Figure 3-28 shows what the display should look like when the contrast is correctly adjusted.

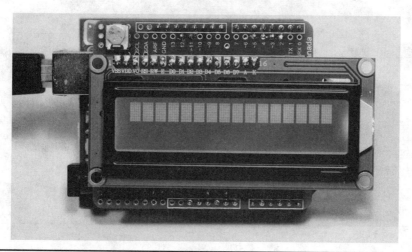

FIGURE 3-28   Adjusting the display for proper contrast.

Software to test the display is included with the Arduino IDE. Start the Arduino IDE and use the File → Examples → LiquidCrystal → HelloWorld menu sequence to locate "LiquidCrystal" and select the entry "HelloWorld." This loads the program shown in Listing 3-1 into the Arduino IDE.

The HelloWorld program distributed with the Arduino IDE contains the following statement:

```
LiquidCrystal lcd(12, 11, 5, 4, 3, 2);
```

The numbers in the statement are pin assignments and are quite arbitrary. While we might prefer different pin assignments, we are sticking with the same pin assignments provided in the distribution software. However, some pins have special purposes. For example, pins 2 and 3 can be used to service external interrupts in addition to normal digital IO functions. Pin 5 can be used as an internal counter/timer. When conflicts arise, as might be the case for a program requiring an external interrupt, pin assignments may need to be changed. The software must reflect those changes.

Once you see the source code for the program appear in the IDE source code window, edit the *lcd* definition statement, then click the check mark icon (just below the File menu option) to compile the program. Fairly quickly a message appears on your PC near the bottom of the screen telling you the compile process is complete. The message also tells you the program's size in bytes and the remaining unused program space. Now click the right-pointing arrow icon (below the Edit menu option) to copy the compiled program into the Arduino's memory via the USB cable. (As an alternative, you can ignore clicking the check mark and directly click the right-pointing arrow. Doing so recompiles the source code and uploads the compiled program with a single click.)

After the upload completes, the IDE tells you the upload is complete and you can now see your handiwork come to life! Your results should look just like Figure 3-29. An explanation of the program follows source code Listing 3-1.

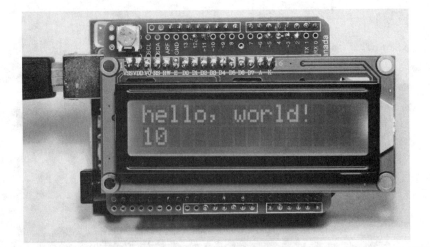

FIGURE 3-29 "Hello World!"

```
/*
 LiquidCrystal Library - Hello World

 Demonstrates the use a 16x2 LCD display. The LiquidCrystal library works with
 all LCD displays that are compatible with the  Hitachi HD44780 driver. There
 are many of them out there, and you can usually tell them by the 16-pin
 interface.

 This sketch prints "Hello World!" to the LCD and shows the time.

  The circuit:
 * LCD RS pin to digital pin 12
 * LCD Enable pin to digital pin 11
 * LCD D4 pin to digital pin 5
 * LCD D5 pin to digital pin 4
 * LCD D6 pin to digital pin 3
 * LCD D7 pin to digital pin 2
 * LCD R/W pin to ground
 * 10K resistor:
 * ends to +5V and ground
 * wiper to LCD VO pin (pin 3)

 Library originally added 18 Apr 2008
 by David A. Mellis
 library modified 5 Jul 2009
 by Limor Fried (http://www.ladyada.net)
 example added 9 Jul 2009
 by Tom Igoe
 modified 22 Nov 2010
 by Tom Igoe

 This example code is in the public domain.

 http://www.arduino.cc/en/Tutorial/LiquidCrystal
 */

// include the library code:
#include <LiquidCrystal.h>

// initialize the library with the numbers of the interface pins
LiquidCrystal lcd(12, 11, 5, 4, 3, 2);

void setup() {
  // set up the LCD's number of columns and rows:
  lcd.begin(16, 2);
  // Print a message to the LCD.
  lcd.print("hello, world!");
}
```

LISTING 3-1   The HelloWorld program.

```
void loop() {
  // set the cursor to column 0, line 1
  // (note: line 1 is the second row, since counting begins with 0):
  lcd.setCursor(0, 1);
  // print the number of seconds since reset:
  lcd.print(millis() / 1000);
}
```

LISTING 3-1   The HelloWorld program. (*continued*)

A "Code Walk-Through" of the "HelloWorld" Sketch

Programmers use code walk-throughs as a more formal way to inspect the source code of a program. When Jack had his software company, he would assign clearly defined tasks to various programming teams. Most of these programming tasks had one- to two-week time frames, and all tasks were presented to all of the programmers on a Friday morning code walk-through, usually on staggered Fridays. The idea was for a different set of eyes to read and comment on the code. If the code "passed muster," the company bought pizza for lunch and *all* of the programmers got the afternoon off. The code walk-throughs provided several benefits.

First, when you write code, often times you are "too close" to the code to see a bad design or a potential problem. It's sort of like someone pointing out that your firstborn has a big nose; it's the elephant in the room but you don't see it. The code walk-through helps uncover potential problem areas before they actually become problems. Second, by making all programmers attend the meeting, you get a fresh perspective on the task at hand and the code designed to solve it. Often a non–team member came up with a design change that significantly improved the code. A third benefit is that everyone is at least somewhat familiar with what the other teams are doing. This was significant to Dr. Purdum's company because it was always a small company and when someone took a week's vacation and someone else had to fill in, at least those substitutes weren't starting from ground zero. Finally, the code walk-throughs fostered an *esprit-de-corps* within the company because of the pizza-and-half-day-off carrot if the team succeeded. It reinforced the idea that one team's success means everyone succeeds. You'd be amazed how often all of the programmers were at the office late Thursday night to help a team prepare for the code walk-through the next day even if it wasn't their project that had the pending walk-through.

While you may not benefit in the same way from a code walk-through, we encourage you to read the discussion of the software issues even if you're not interested in doing your own programming. If nothing else, it helps you understand what the software is doing and how it is done. Indeed, it may be that you can design a better hardware solution by understanding the software. Simply stated, the discussion of each project's software element is like a code walk-through and helps you to better understand what the project is all about.

The HelloWorld program consists of several parts. As you already learned, the text between the first "/*" and "*/" is a multiline comment and everything between these two comment pairs is ignored by the compiler. In this particular example, these comments describe the circuit as you have built it and the names of the programmers who developed the LCD library. The symbol "//" is another comment, in this case a single line.

The *#include* in the line:

```
#include <LiquidCrystal.h>
```

is called a preprocessor directive. Unlike C statements, *preprocessor directives are not terminated with a semicolon.* Only one preprocessor directive can appear on a line. The *#include* directive instructs the compiler to look in the appropriate library directory for the LiquidCrystal.h header file. If the line were written:

```
#include "LiquidCrystal.h"
```

(note the double quotation marks instead of the angle brackets), the compiler looks in the current working directory for the header file rather than the library directory.

The code found in the LiquidCrystal library contains a number of functions that can be called from the program, greatly simplifying the amount of code we have to write. Basically, the hard work has already been done for us.

The next statement:

```
LiquidCrystal lcd(12, 11, 5, 4, 3, 2);
```

is doing a lot of work for you using Object Oriented Programming (OOP) techniques. Think of the Arduino library as a room with a huge wall covered with hundreds of cookie cutters. One of those cookie cutters has the label LiquidCrystal etched on it. Another one has Stepper etched on it. Yet another has WiFi etched on its surface. Each cookie cutter represents a class that is available for you to use in your programs. You can think of a *class as a container that holds information about some kind of object*, an LCD in this example. When you write a statement that begins with a class name (LiquidCrystal) followed by a space, and then another variable name (*lcd*), you are actually calling a special C++ function called a class constructor whose job it is to carve out a chunk of memory according to the specifications of the class you are using.

In the statement above, you are taking the LiquidCrystal cookie cutter from the wall, pressing it into some cookie dough (actually, the dough is the Arduino memory), removing the new cookie (called an object of the class) and putting the label *lcd* on the object. The process of creating a class object is called *instantiation*. The numbers within the parentheses tell the class constructor how you want to initialize certain variables that are buried within the LiquidCrystal class definition. (The long comment at the top of Listing 3-1 gives you some insight as to the meaning of the numbers. We provide some additional information about these numbers in the next paragraph. We also discuss OOP techniques in greater detail in Chapter 7.)

During the process of executing the statement, you have used the LiquidCrystal class to create an instance, or object, of the LiquidCrystal class named *lcd*. If you had used the statement:

```
LiquidCrystal lcd();
```

to call the LiquidCrystal constructor, all of the members of the LiquidCrystal class would have been initialized with default values; zero for numeric variables and null ('\0') for character arrays or other objects. However, since you chose to create the *lcd* object with six parameters, those values are stuffed into the appropriate member variables of the LiquidCrystal display object, *lcd*. If you look at the lengthy comment at the start of the program, you find:

```
 The circuit:
 * LCD RS pin to digital pin 12
 * LCD Enable pin to digital pin 11
 * LCD D4 pin to digital pin 5
 * LCD D5 pin to digital pin 4
 * LCD D6 pin to digital pin 3
 * LCD D7 pin to digital pin 2
 * LCD R/W pin to ground
```

which gives you a pretty good idea of what members of the Liquid Crystal display call mean.

You learned about *void setup()* in Chapter 2. The statement:

```
lcd.begin(16, 2);
```

actually goes to the *lcd* object you created in memory, and calls the *begin()* function that is defined within the LiquidCrystal class. Because the LiquidCrystal class can cope with several different display types, the parameters of our display (e.g., 16 characters in 2 rows) uses the *begin()* class method call to pass our display's information to the class. (If you listen to a conversation among a cadre of OOP programmers, you will likely hear them refer to class methods. *Methods* are almost the same as functions, but they are buried within the class object. We may use the terms *methods* and *functions* interchangeably throughout this text since they serve a similar purpose.)

And now to output some text. The *print(value)* function (or method) call sends whatever is stored in the variable named *value* to the display. In this case, a sequence of characters (also called a string) is displayed, starting at the default cursor position, row 0, column 0. This call:

```
lcd.print("hello, world!");
```

actually sends the text string to the display object. While a simplification, you can think of the statement as going to the *lcd* object in memory, then looking for the memory address of where the *print()* method is stored within the object, and then have program execution branch to that memory address and execute the code found there. Note that, as explained in Chapter 2, since this first set of instructions is part of *setup()*, it is only executed once at program start-up. The type of data held in *value* and sent to the display can be either a text string or a numerical value. The double quotation marks ("") symbols denote textual data. In the case of a numerical value, it can be the result of another call as in the second instruction below or just a numerical value.

The *setCursor()* function call allows you to specify where information is to be written to the LCD display. The instruction:

```
lcd.setCursor(0, 1);
```

places the cursor at the first position in the second row. Remember that row–column coordinates are zero-based values, so the *setCursor()* function call above sets the column at position 0 and row 1 is actually the first column in the second row of the display.

```
lcd.print(millis() / 1000);
```

The *lcd.print()* function call displays an incrementing value at that position. The value is incremented once per second.

It is important to note that the function calls:

```
lcd.print("hello, world!");
lcd.setCursor(0, 1);
lcd.print(millis()/1000);
```

are using functions (methods) that are buried inside the *lcd* object. In other programs, you might see something like:

```
Serial.print("hello world:");
```

In this case, the statement is using the *print()* method that is buried inside the *Serial* class object. The two *print()* methods are quite different. Indeed, they have to be because our *lcd* object must configure the data for display on the LCD, while the *Serial.print()* method must configure the data to first travel over the USB connection and then display the data on your PC's monitor.

Just keep in mind that the period, called the *dot operator* in C++, you see between the object name and the member name or the function name serves the purpose of an "information booth" for the object. For example, you can think of the statement:

```
lcd.setCursor(0,1);
```

as telling the program: "Go to the *lcd* chunk of memory and find its information booth. When you get there ask the person in the information booth for the memory address for the *setCursor()* method. Once you get to the memory address the information booth gave you, dump your data (e.g., 'HelloWorld') in their lap and they will take it from there."

At this point it is worth the time to "play around" a little bit with the LCD and the associated software. Take some time to explore the Web link given in the listing:

http://www.arduino.cc/en/Tutorial/LiquidCrystal

This reference contains detailed explanations of the methods described above as well as the other methods that have not been used yet, but are part of the LiquidCrystal library. Now that you have a piece of hardware that you built yourself up and running, try making modifications to the example program source code in Listing 3-1. For instance, try placing the text in a different position or even output a second line of text rather than the counter. Performing little experiments like this helps make you more comfortable with making changes to the existing code.

Explore the Other Examples

You probably noticed a number of other example programs within the LiquidCrystal library. Take the time to load these and run them. Again, you can experiment with the examples and see how changing various values or parameters change things. If the code doesn't run the first time, try to figure out why. (Did you remember to adjust the pin assignments for your display?) We use some of these additional *lcd* library methods in subsequent projects.

Congratulations! You have successfully completed your first Arduino shield and should now have a working LCD shield! In addition, you learned about the LiquidCrystal library function calls and how they work.

Using Your LCD Display with the TEN-TEC Rebel

In this section we want to show you how easy it can be to interface your LCD display into an existing piece of equipment. The piece of equipment we chose is the Rebel from TEN-TEC. This QRP rig has a long list of nice features for a two-band transceiver. However, the thing we *really* like is that it is Open Source software that helps make things tick inside of it. The beauty of this approach is that, if TEN-TEC or someone else comes out with a nice feature, you don't have to shoe-horn a new board inside the rig. You just change the software. That's what we do in this section: We change the software inside the Rebel so it can use your new LCD display.

The first step is to download the software you'll need to modify the Rebel code. The Rebel source code can be downloaded from:

https://groups.yahoo.com/neo/groups/TenTec506Rebel/files/TEN-TEC%20Repository/Official%20TEN-TEC%20Code%20Repository/

The name of the file is Rebel_506_Alpha_Rev02.pde. This is the origin source code from TEN-TEC. Note that you should keep a copy of this code in its original form in case you want to reload it at some time in the future. Obviously, TEN-TEC isn't going to be responsible for what you do once you overwrite their software. After you download the source code and unzip it, you'll notice that it is a *.pde file, not an Arduino-compatible *.ino file. The different file type segues us into the second step.

The second thing you need is the IDE for the chipKIT UNO32 µC that drives the Rebel. Strictly speaking, the chipKIT UNO32 is not a compatible Arduino µC. It is considerably more powerful and has features like 128K of Flash memory, 16K of SRAM, 42 I/O pins, and lopes along at a much faster 80 MHz clock speed. Despite those differences, we can make it talk to your LCD display. You can download the IDE at:

http://chipkit.net/started/

The Digilent IDE, called MPIDE, is available for Windows, Mac OS, and Linux. Make sure you put MPIDE in its own directory. Their IDE looks and feels like the Arduino IDE and you want to keep the two separate.

Under the Rebel Hood

TEN-TEC obviously made the Rebel to be "played" with, as you can see in Figure 3-30. They provide you with direct access to the µC via nicely labeled headers on the top of the main board. Notice that, unlike the Arduino shields, most of the headers are double rows. This is because of

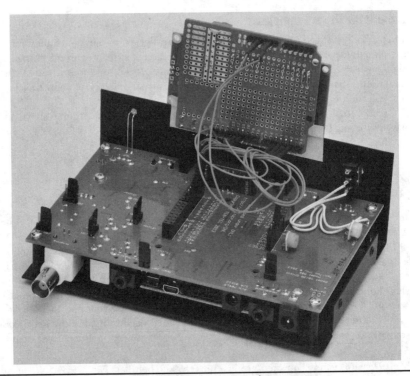

FIGURE 3-30 The rebel with cover off and LCD shield connected.

the higher number of I/O pins available on the UNO32. A USB connector exits the back of the transceiver and, just like an Arduino board, is where you connect the UNO32 to your PC for programming using a USB cable.

As you can see in Figure 3-30, we simply used jumper wires to connect the LCD display to the Rebel temporarily. A more permanent connection could be made using ribbon cable to connect the power and I/O lines from the Rebel to the LCD display. We bought an 18 in. old hard disk drive cable with 40 lines at a flea market for a dime. You could strip an 8 conductor section from that cable (i.e., 2 for power to the shield and 6 control lines) to make an inexpensive way to tie the rig and display together.

Software Modifications

Load the Rebel source code into the MPIDE just like you would an Arduino sketch. The name of the source code we had when this book was written was Rebel_506_Alpha_Rev02.pde. It's quite likely that the name has changed in the intervening months. The line numbers we use in the following paragraphs, therefore, must be viewed as approximations because of changes TEN-TEC may make to later versions of the software.

Now do a search (i.e., Ctrl-F) for "LCD." You will likely end up somewhere around line 63 and see:

```
* LCD RS pin to digital pin 26
* LCD Enable pin to digital pin 27
* LCD D4 pin to digital pin 28
* LCD D5 pin to digital pin 29
* LCD D6 pin to digital pin 30
* LCD D7 pin to digital pin 31
* LCD R/W pin to ground
```

These comment lines provide details on the control and data pins the Rebel expects to be used for an LCD display. Jot these numbers down on a piece of paper or, if you're young, memorize them.

Which Wires to Use to Connect Your LCD Display

Around line 247 you will find the following lines:

```
#include <LiquidCrystal.h>                      // Classic LCD Stuff
LiquidCrystal lcd(26, 27, 28, 29, 30, 31);      // LCD Stuff
```

The lines of code above create a *LiquidCrystal* object named *lcd* for use by the Rebel. More importantly, what this tells you is how to connect your LCD display to the Rebel. In Listing 3-1 presented earlier in the chapter you can find the line:

```
LiquidCrystal lcd(12, 11, 5, 4, 3, 2);
```

which is the code that initializes your LCD display. From the two constructor calls that create the *lcd* object, for our display versus what the Rebel expects, we can construct a wiring table for your display, as seen in Table 3-2. The last two lines provide power from the Rebel to your LCD shield. You should make the connections shown in Table 3-2 to your LCD shield. No change to the code is required at this point. The software is still going to instantiate the *lcd* display object. However, since you are attaching jumpers from their pins to your display, the Rebel could care less who it is actually talking to and outputs the data to your new LCD display.

Rebel Pin #	LCD Shield Pin #
26	12
27	11
28	7
29	6
30	5
31	4
5V0	5V
GND	GND

TABLE 3-2 Rebel-to-LCD Pin Connections

At approximately line 435 you will find the statement:

```
lcd.begin(16, 4);
```

which informs the Rebel that a 16 column by 4 row LCD has been connected to the Rebel. Well, that's not true for our display, so you should change this line to:

```
lcd.begin(16, 2);
```

Your Own Splash Screen

If you would like to add your own custom "splash" screen, you can add lines similar to the following lines immediately after the *lcd.begin(16, 2)* call made in *setup()*:

```
lcd.setCursor(1,0);                       // column 1, row 0
lcd.print(" TEN-TEC REBEL");
lcd.setCursor(1,1);                       // Put cursor on second line
lcd.print("Jane's Display");
delay(1500);                              // Brag for a second and a half...
lcd.clear();      // Clear Display        // ...and then get over it.
```

The code above displays a short splash message about the display. If your name isn't Jane, edit that line and insert your own name (or whatever), but make sure it doesn't contain more than 16 characters. If the line becomes much shorter than 16 characters, you can edit the *lcd.setCursor(1, 1)* line and center the text in the second display line if you wish using:

```
lcd.setCursor((16 - messageLength) / 2, 1);
```

where *messageLength* is the number of characters in your message.

Now find the *loop()* function in the source code. You should find *loop()* near line 503 in the file. Now add the new statement as indicated below to the file after the opening brace for *loop()*:

```
void loop()
{
    static long oldFrequency = 0L;   // Add this new statement
```

Chapter 3: The LCD Shield Project

The keyword *static* allows *oldFrequency* to retain its previous value on each pass through *loop()*, even though it is defined within *loop()*. A little farther down (e.g., line 515) you should see the statement:

```
frequency_tune  = frequency + RitFreqOffset;
```

Add the following new lines immediately below the line above:

```
if (frequency_tune != oldFrequency) {
  LCDFrequencyDisplay();
  oldFrequency = frequency_tune;
}
```

The new lines of code compare the old frequency to the current frequency. If they are not the same, the LCD display needs to be updated with the new frequency (i.e., the user is tuning across the band) via the LCDFrequencyDisplay() function call. We then update *oldFrequency* to the current frequency. On the other hand, if the two frequencies are the same, there is no reason to call the *LCDFrequencyDisplay()* function. If we called the display on every pass through the loop, we could introduce a little bit of flicker in the display.

The code for the *LCDFrequencyDisplay()* function appears in Listing 3-2.

```
/*****
  This method uses the LCD display from Chapter 3 of the Purdum-Kidder book to
  display the Rebel receiving frequency. Jack Purdum, W8TEE, 9/20/2013

  Parameters:
    void

  Return value:
    void
*****/
void LCDFrequencyDisplay()
{
  char row1[17] = {'R', 'X', ':', ' '};
  char row2[17] = {'T', 'X', ':', ' '};
  char temp[17];
  char tail[] = {' ', 'M', 'H', 'z', '\0'};

  if (bsm == 1) {          // 20 meters
    row1[3] = '1';         // Display row 1
    row1[4] = '4';
    row1[5] = '.';         // Make sure we can treat as a string
    row1[6] = '\0';
    itoa((frequency_tune + IF), temp, 10);
    strcat(row1, &temp[2]);
    strcat(row1, tail);
    row2[3] = '1';         // Display row 1
```

LISTING 3-2 Modified LCD display program.

```
      row2[4] = '4';
      row2[5] = '.';              // Make sure we can treat as a string
      row2[6] = '\0';
      itoa((frequency + IF), temp, 10);
      strcat(row2, &temp[2]);
      strcat(row2, tail);
   } else {                        // 40 meters
      row1[4] = '7';               // Display row 1
      row1[5] = '.';
      row1[6] = '\0';              // Make sure we can treat as a string
      itoa((frequency_tune + IF), temp, 10);
      strcat(row1, &temp[1]);      // Ignore the leading '7'
      strcat(row1, tail);
      row2[4] = '7';               // Display row 2
      row2[5] = '.';
      row2[6] = '\0';
      itoa((frequency + IF), temp, 10);
      strcat(row2, &temp[1]);
      strcat(row2, tail);
   }
   lcd.setCursor(0, 0);
   lcd.print(row1);
   lcd.setCursor(0, 1);
   lcd.print(row2);
}
```

LISTING 3-2 Modified LCD display program. (*continued*)

We placed the new function in Listing 3-2 just before the *loop()* function (around line 500) and just after the definition of their *Default_Settings()* function. Simply stated, the code displays the transmit and receive information on the LCD display. Save your work with a name that is different from the original source code, perhaps something like RebelSourceWithLCDDisplay.pde. Now compile and upload the file just like you would an Arduino sketch.

Disconnect the USB cable and power up the Rebel. You should see your splash screen for a second and a half, and then the default start-up frequency for the Rebel. It should look similar to Figure 3-31. As you tune around the band, the display updates accordingly. If it doesn't, you're on your own.

Naw ... just kidding. Simply reload the software into the IDE and go through the steps again. Make sure you are changing the code at the correct code sections as some of the *#define*s appear at multiple places in the code.

Other hams have made similar changes to their Rebel, and have routed a multiline cable from the Rebel to another case that houses the shield and display. This makes the display more permanent and professional looking. Still, having the case open with wires all over the place like you see in Figure 3-30 is kinda cool in a geeky sort of way. Your choice ...

FIGURE 3-31   Rebel display.

Conclusion

In this chapter you built an LCD shield that can be used for a variety of uses, some of which are detailed in later chapters. The construction techniques used to build the LCD display are used for building other shields in other projects. For now, we encourage you to experiment with the shield and perhaps try your hand at modifying the code presented in Listing 3-2. The changes don't have to be earth-shaking, just something that makes a small difference in the way the program functions. You can always start over, so there's no reason not to give it a try.

CHAPTER 4
Station Timer

In this chapter we use the LCD shield you built in Chapter 3 to create a station ID timer. Section 97.119(a) of the FCC Rules and Regulations that govern amateur radio operation states that:

> *(a) Each amateur station, except a space station or telecommand station, must transmit its assigned call sign on its transmitting channel at the end of each communication, and at least every 10 minutes during a communication, for the purpose of clearly making the source of the transmissions from the station known to those receiving the transmissions. No station may transmit unidentified communications or signals, or transmit as the station call sign, any call sign not authorized to the station.*

We have all been guilty of breaking this rule at one time or another, usually because we simply lost track of time. Technically, it appears that those little transmission breaks often reserved for a "Yes" or "No" from the other operator are also in violation of 97.119(a). While we know of no incidents where the FCC has gone after a ham in violation of this section of the Rules, our license remains an earned privilege and there's no reason to tempt Fate when it's so easy to stay in compliance.

In this chapter, we present two station ID timers. The first timer is purely a software-defined timer. That is, you use the program to determine the time lapse between station ID times by reading one of the Atmel timers buried within the μC chip. The software uses the LCD display from Chapter 3 to create a countdown timer. When the timer determines that 10 minutes have passed, the display shows a warning message. A simple reset button recycles the clock.

The second station ID timer is a combination of software and hardware, using the DS1307 Real Time Clock (RTC) chip (see Figure 4-1). The RTC module can be purchased online for less than $5, or you could build your own from the chip. However, if you value your time at more than 10 cents an hour, these small modules you can purchase are a better choice. The RTC module uses the *Inter-Integrated Circuit interface* (I2C), which is a serial bus invented by Philips for connecting various modules to an electronic device. The I2C interface is easy to use and there are a ton of sensors and other devices that use the interface. It's an interface worth knowing. The Arduino IDE and supplemental libraries provide most of the software used by the RTC timer.

FIGURE 4-1 Real time clock/timer shield.

Software Version of ID Timer

In this section we present the station ID timer program that you can load and run with your LCD shield. The program displays an initial count of 10 minutes and counts down to 0 at which time it displays a message stating it's time for you to identify yourself on the air.

The purpose of this section of the chapter is to show you some software design tips and techniques that you can use to improve your programming skills. If you are not interested in programming, you can skip the following sections of this chapter and proceed to the RTC section of the chapter. We recognize that not everyone is interested in the software that drives any given project ... that's just the way it is. What follows, however, is a program that works fine as written in its first draft mode, but can be improved with just a few minor tweaks. It's those tweaks that we concentrate on in the rest of this section of the chapter.

The LCD timer program is presented in Listing 4-1. Version 1.0 of the program was written around midnight on a Saturday night just to see if we could get it working before we hit the sack. We did.

```
/*****
 * Station Identification Timer, set for a 10 minute countdown.
 *
 * Version 1.0, July 20, 2013
 * Dr. Purdum
 *
 * CAUTION: As presented here, the timer triggers after 10 minutes, which
 *     essentially means you are in violation of 97.119(a)
 *
 *****/
```

LISTING 4-1 The LCD timer program, version 1.0.

```
#include <LiquidCrystal.h>
// initialize the library with the numbers of the LCD interface pins
LiquidCrystal lcd(12, 11, 5, 4, 3, 2);

unsigned long current = 0;    // Holds the current millisecond count
unsigned long previous = 0;   // Holds the previous millisecond count
int i;
int minutes = 10;             // Set for ten minutes and ...
int seconds = 0;              // ...no seconds
char buffer[10];              // Working buffer
void setup() {
  // set up the LCD's number of columns and rows:
  lcd.begin(16, 2);
                              // Print setup message to the LCD.
  lcd.print("Ten Minute Timer");
  lcd.setCursor(5,1);
  sprintf(buffer, "%d:%02d", minutes, seconds); // Stuff 10 minutes and no seconds
                                                // into the buffer
  lcd.print(buffer);   // Show the buffer
}
void loop() {
  current = millis();                 // Read the millisecond count
  if (current - previous > 1000) {    // If a second has passed...
    if (previous == 0){               // ...and if first pass through this code...
      lcd.setCursor(5,1);             // ...blank out the "1" character for the "10"
      lcd.print(" ");
    }

    if (seconds == 0 && minutes > 0) {  // If seconds = 0 but minutes are left...
      seconds = 59;                     // ...reset the seconds...
      minutes--;                        // ...decrement the minutes
    } else {
      if (seconds) {
        seconds--;
      }
    }

    if (minutes == 0 && seconds == 0) {
      warnID();
    } else {
      lcd.setCursor(6,1);
      sprintf(buffer, "%d:%02d", minutes, seconds);
      lcd.print(buffer);
      previous = current;
    }
  }
}
```

LISTING 4-1 The LCD timer program, version 1.0. *(continued)*

```
/*****
    This function alters the user that time has expired on the ID timer.
    Parameter List:
        void
    Return Value:
        void
*****/
void warnID() {
  static bool flag = true;
  if (flag) {
    current = millis();                    // Read the millisecond count
    if (current - previous > 500) {        // If a second has passed...
      lcd.display();
      delay(500);
      lcd.setCursor(0, 0);
      lcd.print("    IDENTIFY    ");
      lcd.setCursor(0,1);
      lcd.print("      NOW       ");
      flag = false;
    }
  } else {
    delay(500);
    lcd.noDisplay();
    flag = true;
  }
  previous = current;
}
```

LISTING 4-1 The LCD timer program, version 1.0. (*continued*)

As mentioned earlier, the code worked the first time after correcting a few minor hiccups. When compiled on an ATmega168, the compiled program occupied 4526 bytes of program space. All in all, not too bad and, as the old saying goes: "If it ain't broke, don't fix it."

Well, maybe … maybe not.

Magic Numbers

Good programmers don't like magic numbers. So, exactly what is a magic number? Simply stated, *a magic number is a number that appears in a program yet you have no clue what it means or why it's there.* There are lots of magic number statements in Listing 4-1, some bad, others not so bad. Extracting those magic numbers and classifying them, we might get something like the following list:

List of Bad Magic Number Statements

```
LiquidCrystal lcd(12, 11, 5, 4, 3, 2);
lcd.begin(16, 2);
if (current - previous > 1000) {   // If a second has passed
if (current - previous > 500) {    // If a second has passed...
delay(500);
```

List of Not So Bad Magic Number Statements

```
lcd.setCursor(5,1);
seconds = 59;
```

While researching for this book, we found this statement in a program all by its lonesome:

```
delay(247);
```

Really? 247? And not a hint as to why it's 247. We know the *delay()* function is passed a parameter expressed in milliseconds. Therefore, if it were 250 (i.e., a quarter of a second) perhaps we might make some informed guess about its value, but 247? At the present time, it's even money that it's either some kind of adjustment for the number of clock cycles to execute the delay or it's the programmer's parking spot number.

Bad magic numbers are so classed because we really don't know why they have the value that they do. The first bad statement (LiquidCrystal lcd(12, 11, 5, 4, 3, 2);) is actually explained in the comment at the top of Listing 3-1 in Chapter 3, so it's bad simply because we got lazy and didn't copy the program comment header. The second bad statement is also explained in Listing 3-1, so again, it's our laziness. The next three statements are bad simply because we have to pause a few nanoseconds longer than we should to figure out what the statement does. Note that, even though two of the lines have comments, the comments are the same, but the numbers are different. Sometimes a misleading comment is less helpful than no comment at all. Duplicate comments are especially common when doing cut-and-paste type of editing.

The List of Not So Bad statements are classed as such because we can pretty quickly figure out what they mean because of the use of a good method name (e.g., *setCursor()*) or a good variable name (e.g., *seconds*). The remaining bad magic numbers, 1000 and 500, need to be fixed.

Preprocessor Directives

There are a number of statement lines that can benefit from further explanation. The first time you saw examples of the *#include* preprocessor directive was in Chapter 3. Because we are using the shield you built in Chapter 3, we must include the *LiquidCrystal* header file that links us into that library in the program.

While header files and their associated library files are useful, they still don't solve our magic numbers problem.

Fixing Bad Magic Numbers: *#define*

While you already understand what the *#include* preprocessor directive is, we still have to deal with our bad magic numbers. The fix involves a technique that uses the C preprocessor.

Clearly, the value 1000 is being used in the program to compare millisecond counts. Because 1000 milliseconds is the same as 1 second, we write the following statement near the top of the program, just after the include statement (preprocessor directives must appear in the source code file before any executable program statements):

```
#include <LiquidCrystal.h>
#define ONESECOND 1000
```

Because the *#define* begins with a sharp symbol (#), we know it, too, is a preprocessor directive. The *#define* preprocessor directive causes the compiler to search through the program source code and, every time it finds the characters ONESECOND, it substitutes the value 1000 in its place.

Your response at this point might be: So what? Well, which would you rather read if you're trying to fix a program:

```
if (current - previous > 1000)
```

or

```
if (current - previous > ONESECOND)
```

If we add another #define:
```
    #include <LiquidCrystal.h>

#define ONESECOND 1000
#define HALFSECOND 500
```

then the statement:

```
delay(HALFSECOND);
```

becomes that much easier to read and understand. If you use the *#define* with a little thought, it helps document what your program is doing. We should also mention that you do not have to use capital letters, but most programmers use the convention of capitalizing the letters for a *#define*. But that's not the only advantage to using *#define*s in your programs.

Many years ago Jack was working on code that figured out fines for speeding tickets. The program had three different base fines for cars ($50), light trucks ($65), and tractor trailer trucks ($100) plus additional sums for each mile per hour over the speed limit. At the time, the speed limit for cars and light trucks was initially 55 mph and 50 for big trucks. The program contained codes like the following code fragment all over the place:

```
if (car && speed > 55)
   fine = 50;
// a bazillion more similar statements...
if (lightTruck && speed > 55)
   fine = 65;
```

Then the laws changed and speeds were increased for cars to 70 mph, 65 for light trucks, and 60 for big trucks. Now what do you do? Well, really lazy programmers simply do a global search and replace 55 with 70. This is bad because phone numbers like 555-1234 (and similar numbers) get changed to 705-1234. (Always remember: Computers have the native intelligence of a box of rocks.) A slightly better plan is do a global search for 55 and see if the statement found pertains to a car or light truck (or even something else) and decide how to change the statement. Either way, this search-and-replace approach is a very error-prone process.

Instead, what if the programmer had avoided the magic numbers from the outset and used the following instead:

```
#define CARSPEED 55
#define LIGHTTRUCKSPEED 55
```

```
#define SEMISPEED 50
#define CARBASEFINE 50
#define LIGHTTRUCKBASEFINE 65
#define SEMIFINE 100
```

After the new law change, our smarter programmer takes about a minute and changes the *#define*s to:

```
#define CARSPEED 70
#define LIGHTTRUCKSPEED 65
#define SEMISPEED 60
```

recompiles the program, and ... Shazam! Everything is fixed. If the law happened to change the fines as well as the speeds, it would be equally easy to fix those because the code was written as:

```
if (car && speed > CARSPEED)
    fine = CARBASEFINE;
```

The *#define* preprocessor directive should be used for most magic numbers, especially if there's a chance it may change in the future. Creating symbolic constants using *#define*s makes program code easier to read, program changes faster to implement, and employs much less error prone process for change.

A Second Way to Remove Magic Numbers: *const*

As a general rule, if a *#define* is used to replace a numeric constant, it has no effect on the program size. Indeed, our test program remains at 4526 bytes. However, a *#define* preprocessor directive is not the only way to get rid of a magic number. We should also mention that you could also define a constant with the same effect, such as:

```
const int CARSPEED = 55;
```

The *const* keyword tells the compiler to flag the variable named CARSPEED as a constant and don't let its value change anywhere in the program. As a general rule, you can use the *const* keyword before any regular data type (e.g., *byte*, *int*, *long*). This approach brings the same advantages to the table as does the *#define* preprocessor directive.

So, which method is better? When it comes to memory usage, both produce identical program sizes. If you use capital letters for constants, too, both stand out in the program equally well. However, if the Arduino IDE ever gets a true interactive debugger, using *const* does produce an entry in the symbol table where a *#define* does not. Also, you can perform type checking on a *const* variable but *#define*s are normally typeless. Because of these advantages, many programmers prefer to use the *const* keyword for constants, using capital letters to flag them as symbolic constants. However, after using the C preprocessor for symbolic constants for more than 30 years, these old dogs will likely stick with the *#define*.

Fixing Flat Forehead Mistakes

If you look carefully at Listing 4-1, you can see the statement:

```
int i;
```

a few lines after the *#include* preprocessor directive. The statement defines a variable named *i* for use in the program. Truth is, we thought we would need a loop counter, so we added the data definition during the initial code writing stage. However, if you look closely through the code, you discover the variable *i* is never used in the program. This is what we call a "flat forehead mistake." You know, the kind of mistake where, once discovered, you pound the heel of your hand into your forehead and ask yourself how you could be so careless. Relax, we all make these mistakes. As you get more and more programming experience, the flat forehead mistakes all but disappear. (Alas, you often move on to bigger, more sophisticated mistakes.)

So, what's the harm of one defined, albeit unused, variable in a program? Actually, not much. In fact, the Arduino compiler is smart enough to notice that we never used the variable so it never generated the code to create or manage the variable. The program size remains at 4526 bytes even after removal of the unused definition statement for variable *i*. That's not to say that some other type of unused data definition won't affect program size. Still, it's sloppy to leave such unused data definitions in the code and they should be removed if for no other reason than they add unnecessary clutter to the program.

Encapsulation and Scope

Simply stated, *encapsulation is the process of hiding your data from outside forces* as much as possible. For example, consider the following statements that appear near the top of Listing 4-1:

```
LiquidCrystal lcd(12, 11, 5, 4, 3, 2);
unsigned long current = 0;    // Holds the current millisecond count
unsigned long previous = 0;   // Holds the previous millisecond count
int i;
int minutes = 10;             // Set for ten minutes and …
int seconds = 0;              // ...no seconds
char buffer[10];              // Working buffer
```

Because these data items are defined outside of any function or other statement block, *they are accessible and may be used by any program statement at any point in the program.* There are several different types of scope associated with data definitions. Data that are useable everywhere in a program are said to be *global data* and have *global scope*. Data defined outside a function body is given global scope by the compiler. It's not too much of a stretch to think of "visibility" or "accessibility" as synonyms for scope. In other words, a variable with global scope can be seen and used by virtually any other statement in the program.

In Chapter 2 you learned about the *setup()* and *loop()* functions. If you define a variable inside the *setup()* function, like:

```
void setup() {    // The opening brace for setup()
   int count;
   // more program statements
}                 // The closing brace for setup()
```

you can see that the variable named *count* is defined within the opening and closing brace of the *setup()* function. This means that the variable *count* has function scope. *Function scope means that any data item defined within a function body can be used only within the confines of that function.* Think of the braces as creating a black box named *setup()* and anything outside of *setup()* has no clue what's going on inside the black box. Outside the *setup()* function, for all intents and purposes, *count* doesn't even exist. If you tried to access count outside of *setup()*, the compiler would issue an error message saying the variable was "not declared in this scope." All of the data defined within *setup()* function body, as marked by the opening and closing braces, are invisible outside that function. You can access *count* everywhere within *setup()*, but not outside of *setup()*'s definition.

So why bother encapsulating, or hiding, your data? You hide your data for the same reasons medieval kings hid their daughters in the castle tower … to keep other people from messing around with them. If all of your data have global scope and all of a sudden one variable has a screwy value, you have no idea where to begin looking for the program error because it could be at any point in the program. However, if something goes haywire with variable *count* in the example above, you know it has to be something within the *setup()* function that's causing the problem. You know this because no force outside of *setup()* even knows the variable exists. Because you have encapsulated variable *count* within a function, your search for the error is narrowed from the entire program to a single function. While encapsulation may not seem a big deal in a program as simple and short as that in Listing 4-1, we have worked on programs with over 800,000 lines of code and any help we can get from a good design is welcomed.

Therefore, as a general rule, limit the scope of your data as much as possible. For Listing 4-1, it means moving the following data definitions from outside any program function:

```
unsigned long current = 0;     // Holds the current millisecond count
unsigned long previous = 0;    // Holds the previous millisecond count
int minutes = 10;              // Set for ten minutes and ...
int seconds = 0;               // ...no seconds
```

so they appear inside the *loop()* function. (We deleted the definition of *i* because we never used it.) Upon recompiling the program after the data definitions have been moved, the program size dropped to 3962 bytes for a decrease in program size of 564 bytes, or a little more than 12%.

That's the good news.

The bad news is that our program no longer works correctly. It starts as before, showing 10 minutes of time and counts down to 9:59 and then never changes the display after that. When a program doesn't do what it's designed to do, we say we have a *program bug*.

We have a program bug.

Fixing Our Program Bug

Recall from Chapter 2 that the *loop()* function is called over and over until either the power is removed from the board or there is some kind of component failure. If you look at the statements in the *loop()* function in Listing 4-1, you can find the statement:

```
if (current - previous > ONESECOND) {  // If a second has passed...
```

This expression means we need to use the *previous* and *current* millisecond counts and compare the difference between the two variables to *ONESECOND*. However, because *loop()* now looks like:

```
void loop() {
  unsigned long current = 0;     // Holds the current millisecond count
  unsigned long previous = 0;    // Holds the previous millisecond count
  int minutes = 10;              // Set for ten minutes and ...
  int seconds = 0;               // ...no seconds
```

```
      current = millis();              // Read the millisecond count
   if (current - previous > ONESECOND) {  // If a second has passed...
      if (previous == 0){   // ...and if this is the first pass through this code...
         lcd.setCursor(5,1);         // ...blank out the "1" character for the "10"
         lcd.print(" ");
      }
      // the rest of the loop code...
}     // closing brace for loop() function
```

each time the statements within *loop()* are executed on each pass through the loop, variables *current* and *previous* are re-initialized to 0, so the statement:

```
   if (current - previous > ONESECOND) {  // If a second has passed...
```

can never be *true*. Not good. We want to encapsulate *current* and *previous* inside of *loop()*, but doing so prevents the variables from ever having any value other than 0 on each pass through the loop. We can go back to global scope and sacrifice the benefits of encapsulation, or we can encapsulate the data and have a program that doesn't work properly. A true dilemma ... two choices, both bad.

The *static* Data Type Specifier

Let's make a small change to each data definition within *loop()* and see what happens. Place the word *static* at the front of each data definition in *loop()*, so the data definitions look like:

```
   static unsigned long current = 0;    // Holds the current millisecond count
   static unsigned long previous = 0;   // Holds the previous millisecond count
   static int minutes = 10;             // Set for ten minutes and ...
   static int seconds = 0;              // ...no seconds
```

and recompile, upload, and run the program. Taa-daa! The program works as it should again ... but why? The reason is because the keyword *static* causes the compiler to treat the data definitions differently.

When the compiler sees the keyword *static* in a data definition, it's as though you placed the definition of the variable at the very beginning of the program ... even before the statements in the *setup()* function are executed. If the variables are defined with initial values, as we have done here, the compiler assigns those values into the respective variables. Once the compiler has defined and initialized a *static* variable, *it never redefines or initializes a static variable again*.

What this means is that *static* definition statements are processed at program start and are then ignored from that time on. In our program, therefore, it's like the definitions for *current*, *previous*, *minutes*, and *seconds* are defined outside of *loop()*, but their scope is limited to being known only within *loop()*. We now have the best of both worlds: we have encapsulated those variables inside of *loop()* ... no other part of the program has direct access to them, but they retain their previous values on each new pass through the loop.

We also simplified the *WarnID()* function to just use a *delay()* function call. The *delay()* function is not the best choice in most cases because it literally shuts everything down on the Arduino board during the delay period. However, in this simple program example, we might use it because once the flashing message is read, the operator must hit the Reset button to restart the program anyway. The final version of the code, which is now 4382 bytes, is presented in Listing 4-2.

```
/*****
 * Another program to test the LCD shield
 *
 * Version 1.0, July 20, 2013
 * Dr. Purdum
 *
 * No restrictions
 *
 *****/
#include <LiquidCrystal.h>

#define ONESECOND 1000
#define HALFSECOND 500

// initialize the library with the numbers of the interface pins
LiquidCrystal lcd(12, 11, 5, 4, 3, 2);

char buffer[10];                    // Working buffer

void setup() {
  lcd.begin(16, 2);                 // set LCD's number of columns and rows:
  lcd.print("Ten Minute Timer");    // Print setup message to the LCD.
  lcd.setCursor(5,1);
  sprintf(buffer, "%d:%02d", 10, 0);// Stuff 10 minutes 0 seconds into buffer
  lcd.print(buffer);                // Show the buffer
}

void loop() {
  static unsigned long current = 0;   // Holds the current millisecond count
  static unsigned long previous = 0;  // Holds the previous millisecond count
  static int minutes = 10;            // Set for ten minutes and ...
  static int seconds = 0;             // ...no seconds

  current = millis();                         // Read the millisecond count
  if (current - previous > ONESECOND) {       // If a second has passed...
    if (previous == 0){                       // ...and if 1st pass through this code...
      lcd.setCursor(5,1);                     // ...blank out "1" character for the "10"
      lcd.print(" ");
    }

    if (seconds == 0 && minutes > 0) {   // If seconds are 0 but minutes left...
      seconds = 59;                      // ...reset the seconds...
      minutes--;                         // ...decrement the minutes
    } else {
      if (seconds) {
        seconds--;
      }
    }
```

LISTING 4-2   The LCD timer program, version 1.1.

```
        if (minutes == 0 && seconds == 0) {   // Time to ID ourselves...
          warnID();
        } else {                               // ...otherwise show time that's left
          lcd.setCursor(6,1);
          sprintf(buffer, "%d:%02d", minutes, seconds);
          lcd.print(buffer);
          previous = current;
        }
      }
    }

    void warnID() {
      while (true) {
        lcd.setCursor(0, 0);     // Show the message...
        lcd.print("    IDENTIFY    ");
        lcd.setCursor(0,1);
        lcd.print("       NOW      ");
        delay(HALFSECOND);
        lcd.clear();             // Clears the display
        delay(HALFSECOND);
      }
    }
```

LISTING 4-2   The LCD timer program, version 1.1. (*continued*)

The code presented in Listing 4-2 is our final version of the software ID timer program. For now, we encourage you to experiment with the code presented in Listing 4-2. The changes don't have to be earth-shaking, just something that makes a small difference in the way the program functions. You can always start over, so there's no reason not to give it a try. One change that probably should be made is to issue a warning message when there are only 30 seconds or so left in the 10-minute countdown. After all, the user needs some time to make the station identification. Another addition might be to add a small buzzer to the circuit and activate it when time runs out, or just blink the display.

Using a Real Time Clock (RTC) Instead of a Software Clock

The software countdown timer as presented in Listing 4-2 works, but we can improve on it a little. It would be nice, for example, to have the LCD display serve both the function of a station ID timer as presented in Listing 4-2, but show the current date and time, too. Making these improvements does cost a little more (i.e., less than $5), but does augment the functionality of the timer. Plus, you'll get to learn a few new things along the way ... always a good thing.

The Inter-Integrated Circuit (I²C or I2C) Interface

The first new thing we want to discuss is the I2C interface. The I2C interface was developed by Philips Semiconductor (now known as NXP) back in the 1990s to allow low-speed peripherals to be attached to an electronic device. The interface was embraced by other companies that produced electronic sensors and other devices. While use of the interface has been license-free since 2006,

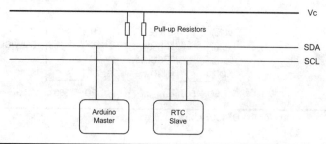

FIGURE 4-2   The I2C bus.

NXP still licenses I2C slave addresses (explained below) used by the interface. The I2C interface is also called the Two Wire Interface (TWI), for reasons that become clear in a few moments.

The interface is actually pretty simple. I2C uses two wires to form a "bus" between electronic devices. The wires are referred to as a "bus" because it can connect more than just two devices together. One device serves as the "Master Device" and is responsible for establishing the communication on the bus. One or more "Slave Devices" are attached to the bus. Each of these slave devices is assigned a slave address that identifies the device on the bus. While both 8- and 16-bit addressing is available, most μC devices use the 8-bit addressing. The slave address for our RTC using the DS1307 chip is 0×68. Therefore, any time the address byte on the bus is 0×68, we know that it is the RTC that is involved in the communication that is taking place on the bus.

Both the master and slave have the ability to transmit and receive data on the bus. The data are sent on the SDA line shown in Figure 4-2. The speed, or pace, of the communications on the bus is dictated by the clock used on the bus. (Line SCL in Figure 4-2.) While there are several different clock speeds defined for the I2C bus, the most common are the 10kbits/s (low speed mode), the 100kbit/s (standard speed mode), and the 400 kbits/s (fast mode). Because of the way the clock is used to strobe the data on the line, actual data transfer rates are less than the mode speeds would suggest. The address space and bus capacitance limit the effective length of the bus to about 5 ft or so. (μCs have WiFi shields available that allow much greater communication links if needed.)

As a general rule, the SDA and SCL lines each need a "pull-up" resistor on them. The purpose of a pull-up resistor is to pull the voltage up to the logic high state (5 V for the Arduino) needed for the I2C interface to work properly. The actual resistor values are not critical, with values between 4.7K and 20K being common. Our circuits use 10K resistors because we had a bunch of them lying around on the bench. Figure 4-2 shows how these pull-up resistors are used in the interface. Note that only one pair of pull-up resistors are needed on the bus, not a pair for each slave device.

While there are a lot of additional details about the I2C bus and how it works, what we have presented here is enough for us to work with. (You can find a ton of additional information about the interface online if you are interested.) The actual hardware used is not much different from what is shown in Figure 4-2. The only exception is the connection of the LCD display.

The I2C and the DS1307 RTC Chip

We chose to use the DS1307 from Maxim/Dallas Semiconductor as the heart of our RTC. The chip is relatively cheap, uses low power, and most modules for the Arduino environment include a small button battery on board to maintain the date and time data if power is lost. Also, there are a number of good libraries for the chip. Figure 4-3 shows the pin-outs for two versions of the chip.

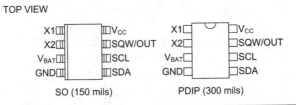

FIGURE 4-3 Pin-outs for the DS1307 chip.

The small version of the chip on the left in Figure 4-3 is a surface mounted device (SMD) and is what most commercial modules use. If you wanted to "roll your own," you might use the larger dual inline package (DIP) shown on the right. Note the SCL and SDA pins that correspond to the clock and data lines shown in Figure 4-2.

We mentioned earlier that the NXT does not charge a royalty fee for using the I2C protocol, but does charge a fee for registering an I2C device address. The DS1307 has been assigned the device address of 0×68, which identifies the chip whenever it and the Master need to communicate with each other.

Buried within the chip are a series of registers that maintain the data for the clock. The Timekeeper Registers are shown in Table 4-1. Note the register addresses and the function corresponding to that address. For example, register 0×05, is the register that holds the Month component of the date while register 0×00 holds the Seconds component of the time.

BCD and the DS1307 Registers

In the DS1307, each register is an 8-bit entity, but values are stored in a format known as Binary Coded Decimal, or BCD. What this means is that each 8-bit register is treated as though it is partitioned into two smaller 4-bit registers. Such *4-bit entities are referred to as nibble (... honest!)*

ADDRESS	BIT 7	BIT 6	BIT 5	BIT 4	BIT 3	BIT 2	BIT 1	BIT 0	FUNCTION	RANGE
00h	CH	\multicolumn{2}{c}{10 Seconds}		\multicolumn{3}{c}{Seconds}			Seconds	00–59
01h	0	\multicolumn{3}{c}{10 Minutes}		\multicolumn{3}{c}{Minutes}		Minutes	00–59
02h	0	12	10 Hour	10 Hour	Hours	Hours	Hours	Hours	Hours	1–12 +AM/PM 00–23
		24	PM/AM							
03h	0	0	0	0	0	DAY	DAY	DAY	Day	01–07
04h	0	0	\multicolumn{2}{c}{10 Date}	\multicolumn{3}{c}{Date}			Date	01–31
05h	0	0	0	10 Month	Month	Month	Month	Month	Month	01–12
06h	\multicolumn{4}{c}{10 Year}		\multicolumn{3}{c}{Year}			Year	00–99
07h	OUT	0	0	SQWE	0	0	RS1	RS0	Control	—
08h–3Fh									RAM 56×8	00h–FFh
0 = Always reads back as 0.										

TABLE 4-1 The DS1307 Registers

Bits:	7	6	5	4	3	2	1	0
Power:	8	4	2	1	8	4	2	1
Binary	0	0	1	0	1	0	0	1
Value:			2		8			1
	<------------------- High Nibble -------------------->	<------------------- Low Nibble -------------------->

TABLE 4-2 *BCD* Conversion to Binary

Now look at the Seconds register at address 0x00 in Table 4-1. The low nibble encompasses bits 0-3. Because any computer register ultimately stores its data in binary, the largest number that nibble can store is 1111 in binary, or 15 in decimal. However, since the Seconds register must be capable of storing values up to 60, it uses bits 4-6 to store the "tens of seconds." For example, if you wanted to store 29 seconds in the Seconds register, the "low" nibble (i.e., bits 0-3) would contain the "units of seconds," or 9 while the high nibble (i.e., bits 4-6, and ignoring the 7th bit) would contain the "tens of seconds," or 2. Therefore, the BCD value in the 0x00 Seconds register would be 0010 1001. Think about it.

Table 4-2 should help you see how the BCD values are determined. Just remember that each binary digit, or bit, is a power of 2. Because each nibble is treated as an individual unit, the binary powers for the High Nibble are the same as those for the Low Nibble. That is, the power multipliers for each nibble are the same (i.e., the range of powers is 1 through 8 for each nibble). Therefore, in Table 4-2, there are 2 "10 Seconds," or 20 in the High Nibble. (This is why Table 4-1 refers to bits 4-6 as "10 Seconds.") In the Low Nibble, you find 8 + 1 = 9 "Seconds." Because BCD treats each byte as a unit, you can see that the byte 00101001 in BCD translates to 29 seconds as a BCD value.

So, why do you need to know all of this BCD stuff? The reason is because the DS1307 registers are fluent in BCD and as dumb as a box of rocks in anything else. Therefore, to make the DS1307 do anything useful for us, we need to send and receive the register data in BCD format. The good news is that writing functions to convert BCD data the registers like to decimal numbers you and I understand is pretty trivial.

One more detail that should help you understand what's going on inside the DS1307 when data are being placed on the I2C bus. The DS1307 maintains an internal register pointer that is automatically incremented each time a byte to read into or out of the chip. Therefore, unless you tell it otherwise, the DS1307 is going to store the first data byte it receives in the Seconds register (address 0x00); the internal register pointer is then incremented so the second byte goes into the Hours register. This sequential process continues until all data bytes have been transferred. This is why the *read()*s and *write()*s to the DS1307 are in the order that you see in Listing 4-3. Using some other order would mess up the data. Fortunately, the libraries that we use take care of most of the details for us. If you're interested, you can examine each of the .cpp files for the libraries to figure out how they talk to the DS1307 across the I2C bus.

Constructing the RTC/Timer Shield

The completed RTC/Timer shield is shown in Figure 4-4. It is constructed on the LCD shield built in Chapter 3 using removable headers to mount the display to the shield.

To construct the RTC/Timer, you can use some of the empty nano-acres of real estate on the LCD shield you built in Chapter 3. We added a RTC module with its associated circuitry and a switch to reset the timer. The circuit for the RTC/Timer shield is shown on Figure 4-5. For the sake of clarity the circuitry associated with the LCD shield is not shown. Adding circuitry to the LCD

82 Arduino Projects for Amateur Radio

FIGURE 4-4 The completed real time clock/station timer shield.

shield is one reason we discussed using a removable header for the LCD display unit. We can unplug the display and add the circuitry underneath. Alternatively, the RTC/Timer can be built on its own shield using stackable headers to plug in the LCD shield.

The RTC module is connected to the shield using a seven-pin header. The stubby ends of the header pins are soldered to the shield. The module should be mounted with the LiOn battery facing up. Should the battery need to be replaced at some future time, mounting the RTC module

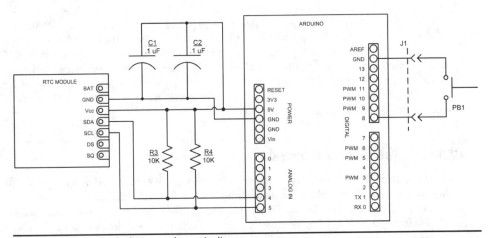

FIGURE 4-5 Real time clock/timer schematic diagram.

FIGURE 4-6   Real time clock breakout module (front and rear).

with the battery facing up makes it a lot easier to access the battery. Do not solder the module to the shield just yet, we first need to install some parts underneath the module. The module that we chose, shown in Figure 4-6, has components mounted on both sides. In particular, there is a quartz crystal (it is the small, shiny, tube-like object with two wires mounted on the side opposite the battery) that must be insulated to prevent shorting out the wiring pads on the shield. We used a small piece of heat-shrink tubing around the crystal, but other insulating materials such as electrical tape would work equally well.

A WORD OF CAUTION: We have learned that some of the RTC breakout modules being sold on eBay are being supplied with a nonrechargeable Li-ion battery, a CR2032. The CR2032 battery can be problematic as some RTC breakout modules are designed to recharge the battery. The RTC breakout modules we are using include a rechargeable Li-ion battery, an LIR2032. The LIR2032 should not cause any problems but make sure that it is fully charged, the DS1307 RTC chip may not function with the battery partially discharged.

Looking at Figure 4-7, you can see how the parts are placed on the shield for assembly. The two 10 kΩ ¼ W resistors (R3 and R4) must be installed before the RTC module can be soldered down. Figure 4-8 shows the placement of the LCD shield with the RTC parts. The two 0.1 μF capacitors (C1 and C2) are added to help control noise on the 5 VDC circuit. This is common practice as noise on the power bus can induce errors if it's severe enough. An ounce of prevention is worth a pound of cure, as they say.

You can mount the timer reset switch, SW1, directly on the shield, but we chose to have it detached so that the entire unit can be mounted in a project box. The switch, PB1, is wired through a plug into J1. Any normally open (NO) pushbutton switch may be used for PB1.

84 Arduino Projects for Amateur Radio

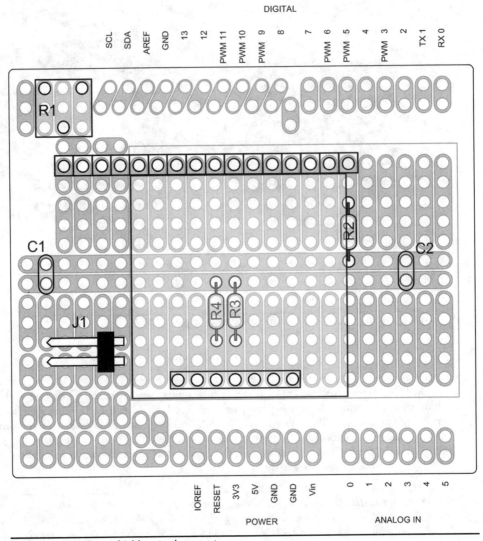

FIGURE 4-7 RTC/Timer shield parts placement.

The RTC/Timer circuitry is wired up using our usual 24AWG solid wire using Teflon tubing added to insulate the wire. Figure 4-8 is a view of the wiring side of the shield. This should make it easier to wire the shield if you are not comfortable yet with schematics. The completed shield is shown in Figure 4-9 with the LCD removed.

An alternate version using a separate LCD shield is shown in Figure 4-9. The alternate design uses the LCD shield that we constructed in Chapter 3. Instead of using regular header pins to attach the RTC/Timer shield to the Arduino, we use stackable headers (see Figure 3-8 for reference). As the LCD is not integrally a part of the RTC/Timer shield, we can eliminate a lot of the wiring shown in Figure 4-8. The wires not needed are shown as dashed lines in the wiring diagram,

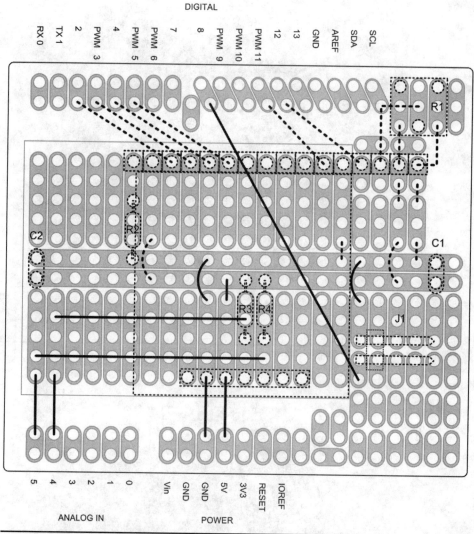

FIGURE 4-8   RTC/Timer wiring diagram (with LCD).

Figure 4-8. Figure 4-9 shows the completed shield with the LCD shield removed and Figure 4-10 shows the wiring side (i.e., Figure 4-8).

The Adafruit RTClib Library

The software we wrote for the RTC uses two additional libraries that we have not used before. The first is the Wire library, which is distributed with the Arduino IDE. The Wire library contains the functions we need to use the I2C bus. The second is the Adafruit RTClib library, which is a free download at:

https://github.com/adafruit/RTClib

FIGURE 4-9 The RTC/Timer shield using the LCD shield.

FIGURE 4-10 The wiring side of the RTC/Timer shield using the LCD shield.

This library greatly simplifies the initialization of the RTC at start-up, as you can see in Listing 4-3.

```
/*****
  This program initializes the RTC module to the time and date on the host PC.
  If the user wishes to use some other date and time, they can hard code the
  values in the SetStartingClockValues() function.

  Feb. 5, 2014
  Dr. Purdum, W8TEE
*/
#include <Wire.h>              //I2C header file
#include <LiquidCrystal.h>
#include <RTClib.h>

#define RTCI2CADDRESS 0x68   // 7 bit device address (without last bit - see
                             // at the datasheet)

#define SECONDS       0    // Offsets into the DS1307 internal registers
#define MINUTES       1
#define HOURS         2
#define DAY           3       // This is a day-of-the-week index register
#define DATE          4
#define MONTH         5
#define YEAR          6
#define MILLENIA      2000

LiquidCrystal lcd(12, 11, 5, 4, 3, 2); // For LCD display from Chapter 3
RTC_Millis rtc;                        // An RTC object
DateTime currentDateTime;              // A date and time object

byte result;
byte timeBuffer[8];
byte flagBit = 0;

void setup()
{
  Wire.begin();        // Initiate Wire library and join the I2C bus as a master

  lcd.begin(16, 2);    // set up the LCD's number of columns and rows

  Serial.begin(9600);  // Initiate serial communication

  rtc.begin(DateTime(__DATE__, __TIME__));
  currentDateTime = rtc.now();
  SetStartingClockValues();
  Serial.println("Clock initialized");
}

void loop()
```

LISTING 4-3 Initialization of DS1307 date and time data.

```
{
  // Nothing to do here...
}

/*****
  This function reads the system clock of the host PC and translates those time
  and date values for use by the RTC. This involves reading the date and time
  values from the PC via the now() method from the RTClib library and copying
  those values into the timeBuffer[] array. Because the DS1307 wants its data
  in BCD format, the timeBuffer[] data are converted to BCD format. The RTClib
  library is available from: https://github.com/adafruit/RTClib.

  Parameter list:
    void

  Return value:
    void

  CAUTION: This function assumes the DateTime variable named currentDateTime
           exists prior to the function call.
*****/
void SetStartingClockValues()
{
  timeBuffer[MONTH]    = currentDateTime.month();                // Month (1-12)
  timeBuffer[DATE]     = currentDateTime.day();                  // Day (1-31)
  timeBuffer[YEAR]     = currentDateTime.year() - MILLENIA;      // Year (0-99)
  timeBuffer[DAY]      = currentDateTime.dayOfWeek();            // Day of week (1-7)
  timeBuffer[HOURS]    = currentDateTime.hour();                 // Hour (0-23)
  timeBuffer[MINUTES]  = currentDateTime.minute();               // Minutes (0-59)
  timeBuffer[SECONDS]  = currentDateTime.second();               // Seconds (0-59)

  Wire.beginTransmission(RTCI2CADDRESS);                 // Select RTC on the I2C bus
  Wire.write(0);                                         // Set internal register pointer
  Wire.write(DecToBcd(timeBuffer[SECONDS]));             // Second
  Wire.write(DecToBcd(timeBuffer[MINUTES]));             // Minute
  Wire.write(DecToBcd(timeBuffer[HOURS]));               // Hour
  Wire.write(DecToBcd(timeBuffer[DAY]));                 // Weekday
  Wire.write(DecToBcd(timeBuffer[DATE]));                // Day
  Wire.write(DecToBcd(timeBuffer[MONTH]));               // Month (with century bit = 0)
  Wire.write(DecToBcd(timeBuffer[YEAR]));                // Year
  Wire.write(0);
  result = Wire.endTransmission();

  if (result)
    Serial.print("RTC Write error occurred");
}
/*****
  This function converts the byte-length decimal value to a Binary Coded Decimal
  value.

  Parameter list:
```

LISTING 4-3   Initialization of DS1307 date and time data. (*continued*)

```
        byte value         A number expressed as a decimal value

   Return value:
      byte              The decimal value expressed as a BCD number

*****/
byte DecToBcd(byte value)
{
   return (value / 10 * 16 + value % 10);
}
```

LISTING 4-3 Initialization of DS1307 date and time data. *(continued)*

Initializing the RTC

If your RTC doesn't come with a battery already in place, or if you need to replace it, you need a way to initialize the registers in the DS1307 chip so they contain valid data. You can do this "by hand," wherein you write specific values into the registers. The program in Listing 4-3 provides a simple way to either use specific data you wish to use or simply copy the date and time data from your PC. We assume that copying the host PC's date and time data is good enough.

The program begins by initializing the I2C interface via the Wire library, which is part of the libraries supplied with the Arduino IDE. The code also creates objects for the RTC, the LCD display, and a communication object *(Serial)*. Having done that, the program performs a *DateTime()* call to the host PC to retrieve its current date and time data, using that data to initialize the RTC object *(rtc)*. The *rtc* object calls the *now()* method to break out that data into *currentDateTime*. The function *SetStartingClockValues()* simply copies the DateTime object's information into the DS1307 registers. The nested calls to the *DecToBcd()* function converts the decimal data stored in into the BCD values the DS1307 wants. The program then displays a message informing you that the RTC is now initialized.

```
/*****
    This program creates a radio station ID reminder using the DS1307 RTC.
    The LCD can also serve as a simple station clock. By default, however, it
    starts in the ID Timer Mode. The Timer Mode is indicated by an asterisk that
    bounces between rows 1 and 2 in column 15. After 10 minutes have passed,
    the LCD display begins to blink and continues to do so until the pushbutton
    is pressed for at least one second.  After one second, the display stops
    blinking, and another 10 minute period begins.You can disable the Timer Mode
    by holding the pushbutton for a period of more than 5 seconds, at which time
    the asterisks disappear from the display.

    Dec. 20, 2013
    Dr. Purdum, W8TEE
```

LISTING 4-4 The RTC program.

```
*****/
#include <Wire.h>                    //I2C header file
#include <LiquidCrystal.h>
#include <RTClib.h>

//#define DEBUG 1                    // Use this to toggle debug stuff

#define RTCI2CADDRESS   0x68         // 7 bit device address
#define LEDPIN          13
#define RESETSWITCHPIN  8

#define IDTIMESLICE     600000  // 10 minutes. Use shorter period for debugging.
//#define IDTIMESLICE   10000    // For debugging, uncomment this one
#define TENSECONDS      10000
#define HALFSECOND      500
#define DEBOUNCEDELAY   20
#define ERRORREPEAT     6       // Blink rate for error LED indicator
#define FLIPDELAY       1       // Alternates LCD asterisk after this many seconds
#define MODESWITCHDELAY 5000    // Hold the push button > 5 seconds, activeTimer
                                // changes state.

#define MILLENIA        2000
#define SECONDS         0       // These are offsets for DS1307 internal registers
#define MINUTES         1
#define HOURS           2
#define DAY             3       // This is a day-of-the-week index register
#define DATE            4
#define MONTH           5
#define YEAR            6

#define RTCWRITEERROR 1 // Unable to write RTC time and date to DS1307
#define RTCREADERROR  2 // Unable to read RTC time and date from DS1307

LiquidCrystal lcd(12, 11, 5, 4, 3, 2); // For LCD display from Chapter 3
RTC_Millis rtc;

byte result;
byte timeBuffer[8];
byte flagBit = 0;

boolean activeTimer;             // Holds state of timer: 1 = active, 0 = inactive

char* weekdayname[] = {"Sun", "Mon", "Tue", "Wed", "Thu", "Fri", "Sat"};
char *timePeriods[10];

long millisStart, millisCurrent;

void setup()
{
```

LISTING 4-4 The RTC program. (*continued*)

```c
  Wire.begin();                         // Initiate the Wire library and join the I2C
                                        // bus as a master
  lcd.begin(16, 2);                     // set up the LCD's number of columns and rows

  pinMode(LEDPIN, OUTPUT);                     // Set error LED
  pinMode(RESETSWITCHPIN, INPUT_PULLUP);       // The reset clock pushbutton

  millisStart = millis();               // Start the count...
  activeTimer = true;                   // Assume they want timer started
}

void loop()
{
  static boolean timesExpired = false;
  boolean featureChanged;
  byte sentBack;
  int i;
  static long switchMode;               // How long the button has been pushed.
  static long count = 0L;
  long timeThatSwitchWasClosed = 0L;

  millisCurrent = millis();
  if (activeTimer == true) {            // If want to use the timer feature and
    if (millisCurrent - millisStart > IDTIMESLICE) {  // Is it time to tell them?
      timesExpired = true;                            // Yep...
    }
    if (timesExpired) {                 // Show them they're passed due for an ID
      lcd.noDisplay();
      delay(HALFSECOND);
      lcd.display();
      delay(HALFSECOND);
      DisplayDateAndTime(timeBuffer);
    }
  }
  if (debounce(RESETSWITCHPIN) == true) {  // Pressed the switch? Yes, so...
    if (activeTimer == true) {
      timesExpired = false;             // ...at least they want to reset timer...
      millisStart = millis();
    }
    switchMode = millis();    // Start measuring how long switch closed...
    count = millis();
    featureChanged = false;             // Assume no change in timer feature...
    while (digitalRead(RESETSWITCHPIN) == LOW) {  // As long as it's LOW...
      count = millis();                           // Yep. Count how long it's LOW...
      timeThatSwitchWasClosed = count - switchMode;
      if (timeThatSwitchWasClosed >= MODESWITCHDELAY) {  // Hold it long enough to
                                                         // toggle mode?
        featureChanged = true;                           // Yep, they did...
        break;
      }
```

LISTING 4-4   The RTC program. (*continued*)

```
      }                   // End while() switch was closed
    }                     // The switch is now opened

    if (featureChanged == true) {       // They want to change clock mode
      if (activeTimer == true) {        // So, if ID timer was running...
        lcd.setCursor(15, 0);           // Clear out asterisks
        lcd.print(' ');
        lcd.setCursor(15, 1);
        lcd.print(' ');
        activeTimer = false;            // ...shut the ID timer feature off
      } else {                          // Otherwise, must want to turn it on.
        activeTimer = true;             // ...turn on the ID timer feature.
        millisStart = millis();         // Start the count
      }
      featureChanged = false;
      timesExpired = false;             // ...at least they want to reset timer...
    }

    Wire.beginTransmission(RTCI2CADDRESS);  // Send RTC device address on the bus
    Wire.write(0);                          // Reset internal register pointer

    result = Wire.endTransmission();
    if (result)                             // Non-zero = could write to the chip
      SetError(RTCWRITEERROR);

    sentBack = Wire.requestFrom(RTCI2CADDRESS, 7);  // Ask for all 7 data bytes
    if (sentBack == 0)                              // It didn't like that request...
      SetError(RTCREADERROR);

    for (i = 0; i < sentBack; i++) {        // Read data from BCD to decimal,and...
      timeBuffer[i] = BcdToDec(Wire.read());// ...stuff into buffer
    }
    DisplayDateAndTime(timeBuffer);         // Show what we got back...
}

/*****
  This function is a pretty standard debounce function so we can determine
that the reset switch has really been pressed.
  Parameter list:
    int pin         the pin that registers the switch state
  Return value:
    boolean         true if the switch was pressed, false otherwise
*****/
boolean debounce(int pin)
{
  boolean currentState;
  boolean previousState;
  int i;

  previousState = digitalRead(pin);
  for (i = 0; i < DEBOUNCEDELAY; i++) {
    delay(1);       // small delay
    currentState = digitalRead(pin);        // Read it again...
```

LISTING 4-4   The RTC program. (*continued*)

```
      if (currentState != previousState)      // If not same, could be bouncing...
      {
        i = 0;                                 // ...so try another pass...
        previousState = currentState;
      }
    }
    if (currentState == LOW)
      return true;
    else
      return false;
}

/*****
  This function displays the date and time returned from the RTC on the LCD
display.
  Parameter list:
    void
  Return value:
    void
  CAUTION: This function assumes that all of the date-time variables have been
read from the RTC before this function is called.
*****/
void DisplayDateAndTime(byte *ptr)
{
    lcd.setCursor(0, 0);                         // First LCD display line
    lcd.print(weekdayname[*(ptr + DAY)]);        // Note: you can use a pointer with an
                                                 // offset, or...
    lcd.print(" ");
    DigitFill(ptr[MONTH]);                       // ...you can use it as an array index
    lcd.print("/");
    DigitFill(ptr[DATE]);
    lcd.print("/");
    DigitFill(ptr[YEAR]);

    lcd.setCursor(4, 1);                         // Second LCD display line
    DigitFill(ptr[HOURS]);
    lcd.print(":");
    DigitFill(ptr[MINUTES]);
    lcd.print(":");
    DigitFill(ptr[SECONDS]);

    if (activeTimer) {
      TimerIndicator(ptr);    // Display asterisk at line end if timing.
    }
}

/*****
  This function determines if the value passed in requires one or two digits
to display. If the number is less than 10, the display is padded with a
leading (tens units) of '0' and the units digit is displayed. Otherwise, the
tens and units digits are displayed.
```

LISTING 4-4   The RTC program. (*continued*)

```
  Parameter list:
    byte digit        A number
  Return value:
    void
*****/
void DigitFill(byte digit)
{
  if (digit < 10)
    lcd.print('0');
  lcd.print(digit);
}

/*****
  This function is used to indicate whether the ID timer is active or not.
When the timer is not active, the LCD acts like a simple clock. When the timer
is active, the LCD display blinks at the end of the ID period. This function
places an asterisk in column 15 of the display only when the timer is active.
The asterisk alternates between rows 0 and 1 every FLIPDELAY seconds.
  Parameter list:
    byte value        A number expressed in BCD
  Return value:
    byte              The BCD value expressed as a decimal number
*****/
void TimerIndicator(byte *ptr)
{
  static byte flip = 0;
  static byte oldSeconds, newSeconds;

  if (activeTimer == false)
    return;
  newSeconds = ptr[SECONDS];
  if (newSeconds == 0) {      // rollover at 0 seconds, leaving oldSeconds at 59,
                              // which would stop the display of the asterisks
    oldSeconds = 0;           // from then on, so reset it.
  }

  if (newSeconds - oldSeconds >= FLIPDELAY) {     // Flip every FLIPDELAY seconds
    if (flip) {
      lcd.setCursor(15, 0);
      lcd.print("*");
      lcd.setCursor(15, 1);
      lcd.print(" ");
      flip = 0;
    } else {
      lcd.setCursor(15, 1);
      lcd.print("*");
      lcd.setCursor(15, 0);
      lcd.print(" ");
      flip = 1;
    }
```

LISTING 4-4   The RTC program. (*continued*)

```
      oldSeconds = newSeconds;
    }
  }
/*****
  This function converts the byte-length Binary Coded Decimal value returned
from the RTC to a decimal number.
    Parameter list:
      byte value         A number expressed in BCD
    Return value:
      byte               The BCD value expressed as a decimal number
*****/
byte BcdToDec(byte value)
{
  return ((value / 16) * 10 + value % 16);   // Think about it...
}

/*****
  This function converts the byte-length decimal value to a Binary Coded
Decimal value.
    Parameter list:
      byte value         A number expressed as a decimal value
    Return value:
      byte               The decimal value expressed as a BCD number
*****/
byte DecToBcd(byte value)
{
  return (value / 10 * 16 + value % 10);   // Think about this one, too.
}

/*****
  This method blinks the Arduino board LED if an error occurs. The error
type is coded into the blink rate, as explained below. There is a 10 second
pause between the error blick rate. Note that you could assign LEDPIN to a
different pin and run the LED "outside" of the Arduino board to make it more
visible.
    Parameter list:
      int                integer that defines type of error. Error values are #define
                         for: RTCREADERROR = 2 and means we could not read the RTC
                         chip, and RTCWRITEERROR = 1 which means we could not write
                         to the chip.
    Return value:
      void
*****/
void SetError(int error)
{
  int i;
  int repeat = ERRORREPEAT;    // Number of times to repeat error indicator
  while (repeat--) {           // Go back to the caller after this many
                               // repeats of blinking the error LED
  {                            // for 10 seconds
```

LISTING 4-4   The RTC program. (*continued*)

```
    for (i = 0; i < error; i++)
    {
      digitalWrite(LEDPIN, HIGH);
      delay(HALFSECOND);
      digitalWrite(LEDPIN, LOW);
      delay(HALFSECOND);
    }
    delay(TENSECONDS);            // wait 10 seconds, repeat up to ERRORREPEAT times.
  }
}

/*****
  This method reads the high bit of the Seconds register of the DS1307 to see
if the clock is running.
  Parameter list:
    void
  Return value:
    byte             1 if the RTC clock is running, 0 if not.
  CAUTION: This function is currently not used, but is present to show you one
way to read a bit value from a register if you so desire (e.g., change from 24
to 12 hour format and implement AM/PM indicator.)
*****/
byte Isrunning(void)
{
  byte sentBack;
  uint8_t secondsRegister;

  Wire.beginTransmission(RTCI2CADDRESS);    // Wake up
  Wire.write(0);                            // Set RTC register pointer to 0
  Wire.endTransmission();

  sentBack = Wire.requestFrom(RTCI2CADDRESS, 1);// Read Seconds register byte
  if (sentBack == 0)                        // It didn't like that request...
    SetError(RTCREADERROR);

  secondsRegister = Wire.read();
  return !(secondsRegister >> 7);           // Get Clock Halt (CH) bit, invert it, and
                                            // return it
}
```

LISTING 4-4 The RTC program. (*continued*)

The software presented in Listing 4-4 is designed to be used in conjunction with the shield schematic depicted in Figure 4-5. We should point out that for the first RTC shield we built, Jack installed the RTC clock module you see in the middle of Figure 4-11 with the battery under the RTC module. While this approach makes it easier to see what the pins on the module are, changing the battery is going to be a problem. This is why the construction details for the RTC installed the RTC module with the battery side up. However, given the projected battery life associated with the DS1307 and using a CR2032 button cell, we have about 10 years before we need to think about changing the battery for the shield Jack built upside down.

Running the Program

When the timer is first turned on, the display shows the current time and causes a counter (*millisStart*) to record the current number of milliseconds as stored in the Arduino via a function call to *millis()* in the *setup()* function. On each pass through *loop()*, the current millisecond value is read via another call to *millis()* and that value is stored in *millisCurrent*. When the difference between *millisStart* and *millisCurrent* is 10 minutes (*IDTIMESLICE*, or 600,000 milliseconds), the code blinks the LCD display. The LCD display continues to blink until the user presses the pushbutton (PB1). Once the user presses the pushbutton, *millisStart* is reset and the 10-minute timer period starts again. We call this mode the Timer Mode.

If PB1 is pushed, but held for more than MODESWITCHDELAY seconds, the RTC clock toggles to its alternative state. For example, if the clock is in Timer Mode and the user pressed PB1 for more that MODESWITCHDELAY seconds, the clock switches to Clock Mode. We chose to call it Clock Mode because ... wait for it ... in Clock Mode, the RTC clock acts as a standard clock. This means the RTC is not acting as a timer. You can tell which mode is active by the presence (Timer Mode) or absence (Clock Mode) of asterisks in column 15 of the LCD display. In Timer Mode, the asterisks bounce between rows 1 and 2 on the display (see Figure 4-1). In Clock Mode, no asterisks are present.

The LCD shield you built in Chapter 3 stacks on top of the RTC shield shown in Figure 4-11. (See Figure 4-12.) The "Arduino sandwich" shown in Figure 4-12 is not uncommon in Arduino projects. Indeed, being able to stack shields is one of the real strengths of the Arduino design, as it adds a tremendous amount of flexibility to the functionality of the Arduino.

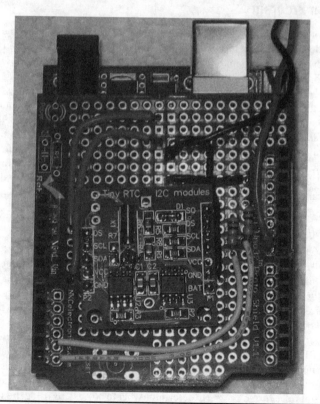

FIGURE 4-11 Jack's initial RTC shield with hidden battery.

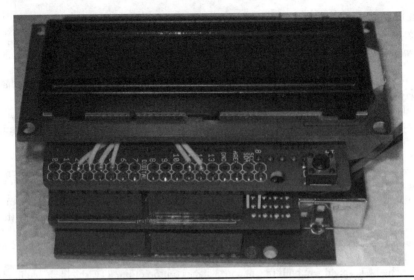

FIGURE 4-12   The Arduino on the bottom, the RTC in the middle, and the LCD shield on top.

The RTC Timer Program

There's not too much new in the code presented in Listing 4-4. Note that the program does assume that you downloaded and installed the RTClib library mentioned earlier in the chapter. The code defines several global variables from the libraries, including *lcd* for the LCD display and *rtc* for the RTClib library.

The *beginTransmission()* method puts the RTC device address (i.e., RTCI2CADDRESS or 0x68) on the I2C bus and writes a 0 on the bus to wake up the RTC module and tell it that some data for it are coming across the bus and reset its internal register pointer. After the data have been sent, the *endTransmission()* method of the Wire object tells the DS1307 that is has all of the data. If everything went correctly, *endTransmission()* sends the value 0 back from the call. As mentioned earlier, the sequence of the *Wire.read()* and *Wire.write()* method calls is important because of the way the DS1307's internal register pointer works. If you observe screwy data coming back from the clock module, this is probably the first thing you should check.

The *loop()* Function

The *loop()* function is designed to simply update the clock on each pass through the loop. It begins with a call to *millis()* and assigns the data returned into *millisCurrent*. If the RTC is in Timer Mode, we check to see if the 10-minute time slice has been exceeded. We can do this because *millisStart* was set in *setup()*, so the difference between *millisCurrent* and *millisStart* is the elapsed time since the clock started timing, in milliseconds. If the 10 minutes has expired, we flash the LCD display and call the *DisplayDateAndTime()* function to update the display.

Next, the code checks to see if the user pressed the pushbutton switch, PB1. As mentioned earlier, PB1 has two functions: 1) to reset the timing period when in Timer Mode, or 2) to switch between modes. The length of the press on PB1 determines which action is taken. If the user holds PB1 longer than MODESWITCHDELAY seconds, *featureChanged* is set to *true*. Subsequent code uses the state of *featureChanged* to determine what to do next. By using the states of *featureChanged* and *activeTimer*, the code toggles the program into the desired mode.

Once the state of the program is known, a call to *Wire.beginTransmission(RTCI2CADDRESS)* is used to reset the internal register pointer of the DS1307 chip. The DS1307 maintains the current state of this pointer internally and if you don't reset it yourself, it simply increments the pointer and reads whatever it happens to be pointing to. The register pointer is smart enough to "roll over" if your *Wire.read()* calls cause the pointer to point past the timer registers. Clearly, forgetting to reset the register pointer can cause the data returned from the *Wire.read()* calls to be out of sync with what you are expecting.

The call to *Wire.requestFrom()* is used to place the device address on the I2C bus and inform the device we want to retrieve 7 bytes of data from it. The method returns how many bytes are ready to be sent, which we assign into *sentBack*. If something went wrong, *sentBack* is set to 0 and our error function is called. Assuming no error occurred, *sentBack* should have the value of 7. We use that value to control a *for* loop to stuff the data from the RTC into the *timeBuffer[]* array. Again, the DS1307 only speaks BCD, so we must convert the data back to decimal before we place it in the *timeBuffer[]* array. This is why the *Wire.read()* calls are nested within the *BcdToDec()* function calls. The call to *DisplayDateAndTime()* simply moves the data from the *timeBuffer[]* array to the LCD display. After the data are displayed, the *loop()* method repeats the process again.

If 10 minutes passes without pressing the pushbutton, the LCD display starts blinking at a half second rate. The LCD display retains the time at which the display started blinking. A quick glance at your wristwatch and the LCD time will quickly tell you how long it's been since you *should* have identified yourself. As a rough approximation, figure one month of jail time for every minute you're late in identifying yourself.

You should pick an enclosure to house the cluster shown in Figure 4-12. Keep in mind that once you disconnect the USB cable, the display darkens, but the RTC clock keeps ticking by drawing power from the battery. Therefore, you should use a small power source (i.e., "wall wart") that plugs directly into a wall socket and supplies between 7 VDC and 12 VDC. The Arduino's onboard voltage regulator adjusts the incoming voltage to the requisite 5 V expected by the shields. Also, keep in mind that you need to mount the pushbutton in a way that makes it easily accessible for the user. You could add a power switch, but we just disconnect the wall wart when we don't want the clock running.

A Software Hiccup

As stated earlier, we initialize the DS1307 registers with the date and time data from the host PC. (You can hard code the data, but that gets very tedious after recompiling the code several dozen times.) Obviously, initializing via the PC means we need to establish a *Serial* object in the code to handle the communications across the USB cable. When we uploaded the compiled code to the Arduino, the LCD display worked like a champ and displayed the proper time. No problem.

In theory, when we disconnect the Arduino from its power source (e.g., the USB cable), the onboard battery kicks in and maintains the RTC clock via battery power. So, we disconnected the USB cable and let the board sit for 30 minutes or so, then plugged a wall wart into the Arduino's external power plug and observed the time on the LCD display. The time that was displayed was the time when we disconnected the USB cable from the Arduino, *not* the current time. Really? It appears that the DS1307 was storing the time and date data, probably somewhere in the 56 bytes of nonvolatile RAM that the chip has. This was puzzling.

We knew that the code should reinitialize the DS1307 each time we compiled and uploaded the code. However, the code should be set-it-once-and-forget-it ... at least for 10 years or so. While debugging the code, we used the standard technique of toggling the debug code into and out of the program using the following type of preprocessor directive

```
#ifdef DEBUG
// Debug statements
#endif
```

which we used to encapsulate all of the *Serial.print()* calls, so commenting out the line:

```
#define DEBUG 1
```

at the top of the program and recompiling has the effect of removing all of our debug statements using the *Serial* object from the program.

Well … sort of.

As it turns out, we made yet another flat forehead mistake: We forgot to surround the:

```
Serial.begin(9600);
```

statement that is called in *setup()* with our *#ifdef* preprocessor directives. We are embarrassed to admit that we chased this "undocumented feature" of the program for over a day before we realized that the *Serial.begin()* method call causes the bootloader to perform a reset on the board. Evidently, the reset causes the DS1307 to cache the date/time data and shut down the chip's internal counter. When power is reapplied, it restarts the internal counter, but using the values that were stored when the reset was sensed.

So what's the solution? One solution is to simply compile and upload the code twice. The first time you maintain the link to the host PC with the *Serial* object so you can initialize the DS1307 registers with the current date and time. The second time, comment out the *all* of the code that uses the *Serial* object that fetches the date/time data from the PC or sends debug data to the host PC. Then simply recompile and upload the "No PC Talk" version of the code to the Arduino. Once we followed these procedures, the RTC performed as designed.

Instead of this compile-edit-compile process, we broke the initialization of the clock out into a separate program. This initialization program is the code presented in Listing 4-2. The code in Listing 4-4 is then used for the next 10 years.

Conclusion

In this chapter you have examined both a software and a hardware approach to creating a station ID timer. There are many areas where you can improve upon the design presented in this chapter. To avoid jail time, you could wire in a 200 W claxon (or perhaps a smaller buzzer) to sound instead of the blinking display. It's easy to miss the blinking display: A 200 W horn … not so much. You could also try your hand at building an RTC module of your own from the chip. There are also a variety of ways that you can implement the Reset button once the 10-minute time period has passed. You might also want to blink the display at 9 minutes and 30 seconds to give the operator some time to end the current transmission. (Then fire off the claxon at 10 minutes because they ignored your warning.) To test your understanding of the DS1307 registers, try changing the software so the clock uses a 12-hour format rather than the 24-hour format we prefer. With a little imagination, there are a lot of modifications you might make to the hardware and software, if you are so inclined. If you do so, we hope you'll post those on the book's web site.

Have fun and experiment … it's a great way to learn.

CHAPTER 5

A General Purpose Panel Meter

As we mentioned in Chapter 2, one important aspect of any µC system is to be able to display data. Panel meters are found in many "homebrew" projects. Any radio amateur who's been around a while is bound to have a few old meters hanging around the workshop. We find them at the swap meets or remove them from some old piece of gear we've parted out. Chances are good, however, that the one you found in your junk box has unknown characteristics and a scale that needs to be replaced. Although replacing a scale is not too difficult—there is software available to print a new scale—you run the risk of damaging the delicate meter movement while swapping the scales.

This chapter's project creates a general purpose panel meter that is designed to read 0–1 mA full scale. The project presents both digital and analog displays on the two-line LCD constructed in Chapter 3. Changing the range of the meter is easy using external resistors, and we provide examples of how this is done. Because the display is generated by software, it becomes trivial to change the displayed range; it's just a text string. We also show a simple means of calibrating the meter so that it reads correctly. Figure 5-1 shows the meter in action.

FIGURE 5-1   The general purpose panel meter in action.

Circuit Description

The panel meter is modeled after a typical D'Arsonval 1 mA meter. In this case, the meter presents approximately a 50 W load with 1 mA of current flowing. The actual value of the input resistor (R1 in Figure 5-2) is not critical, but it is needed to calculate any external scaling that is utilized. We used a value of 51 Ω in the current version. Ohm's law tells you that 1 mA of current through a 50 Ω resistor produces a voltage of 50 mV. That 50 mV is then read by the Arduino analog input.

The analog input uses an analog-to-digital converter, or "ADC," to represent the analog voltage as a digital value. The digital value is 10 bits in length or, to look at it in a different way, the analog input is represented by 1024 digital values (0000 through 1023). However, the input range of an Arduino analog pin is from 0 V to 5 VDC (5 VDC = 1023) and this presents a problem. Our voltage across the input resistor is only 50 mV. If 50 mV is fed to the ADC, the highest value from an analog read is 0010 and the accuracy and resolution of the incremental values would be very poor. We would be able to display only 11 discrete values. This would not be a very good panel meter.

Instead, we use an operational amplifier (op amp) (U1a in Figure 5-2) to amplify the DC input voltage to a higher value. We chose an LM324 op amp for this project. The LM324 is a quad op amp, meaning that there are four identical op amps in a single IC package. There are single op amps available, such as an LM741. However, we chose to go with the LM324 for several reasons. First, it operates off of a single voltage supply without additional biasing. Second, they are readily available and inexpensive, usually less than a $1 each. Third, we plan on adding some additional circuitry to the shield in future projects and will make use of those additional, unused, op amps later on.

Because the op amp is running on 5 VDC supplied by the Arduino, the maximum output voltage from the op amp is about 3.5 VDC—a function of the design of the op amp. While not perfect, the LM324 provides better range than a maximum input voltage of 50 mV. Resistors R2, R3, and R4 in Figure 5-2 set the gain of the op amp (GAIN = 1 + ((R3 + R4) / R2)). R3, a 10-turn potentiometer, provides the means to make fine adjustments to the gain of the amplifier. With the gain set to 70, we would have roughly 716 values giving us greater resolution of the meter's input current.

NOTE: According to the LM324 datasheet, the maximum output voltage would be

$$VCC - 1.5 = 5 - 1.5 = 3.5$$

If we apply the GAIN formula using the values suggested in Figure 5-2, then:

$$GAIN = 1 + ((R3 + 270K) / 4.7K)$$

and the range of values for GAIN is a minimum of about 57 to a maximum of almost 79.

The bad news is that adding the op amp is still not optimum because we would like to have our maximum input provide a full count of 1023 in the ADC. However, the good news is that there is an easy cure! The Arduino includes a little feature on a pin called "AREF." AREF, or "analog reference," allows us to provide an external reference voltage to the ADC. The ADC samples the voltage present on the analog input pin using the analog reference for comparison. The voltage used as the analog reference represents the largest voltage that can be measured on the analog input. The Arduino defaults to the internal reference voltage of 5 VDC, which means that, in order to have a reading on the analog input of 1023 counts, we would need 5 VDC on the analog input. Because our op amp circuit can only produce about 3.5 VDC maximum, we can configure the Arduino to use an external reference, with 3.5 VDC applied to AREF, in order to obtain a count of 1023 when the op amp is producing the maximum output. So, how to produce 3.5 VDC?

One nice feature of using a quad op amp, such as the LM324, is that not only are they in the same package, all four devices share the same substrate material on the integrated circuit die; therefore, their electrical characteristics tend to be almost identical. We can take advantage of this "feature" to create a voltage of 3.5 VDC to use as the external analog reference. With an unused section of the LM324, we create a 3.5 VDC reference by configuring the op amp as a voltage follower (a non-inverting amplifier with unity gain) and pull the non-inverting input to the positive voltage rail. With 5 VDC on the non-inverting input, the output of the op amp is 3.5 VDC and the same as the maximum voltage that is produced by the DC amplified section of our circuit. Now when we apply the maximum input value to our meter circuit, the ADC provides a count of 1023 rather than 716 as we described earlier when using the internal reference.

Note how the output of the LM324 feeds into the AREF pin on the Arduino. This means that the AREF pin receives the 3.5 VDC, which serves as the internal reference voltage for the ADC circuit of the Arduino. The line in Listing 5-1 (presented toward the end of the chapter):

```
analogReference(INTERNAL);    // set the A/D to use the AREF input
```

is responsible for establishing the reference voltage for the Arduino. Note that you can change the symbolic constant INTERNAL to EXTERNAL for debugging purposes without having to use the reference voltage. To use an external reference voltage on the AREF pin, you must call *analogReference(EXTERNAL)* before calling *analogRead()*. **CAUTION:** Do not apply less than 0 V or more than 5 V to the AREF pin, otherwise you might "brick" your Arduino. (The term "brick" means to transform your Arduino board from a useful electronic μC into a silicon brick.) You can use *analogReference(INTERNAL)* to generate "fake" analog readings (i.e., *analogIn* in Listing 5-1) for testing purposes. This proved helpful while the hardware was being built in California but the software was being written in Ohio!

As can be seen in Figure 5-2, section U1b of the LM324 is configured as a voltage follower with the non-inverting input connected to 5 V through a 100K resistor (R5), driving the output of the follower to the maximum voltage or about 3.5 VDC. A parts list for the panel meter is presented in Table 5-1.

Another requirement of a digital panel meter is to indicate when the input value exceeds the maximum range. Unlike a mechanical panel meter whose indicator can move past the full-scale

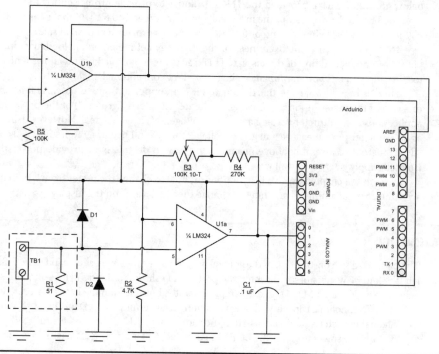

FIGURE 5-2 Schematic of the general purpose digital panel meter.

C1	0.1 µF 50V Monolithic capacitor
D1, D2	1N4001, 50V/1A rectifier
R1	51 Ω, ¼ W, 5% resistor
R2	4.7 kΩ, ¼ W, 5% resistor
R3	100 kΩ, 10-turn potentiometer
R4	270 kΩ, ¼ W, 5% resistor
R5	100 kΩ, ¼ W, 5% resistor
TB1	2 Pins, 5mm Pitch AC 250V 16A Block Terminal Connector
U1	LM324, Quad op-amp and 14-pin DIP socket
Misc	prototype shield, LCD shield, Arduino, 26-28 AWG Hookup wire

TABLE 5-1 General Purpose Panel Meter Parts List

value (or becomes "pegged"), the digital panel meter reaches the full-scale value and goes no higher. It is the nature of a digital panel meter design that, as the input voltage increases above the 50 mV full-scale value, the displayed value does not increase. The op amp is "saturated" and can't produce a higher output voltage for the ADC to read. Because the ADC can count to 1023, and we want to be able to detect the overrange condition, we arbitrarily set a count of 1000 to equate to 1 mA, giving us the range of 1001 through 1023 to indicate the overrange values. So, what we want is to have 1 mA input current equal to something less than count of 1023. The beauty of using the AREF input is that, if we set the gain of the op amp so that 1 mA input current equals a count of 1000, the output voltage is less than the maximum, giving us the extra "headroom" to detect the overrange values. Anything over 1000 and the meter is "pegged" and displays the overrange message.

Because the LM324 input has a maximum rating that cannot exceed the supply voltage or below ground, we have provided the 1N4001 diodes to provide protection for the op amp input circuits. The diodes do "clamp" the input voltage approximately 0.6 VDC above the positive supply voltage or 0.6 VDC below ground or 0 VDC. The diodes provide a margin of safety for the op amp.

To further improve the usefulness of the digital panel meter, an analog bar graph has been placed on the second line of the display. This tracks the input quickly and allows a nice indicator for fine adjustments you may be making, much like a conventional analog meter.

There is one drawback to this design and it is an important one to note. As it is constructed, one side of the meter input is connected to the negative lead ("ground") of the power source. For applications as a voltmeter reading a positive voltage, this is no big deal. However, using the meter as an ammeter or to measure a negative voltage, the ground becomes an issue. The solution is to "float" the entire assembly above ground. One way to do this is to provide power through an isolated supply, such as a "wall wart."

Later in the chapter we give instructions on how to change the meter from a milliammeter to a voltmeter or an ammeter, and how to change the unit's value on the display.

Construction

The parts used for construction of the general purpose panel meter are shown in Figure 5-3. At the top left, the four ¼ W resistors; top center, the prototyping shield; to the right, a 14-pin DIP socket; and below the socket, the LM324 Quad op amp. Across the bottom from the left: R3, a 100K 10-turn pot; C1, a 0.1 µF monolithic capacitor; the two position screw terminal; and D1, D2, the 1N4001 diodes.

The panel meter shield is constructed on an Omega MCU Systems ProtoPro-B prototyping shield, the same shield used for the LCD shield and RTC/Timer projects. Figure 5-4 shows how we

Chapter 5: A General Purpose Panel Meter

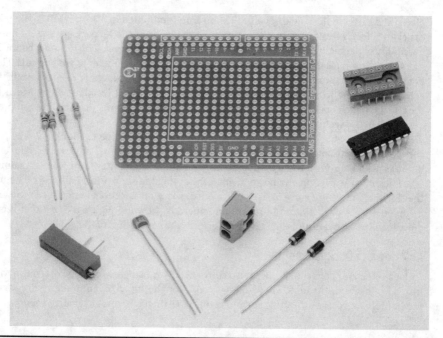

FIGURE 5-3   Parts used for the general purpose panel meter.

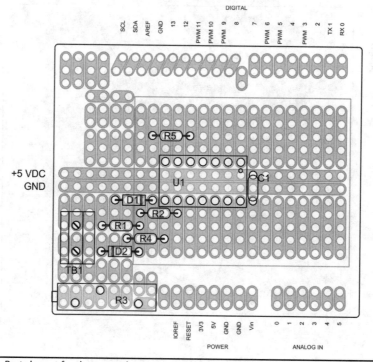

FIGURE 5-4   Parts layout for the general purpose panel meter shield.

placed the parts on the shield. The LM324 op amp is mounted in a 14-pin DIP socket. It is much easier to solder wires to the socket rather than the leads of the IC package and it eliminates the possibility of damaging the IC with too much heat while soldering. The terminal connector block (or "screw terminal") is used to make it easier to attach the panel meter to your circuit. The 10-turn pot should be mounted so that it can be accessed with the LCD shield or display installed, with the adjustment screw to the outside edge of the shield. You need to adjust the 10-turn pot while the panel meter is operating in order to calibrate the meter reading. (The calibration procedure is described in the section on testing and calibration of the meter.)

We use the same wiring techniques described in the preceding chapters, solid 24 AWG, bare wire with Teflon tubing as insulation where needed. Figure 5-5 shows how we wired up this version of the panel meter. We have left ample room for additional circuitry to be added at a future time.

The actual wiring on the bottom side of the shield is shown in Figure 5-6. The top side of the shield is shown in Figure 5-7. There is lots of space for additional circuitry in the future. The completed stack of shields and the Arduino board is shown in Figure 5-8. All that remains is to download the software and then test and perform the calibration of the panel meter.

An Alternate Design Layout

As an alternative, a panel meter was also constructed on a single Mega shield. Using a Mega shield instead of a simple Arduino shield allows room for the input "conditioning" circuitry and also leaves room for future projects that employ the panel meter. In this alternative version, rather than

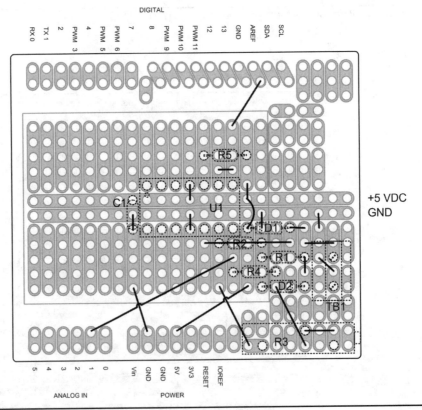

FIGURE 5-5 Wiring diagram for the general purpose panel meter.

Chapter 5: A General Purpose Panel Meter   107

FIGURE 5-6   Bottom view showing wiring of general purpose panel meter shield.

FIGURE 5-7   View of completed general purpose panel meter shield with LCD removed.

FIGURE 5-8 The completed general purpose panel meter.

using the LCD shield constructed in Chapter 3, we chose to mount the LCD display directly to the meter shield using socket headers so that the display may be removed to add other projects. Figure 5-9 shows the top of the shield and component placement. This particular Mega shield has only two buses: one for Vcc (+ 5 VDC) and a second for ground. There are no other interconnected pins that we can use. Connections to the header pins are made in the same manner as shown in

FIGURE 5-9 General purpose panel meter alternative Mega shield layout.

Chapter 5: A General Purpose Panel Meter

FIGURE 5-10   General purpose panel meter alternative Mega shield layout showing wiring.

Chapter 3, Figures 3-16 through 3-19. The wire is wrapped around the stubby end of the pin and soldered. You can see this on the bottom side of the shield in Figure 5-10.

The completed alternative design is shown in Figure 5-11 with the LCD installed. Note that the parts and the circuit are the same, only the position of the parts is different and so is the shield when compared to the first version of the meter.

FIGURE 5-11.   Completed alternative general purpose panel meter.

Loading the Example Software and Testing

The general purpose panel meter software sketch is provided in Listing 5-1. Use the Arduino IDE to compile and upload the sketch into your project.

> **WARNING:** One very important point while uploading, DO NOT upload the sketch for the first time with the panel meter shield in place. To do so can seriously damage your Arduino. As mentioned earlier, the Arduino is designed to allow an external reference voltage for the A/D converters. The general purpose panel meter shield places a reference voltage on the AREF pin and unfortunately, the internal reference voltage is connected directly to the external reference voltage until the AREF pin is assigned in the code as using an external source. The best practice here is to compile and upload the code, disconnect your USB cable from the Arduino, install the panel meter shield, and then reconnect the USB cable. This prevents you from damaging the Atmel processor.

The software written for the general purpose panel meter uses two libraries. One we have used before with the LCD shield in Chapter 3. The second is a new library that we haven't used before and is the LCD Bar Graph library and may be freely downloaded from the Arduino "Playground" at:

http://playground.arduino.cc/Code/LcdBarGraph#Download

The latest version of the library as of this writing is 1.4.

```
/* Simple LCD Panel Meter ver1.0 23 December 2013 D.Kidder W6DQ and
   J.Purdum, W8TEE

Features:
  * Along with associated hardware, displays 0-1 mA fullscale
  * Displays overrange condition when input exceeds 1 mA
  * Displays an analog bar graph on the second line of LCD

The meter is user-configurable by changing the scaling value and scale.
As presented, the associated hardware produces a 0-3.5 VDC output to the
analog input of the Arduino. The associated hardware also provides a
3.5 VDC source to the AREF input (analog reference), thus at 1 mA input,
the A/D count as 1000. The hardware includes over- and under-voltage
protection. The meter hardware presents a 51 Ohm input impedance.

The user may change the scale by changing the value of the "scale" string. The
meter range may be modified by changing the "scaleFactor" variable. Voltage
dividers or current shunts may be added to change the hardware range.

*/

#include <LiquidCrystal.h>   // Standard Arduino LCD library
```

LISTING 5-1   The general purpose panel meter program.

Chapter 5: A General Purpose Panel Meter

```c
/*
 * LCD Analog Bar Graph library (with modifications)
 *
 * Author: Balazs Kelemen
 * Contact: prampec+arduino@gmail.com
 * Copyright: 2010 Balazs Kelemen
 * Use is covered under the GNU GPL
 * This library creates an analog bar graph on the LCD
 */
#include <LcdBarGraph.h>

#define DEFAULTPRECISION  3         // default decimal places
#define MAXANALOGVALUE    1000      // Set valid input to 0 - 1000
#define VALUEWIDTH        6         // Numeric field width, include '-', '.'

#define UPDATEDELAY       100       // Delay for normal reading
#define ERRORDELAY        400       // Delay if error message displayed
#define STARTUPDELAY      3000

#define BARGRAPHROW       1         // Draw graph on second row
#define BARGRAPHCOL       0         // Start graph in column 0

#define LCDNUMCOLS        16        // number of columns on LCD
#define SENSORPIN         A1        // define the analog input as A1

/*
 * The user can modify the following two lines to change scale and scale factor.
 *
 * "scale[]" provides the displayed units of measurement. Replace the string within
 * the quotes to assign the units you require. E.g: "Volts" "Amperes" etc.
 *
 * "scaleFactor" is used to set the maximum displayed range of the measured value.
 * The default value of 1000.0 equates 1 mA input current (a count of 1000 in
 * the ADC) to 1.00 mA displayed. For example, if you wish to display 10 Volts
 * full scale, then 1 mA input current would be an ADC count of 1000 and the scale
 * factor would be 100. The displayed range would be 0. to 10.00.
 *
 */

char  scale[] = "mA";           // scale string for display units
float scaleFactor = 1000.0;                // scaling factor for the meter

LiquidCrystal lcd(12, 11, 5, 4, 3, 2);       // setup the LCD hardware
                                              // interface pins
LcdBarGraph   lbg(&lcd, LCDNUMCOLS, BARGRAPHCOL, BARGRAPHROW);

void setup()
{
  analogReference(EXTERNAL);      // set the A/D to use the AREF input
```

LISTING 5-1 The general purpose panel meter program. *(continued)*

```
  lcd.begin(2, LCDNUMCOLS);           // initialize the LCD

  lcd.print("LCD Panel Meter ");      //sappy stuff, name, version and date
  lcd.setCursor(0,1);
  lcd.print("ver 1.0 23DEC 13");
  delay(STARTUPDELAY);
  lcd.clear();
}

void loop()
{
  int i;
  int analogIn;
  int len;
  char buffer[LCDNUMCOLS + 1];   // Make it big enough for a 4x20 display
  float val;

  analogIn = analogRead(SENSORPIN);       // Read the analog input value

  val = (float) analogIn / scaleFactor;   // Scale the input reading...
  dtostrf(val, VALUEWIDTH, DEFAULTPRECISION, buffer); // Convert and format value

  buffer[VALUEWIDTH] = ' ';               // Append a space...

  i = 0;                                  // ...now add the scaling factor.
  while (scale[i]) {
    buffer[VALUEWIDTH + 1 + i] = scale[i];
    i++;
  }
  buffer[VALUEWIDTH + 1 + i] = '\0';

  lcd.setCursor(0,0);                     // Set up the display...
  lcd.print(buffer);

  if (analogIn <= MAXANALOGVALUE) {       // If analog value within scale...
    lbg.drawValue(val, MAXANALOGVALUE);   // show analog value of input read
    delay(UPDATEDELAY);
  } else {                                // ...otherwise complain about it.
    lcd.setCursor(0, 1);
    lcd.print("   Overrange!   ");
    lcd.noDisplay();                      // flash the display on and off
    delay(ERRORDELAY);
    lcd.display();
    delay(ERRORDELAY);
  }
}
```

LISTING 5-1 The general purpose panel meter program. (*continued*)

Code Walk-Through

There are two libraries used here. The first is one we are familiar with and is the LCD library LiquidCrystal.h that is part of the Arduino IDE. We have added a new library, LcdBarGraph.h, that allows us to have the analog bar graph on the bottom row of the LCD to represent the numeric value displayed on the top row. Follow the steps described in Chapter 4 for RTClib.h to install LcdBarGraph library files.

DEFAULTPRECISION is used to set the number of decimal places that are shown on the LCD. In this case we used 3, which is probably a bit overambitious. If you think about it, a good rule of thumb for a panel meter is that any reading is going to only be readable to half of the smallest scale division. If the typical meter has 50 scale divisions, half of that would be 1/100 of full scale. Depending on how accurate your calibration source is, one might expect that this meter could achieve the same accuracy, so a better choice might be to use "2" for the number of decimal points.

Of course, you have seen the next line before whenever we use the LCD. *LCDNUMCOLS* sets up the number of columns to display on the LCD. This is the first time we have used an analog input and the line:

```
#define SENSORPIN   A1
```

assigns our analog input to be analog pin 1 on the Arduino. (We use A1 instead of 1 to denote the analog pin assignment, as that symbolic designation for analog pin 1 is known by the IDE. We think using A1 better documents the fact that analog pin 1 is being used in the assignment statement.) The analog input uses an analog to digital converter, or ADC, to provide a digital representation of the analog value. The ADC is a 10-bit device meaning that the output provides a count of 2^{10} or a range of values between 0000 and 1023 (decimal).

The next two lines are very important for tailoring the meter to your application. These are the *scale* and *scaleFactor* definitions. The character string defined by *scale* allows you to display the units of measurement of your meter. We have used the string *mA* because our meter is a milliammeter. You can replace this string with anything you would wish to use. Maybe you are measuring speed, so "furlongs per fortnight" might be your desire. Unfortunately, that one won't fit, as there are only 9 positions on the display for the scale (unless you change the default precision to 2, in which case there would be 10 positions), so you might use "flngs/fort" as your units. You may be just as likely to use "Amperes" or "Volts."

The scale factor is equally important in that it determines the full-scale value that you can display. The default is a trivial case where 1000 counts on the ADC is equal to a displayed value of 1.000 mA. But what if you wish to display 20 V as the full-scale value? Take a look at the line a little bit farther down in the code within the "loop" portion of the code. The line:

```
val = (float) analogIn / scaleFactor;
dtostrf(val, VALUEWIDTH, DEFAULTPRECISION, buffer);
```

is the one we are interested in. Ignoring the function call *dtostrf()* for the moment, here we read the analog input on analog pin 1 and divide by the scale factor and assign it into *val*. In this example, we want the count of 1000 to equal 20.00 so the scale factor would be 50 because 1000 divided by 50 would result in a value of 20.00. We discuss scale factor a bit more when we talk about the hardware side of changing the scale and range in an upcoming section.

Instantiating the *lcd* and *lbg* Objects

The line:

```
LiquidCrystal lcd(12, 11, 5, 4, 3, 2);
```

is also one we've seen before and is used to set up the hardware pins used by the LCD. We are creating an object named *lcd* using the class named "LiquidCrystal." The object, *lcd*, uses the object constructor that asks for the pin numbers that we have chosen for the LCD hardware interface. These pin numbers are passed to the constructor as parameters that initialize the object to talk to our LCD hardware.

The next line is new and is associated with the bar graph library. Look at this line:

```
LcdBarGraph    lbg(&lcd, LCDNUMCOLS, BARGRAPHCOL, BARGRAPHROW);
```

Here we are calling the C++ *LcdBarGraph* class constructor to create an object named *lbg*. The constructor initializes certain members of the class to specific values: 1) a reference (i.e., the memory address, or lvalue, as denoted by the "&" operator) to the object *lcd* (that was just instantiated), 2) define the number of columns used for the bar graph, and 3) places the start of the bar graph in column 0 of row 1. This places the bar graph in the bottom row of the LCD (remember that the rows and columns start with 0).

We learned about *void setup()* in Chapter 2. The next line:

```
analogReference(EXTERNAL);
```

is how we tell the Arduino to use the AREF pin as an external reference voltage. You recall that the voltage applied to AREF when set to *EXTERNAL* allows us to have a full count on the ADC (1023) when the analog input voltage is equal to the AREF input.

The next series of statements should also be familiar by now. Each of the statements begins with the object name, *lcd*. Each statement calls the stated function within an *lcd* object as defined within the *LiquidCrystal* class. So *lcd.begin()*, *lcd.clear()*, *lcd.home()*, *lcd.print()*, and *lcd.setCursor()* are all class methods available to perform some specific action on the display. We use *lcd.begin()* to set the number of rows and columns that are to be used by the display. We then use *lcd.print()* to display a "splash screen," which is a way of telling us that the program is starting. We move the cursor to the next row with *lcd.setCursor()* and display the version and date of the program. In order to keep the splash screen on the display long enough to read it, we add a little time delay with *delay()*, in this instance, 3 seconds (3000 milliseconds) and then call *lcd.clear()* to clear the display. (We avoid using the *clear()* method outside of the one call in *setup()* because it is fairly slow.)

The *loop()* Code

The first dozen lines or so of the *loop()* function simply establish the working variables and call *analogRead()* to fetch the current value of the analog input. The value returned in *analogIn* is then scaled and assigned into *val*. The *dtostrf()* function is a standard library function that was added to the Arduino IDE in the post-1.0 era. The *dtostrf()* function converts a *double* data type to an ASCII string using a floating point format that is specified by its arguments. The parameter *VALUEWIDTH* is the total number of characters you want the string to use, including the decimal point and a minus sign (if applicable). Because we only have 16 columns to work with, we arbitrarily set *VALUEWIDTH* to 6. The symbolic constant *DEFAULTPRECISION* is the number of digits you want displayed after the decimal point. We set *DEFAULTPRECISION* to 3. The final argument, *buffer[]*, is where we want to store the result after the conversion. The *dtostrf()* function returns a pointer to *char*, so we could nest the function call in a *print()* (or other) function call if needed.

After the *dtostrf()* function call, *buffer[]* contains the ASCII representation of the value read by *analogRead()*. Because we now want to append the appropriate scale value to the string, we use the following code:

```
buffer[VALUEWIDTH] = ' ';

i = 0;
while (scale[i]) {
   buffer[VALUEWIDTH + 1 + i] = scale[i];
   i++;
}
buffer[VALUEWIDTH + 1 + i] = '\0';
```

If you walk through this code snippet, you should be able to convince yourself that, after the *while* loop finishes, *buffer[]* contains a string with the measured analog value and its scale (e.g., ".123 mA"). The code then moves *buffer[]* to the LCD display. When studying the code, it helps to remember that *VALUEWIDTH* characters were written to *buffer[]* during the *dtostrf()* call and that we added a space character after that number. That hint should help you ferret out what *buffer [VALUEWIDTH + 1 + i]* is all about.

Earlier we discussed the concept of having an overrange value for the analog input. In other words, we want to provide a means of determining that the input value is greater than our full-scale display value of 1.000 mA. We know what the hardware does, but now we have to provide the software to complete the job. One way to look at the notion of overrange is to say that the analog input value exists in one of two "states." One state is within range and the other is out of range. We know that the only possible values for the analog input are within the counts of 0000 through 1023. To break this down a bit further, the values that exist "within range" are from the count 0000 through 1000, and the counts that are "out of range" exist from 1001 through 1023. Because there are only two possible states, a simple *if* statement:

```
if (analogIn <= MAXANALOGVALUE) {
```

determines which state currently exits for the analog input value.

If the analog input value can be displayed properly, the current state is such that we can display the bar graph to depict its value. The bar graph uses the *lbg* object to draw the bar graph using the scaled value, *val*, to draw the bar graph within the range established by *MAXANALOGVALUE*. The code then calls *delay()* to allow the LCD to update the display. (Again, *delay()* is okay to use provided you don't have the Arduino doing something important in the background. The reason is because *delay()* shuts down most of the board's functionality during the delay period. Interrupt service routines, however, can still be serviced.)

If the analog value causes an out-of-range state to exist, the bar graph is not displayed. The *else* condition of the *if* test causes "Overrange!" to be displayed on the second line of the display. The entire display is then blinked by calls to *lcd.noDisplay()* and *lcd.display()* with *ERRORDELAY* pauses between the calls.

Testing and Calibration of the Meter

Once you have completed construction of the meter, three steps remain. First, you need to compile and upload the code presented in Listing 5-1. Second, you need to test the meter to make sure it is working properly. Third, after you know the meter is working, you need to calibrate it. It is the second and third steps that we discuss in this section.

To test the meter, you need do nothing more than provide a reliable 1 mA current source. So, how do you do that? A simple way is to use a 9 V transistor radio battery and a resistor. You can employ Ohm's law ($E = I \times R$) to determine the value of series resistor to produce 1 mA of current. And, oh by the way, in doing so you have created a nice tool to calibrate the meter!

So, 9 V at 1 mA results in a resistor value of 9 kΩ. However, that is a nonstandard value for 5% resistors. One solution is to purchase a precision resistor of 1% or better tolerance. But do you really need to do that? If you know the *real* value of the resistor and the *real* voltage of the battery, you can determine the *real* current. Also, a fresh 9 V transistor radio battery often has a voltage higher than 9 V! A used battery normally is much less than its rated voltage. Hopefully, you have a multimeter that you can use to measure the voltage and the resistor value. If not, Harbor Freight is a good source for an inexpensive digital multimeter! Or, maybe you can borrow one from a friend who does. Other potential sources for a multimeter are another ham, the physics department at your local high school, community college, or university. Most are happy to allow you to bring in your resistor and battery and use their equipment. Once you know the real current, it is a simple matter of adjusting the 10-turn pot to the correct displayed value.

Changing the Meter Range and Scale

You now have a digital panel meter that can read current up to 1 mA. But what if you need to measure the output of a solar charger? Or perhaps the current being delivered by the solar panel? You need to "scale" the meter to the desired range you want to measure. To do this, you add "scaling circuits" to the meter. You can also change the "units" displayed on the LCD by a simple change to the software. This section describes how these changes are made.

Few applications require a meter that only reads 0–1 mA. More often the case is that a meter is needed to measure a voltage, say 0–150 VDC, or a higher current, 0–10 A. Well, there is a solution. You can add series resistors to create a voltmeter or a shunt resistor to create an ammeter. Because you know that 50 mV across the input resistor creates a 1 mA full-scale reading, it is a simple matter using Ohm's law to determine the correct value to use as a scaling resistor. Figure 5-12 shows the basic circuits used for scaling. The dashed lines are the existing circuitry of the panel meter.

Voltmeter

In the voltmeter circuit of Figure 5-12, a series resistor is added to form a voltage divider. You know that the voltage across the input of our meter needs to be 50 mV (or 0.05 V) across R1 for a full-scale reading. You also know that two resistors in series carry the same current and we know that the full-scale current is 1 mA. Therefore if you subtract 50 mV from the full-scale reading you want, you can then use Ohm's law to determine the missing resistor value that must be added into the circuit.

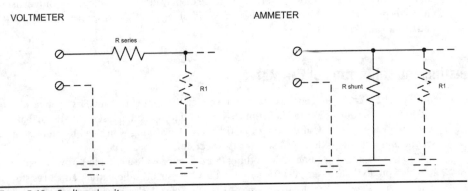

FIGURE 5-12 Scaling circuits.

For example, let's say you want to construct a voltmeter with a full-scale reading of 15 VDC. To determine the value of the series resistor, you subtract 0.05 V from 15 V and then divide the result by 0.001 A (1 mA). Some quick math indicates the resultant value is 14,950 Ω. When we consider all of the accumulated tolerance errors of the resistors in this circuit, a good old 15K Ω resistor should work just fine. Remember that there is a potentiometer in the meter circuit that allows for calibration of the full-scale reading.

One thing that is important to remember for all of these circuits is power dissipation. Any current flowing through a resistor creates heat and the resistor must have a sufficient power rating (P) to handle the heat load generated. $P = I \times E$, or to put it another way, $P = I^2 \times R$. So, in this case, with 0.001 A and 15K Ω, the power dissipated is 0.015 W. So, a ¼ W resistor is more than adequate for this application.

Ammeter

In the ammeter circuit, a shunt resistor is added to "divert" some of the current from the meter. In this case, the important thing to remember is that two resistors in parallel have the same voltage across the two of them, and that the total current is the sum of the current through each resistor. R1 is the meter input resistor and, by design, draws 1 mA at 50 mV. This means you can subtract 1 mA from the total current you wish to measure. Divide 50 mV by that result and you have the value of the required shunt resistor.

For example, let's say we need to measure up to 500 mA DC current. Our total current to be measured is 500 mA and we can then subtract the current through the meter to determine the current through the shunt. So, 500 − 1 = 499 mA or 0.499 A. Divide this by the voltage, 0.05 (50 mV), and we get 0.1002 Ω. The power rating of this resistor would be 0.499 A times 0.05 V or 0.025 W.

Let's take this example into the real world. How do you find a 0.1 Ω resistor with a 0.025 W rating? Well, you don't. But, we can make one using other resistors. Ten 1 Ω resistors in parallel make a dandy 0.1 Ω resistor. And if you use ¼ W or even ⅛ W resistors, there is more than enough power dissipation to meet our requirements. Remember that these resistors are not exactly what they say they are. There is a tolerance rating for them. *Tolerance rating* means that, at a specified resistance, the actual resistance value is somewhere within that tolerance range. If we use 5% tolerance resistors, it means that the final value of a 1 Ω resistor is somewhere between 0.095 and 1.05 Ω. Again, you use the 10-turn pot on the meter shield to adjust your meter for the proper full-scale reading.

Changing the Scale

Obviously, if you change the way the meter scales the readings, you also need to change the scale of the meter. Changing the scale is quite simple. It is merely a matter of editing the character string variable *char scale[]* to the desired value. As coded, the current scale is in milliamps; "mA" in Listing 5-1. Use the Arduino IDE to replace "mA" with the desired character sequence, say, "Volts," or "Amperes," as the case may be.

Conclusion

There are many things that can be "tweaked" in this design. For instance, maybe you want the basic meter to read a different full-scale value other than 1 mA. Well, it is a matter of adjusting the gain of the op amp circuit and possibly changing the scale factor in the code. Such modifications are left as a (kinda fun) exercise for the reader. For the more ambitious hardware types, one possible change is to provide a true differential input that doesn't require a ground reference. If you want to explore the software, maybe adding meter "ballistics." That is, account for the fact that an analog meter pointer has some mass to it and exhibits accelerations and time lags. Another possibility is to add a peak reading indicator. One might explore how to add those features to this project. You could also add a switch that would toggle different resistor combinations for different meter ranges, with software modifications to change the scale automatically.

This project provides a simple to construct and simple to use general purpose panel meter. How it can be used is really up to your imagination.

CHAPTER 6

Dummy Load

Jack remembers the ham who introduced him to amateur radio. His name was Chuck Ziegler, W8RV, and he administered Jack's Novice Class exam when Jack was 11 years old. Chuck not only introduced Jack to ham radio, he was also a mentor who stressed practices that make ham radio the enjoyable avocation it is. One of the things Chuck said was: "Never go swooshing across the band when you're tuning up your transmitter. That's just plain rude!" He's right, but we hear it all the time on the air.

A dummy load allows you to pump your rig's RF power into a dummy (i.e., nonradiating) load while you make any necessary tuning adjustments. Chuck sold Jack a 75-W dummy load that Jack used to tune up his old Heathkit DX-20. The dummy load was about the size of a shoebox, had a finned metal shell to dissipate the RF energy, and weighed slightly less than a Volkswagen.

Figure 6-1 shows the dummy load (DL from now on) we build in this chapter. The DL is based on a design by Ken Kemski, K4EAA, and used extensively in his shop where he repairs ham radio equipment. Essentially, the DL is a group of noninductive (metal film or metal oxide) resistors that

FIGURE 6-1   The finished dummy load.

are bathed in a 1 quart container filled with mineral oil. The construction is sufficiently robust to handle the output of most QRP rigs all day without fear of damaging the resistors. Ken has used the DL with 100 W rigs for short periods of time without problems. Depending upon the depth of your junk box, the DL should cost less than $20 to build and weighs considerably less than a Volkswagen.

Another nice capability that Ken included is the ability to calculate the power being sent to the DL. Ken uses a vacuum tube voltmeter (VTVM) to measure the voltage across the binding posts you see in Figure 6-1, performs a little electronic math on the measurement, and derives the power going into the DL. Because you have the panel meter from Chapter 5, you can use that to display the DL power and let the Arduino do the math for us.

Mechanical Construction

The heart of the DL is the resistor load built from 20 metal film 1K resistors (1% precision) rated at 3 W each. We were able to find them locally for about $0.30 each. You can also find them online (Ken sells them), but they are often sold in 50 piece lots for about $12. (You might find members of your local ham club would also be interested in building a DL and split the resistor cost.) By ganging the 20 resistors in parallel we end up with a 50 Ω load rated at 60 W. The mineral oil bath allows us to run up to 100 W through the DL for short periods of time. As Ken points out, if you measure the resistance across the load and find it noticeably higher than it used to be, chances are you cooked one or more of the resistors. If that happens, you can disassemble the DL and replace the damaged resistor(s).

Figure 6-2 shows how the resistors are placed together in the DL. We purchased a 4 × 4 in. piece of sheet brass and a 1 quart paint can at a local hardware store. First we cut the brass sheet

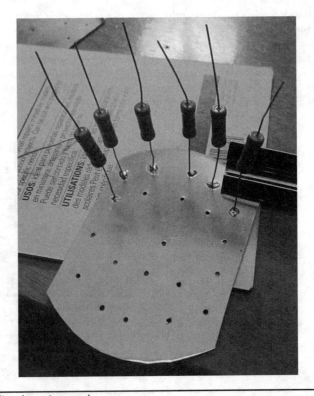

FIGURE 6-2   Building the resistor pack.

in half and trimmed the edges with tin snips so they would fit into the paint can. Next, we stacked the two halves on top of each other and drilled 20 small holes for the resistors and one hole more-or-less centered in the middle of the sheets. We enlarged the center hole of one of the pieces so we could pass a short piece of RG58 coaxial cable through it. (The cable's shield is eventually soldered to the sheet with the larger diameter center hole and the center conductor to the other sheet.)

After the holes are drilled, we made about a ¼ in., 90° bend in each resistor lead, threaded it through one of the holes, and soldered it in place. Figure 6-2 shows how this step is done. Solder all 20 of the resistors in place, leaving the center hole empty for the moment.

Resistor Pack Spacing

Figure 6-3 shows the resistor pack with the top brass sheet in place and the short piece of RG58 passing through the top brass sheet and connecting to the center hole in the bottom brass sheet. You need to give a little thought to the placement of the top brass sheet. Our paint can was just under 5 in. tall, which means that our resistor pack must be shorter than this because we want the resistor pack to "float" in the middle of the can and not touch either the top, bottom, or sides of the can. (One of our Beta testers, Leonard Wong, suggests lining the can with kraft paper!) Therefore, you must place the top brass sheet on the resistor leads in such a way as to give you clearance at all points inside the can. Don't forget that the BNC connector protrudes into the can from the lid, so that length needs to be factored into the spacing between the sheets. The calculation doesn't have to be precise, but obviously nothing in the resistor pack can touch any part of the can. For us, the spacing was about 3 in., but may vary depending upon the size of the can you are using.

To be able to use the DL for power measurement, you need to connect a diode between the bottom brass sheet and the positive (red) binding post. Ken's design calls for a BAV21 for the diode,

FIGURE 6-3   The assembled DL resistor pack.

FIGURE 6-4 Connecting the diodes to the binding post.

but Jack used two 1N4148s because he picked up 50 of them for a dollar at a flea market. (Two diodes were wired in series because the continuous reverse voltage for the 1N4148 is only 100 V while the BAV21 is twice that level.) In Figure 6-4, you can see the two diodes coming from the bottom sheet and connecting to the positive binding post on the lid. If you look closely, you'll see that we cut a small notch in the top sheet to allow the diode to pass the sheet without shorting out. Make sure the diode lead remains free of contact with the top sheet when you place the assembly in the can.

Fabricating the Lid Connections

You need to drill three holes in the lid of the paint can: one for the BNC antenna connector and two for the positive and negative power leads. As you might expect, the metal used for the lid is pretty thin and may "tear" when you drill holes with a bit large enough to accommodate a BNC connector. You can, of course, mount an SO-239 connector in the center, too, and they do make feed-through versions of the connector. However, we tend to run QRP quite often and those rigs usually come with BNC connectors. Select the drill bit size that fits your connector.

We purchased a dual binding post for our power leads. It has both banana jack inputs, which also allow bare wires to be attached to the binding posts (see Figure 6-5). These are available at your local parts store and online. Sometimes the packaging has a template you can use for drilling the holes. If you are using individual terminal posts, they should be mounted fairly close together because we need a capacitor to span the gap between the two posts. You can use the fastening nut to gauge the size of the drill you need for the job. (Banana jack binding posts like those shown in Figure 6-5 typically have a 0.75 in. spacing. Ours came in a plastic package that had a template printed on the back.)

Because the metal lid is prone to tear when drilled, we clamped two pieces of scrap wood snugly to the top and bottom of the lid with C-clamps. Make sure the wood is tight to the lid and not bridging the lip on the lid. Now mark a center hole and the two holes that are appropriate for your terminal posts. *Slowly* drill all three holes (i.e., two for the terminal posts and one for the BNC connector).

We applied a small bead of silicone caulk between the two pieces of the dual binding post and then mounted it into place. Next we attached the BNC (or SO-239) connector in the center hole.

FIGURE 6-5   Dual binding posts.

As you can see in Figure 6-4, we applied a liberal amount of silicone caulk to the holes to keep the mineral oil from leaking out.

Attaching the Lid to the Resistor Pack

After determining the correct distance from the lid to the bottom of the resistor pack, we cut the top of the coax to length. We stripped away enough of the outer insulation to fan the ground braid so it would reach the negative terminal binding post. We attached the center conductor of the small piece of coax to the bottom sheet. The top of the center conductor is attached to the center pin of the BNC connector. We also routed a fairly stiff, single-conductor piece of wire from the ground lug of the BNC connector, wound it around the negative terminal post and then continued the wire from the terminal and soldered it to the top sheet. This helped to mechanically strengthen and stiffen the assembly, too.

Looking at Figure 6-4, it may appear that things look a little messy around the binding posts, the BNC connector, and the lid of the can. The reason is because it is messy! We applied more silicone caulk around each connection to help seal the can. While you could use regular oil as the coolant, mineral oil is used because it is pretty benign stuff if you spill it. In fact, you can drink it, although you may find yourself doing a little of the Green-Apple-Quickstep shortly afterward. All in all, the project is fairly friendly to the environment compared to some possible alternatives for a DL.

Lastly, you need to connect a 0.01 µF disk ceramic capacitor between the two terminal posts. The rating should be at least 200 V. We didn't have that value in the junk box with a sufficient voltage rating, so we connected two 0.0047 µF 200 V disk ceramic caps in parallel. While this is not exactly 0.01 µF, it's close enough for our needs and gives us a safety margin on the voltage rating.

Before filling the can with mineral oil and sealing the lid, check that the connections and resistor pack do not touch the inside of the can, including the lid. Use your VTVM to check for shorts in the system. If you wish to use sheets of kraft paper on the sides and bottom of the can, this is a good time to add them, although they shouldn't be required if you've constructed the resistor pack correctly. As another option, you could "invert" the assembly and tie the upper brass sheet directly to the center pin of the BNC connector and tie brass strips to the lid and to the bottom sheet. This would also make connecting the diodes a much easier process plus it would do away with submerging the coax in the oil. It would also make the assembly more rigid. (This is a good example of *ex post* thought processes.)

When you have finished assembling the resistor pack, measure the resistance between the two brass sheets. If you have constructed the DL correctly, you should get a value that is very close to 50 Ω. Write the resistance value down because you need it for the symbolic constant *MESASUREDDUMMYLOADRESISTANCE* in the program that measures the watts being put into the DL.

Electronic Construction

There really isn't much to do if you have already built the panel meter described in Chapter 5. That chapter also told you how to construct the panel meter as a voltmeter, which is how we use the meter in this project. If you look at Figure 5-12 in Chapter 5, you can see that all you have to do is add a series resistor to the panel meter to make it a voltmeter. The question is: What's the requisite value of the resistor?

Doing the Math

If you recall the discussion from Chapter 5, R1 in Figure 5-12 is the meter's input resistor and is designed to provide 1 mA of current at 50 mV. Let's suppose we want to enable our panel meter to be able to read up to 100 VDC. To do that, we can use Ohms law to figure out the series resistor we need to add to the circuit in Figure 5-12. Therefore:

R = Voltage/Current
R = (100 V − 0.05 V)/0.001 A
R = 99.95/0.001
R = 99950 Ω

Because 100K is a standard resistor value and it's well within our tolerance for the meter, that's what we'll use. However, we still need to consider the power rating for that resistor. Because power is the current squared (0.001 A^2) times the resistance (100K), the power requirement is about 0.10 W, so a quarter-watt resistor is more than adequate. Therefore, we can simply add a ¼ W 100K resistor to the circuit shown in Figure 5-12 and use the panel meter to observe the RF power going to the antenna.

You could also add a switch that would toggle different resistors into the circuit. This way, you could select between low power (e.g., 0–10 W) and high power (0–100 W) measurements. Using 1% resistors would result in accuracy that approaches that of commercial watt meters. Once you have selected the appropriate resistor(s), simply attach the power leads to the panel meter circuit.

We chose to use two small header pins to hold the series resistor. If you look closely at Figure 6-6, you can see a resistor just above the terminal block on the right edge of the panel meter board, directly above the power connector of the Arduino board. Using a 100K resistor makes the full-scale voltage almost 100 V. Because we are employing a 100K series resistor, we plan on using the DL with transmitters of up to about 100 W of power. Because the resistor pack is comprised of twenty 3-W resistors, 60 W is the "real" power limit for the DL. However, the 60 W load, coupled with the mineral oil bath, can sustain the heat for enough time for you to get a quick reading at higher power levels. Still, at these power levels, you shouldn't dawdle while taking measurements. If you're running QRP at 5 W, you don't need to worry too much about the DL handling the power.

You should, however, always be respectful of the voltage that appears across the DL's binding posts. As you will see in a few moments, even a QRP transmitter generates enough voltage at the DL to demand respect. Connect the Arduino and the power leads to the DL before you connect

Chapter 6: Dummy Load 125

FIGURE 6-6 The DL mounted in its display case. (*Display case courtesy of Linksprite*)

the coax from the DL to the transmitter. No sense bumping the transmitter key while you have both power leads in your hands. Always make connections to the DL with the transmitter turned off.

As you can see in Figure 6-6, we mounted the LCD display "off board" and moved it to the window provided in the display case (see Figure 6-7). In Figure 6-7, you can see that the LCD display fits perfectly in the provided window. The case also has premolded spacers for the Arduino boards, special end panels for the power and USB connectors, plus the required hardware. The case is manufactured by Linksprite and makes an attractive way to package the DL wattmeter.

The "dangling pot" that you see in Figure 6-6 is R1 in Figure 3-4 from Chapter 3. If you recall, adjusting R1 controls the contrast on the LCD display. We wanted to determine the resistance that provided the correct contrast for our LCD display, as it can vary for each LCD display module. For our display, about 1.5K provided the contrast we wanted.

The two leads that connect to the terminal block of the power meter board are terminated with two banana jacks. These banana jacks plug into the binding posts connected to the top of the DL, as shown in Figure 6-8, and afford a good solid connection to the DL. The leads then supply a

FIGURE 6-7 The "Window" side of the Linksprite project case.

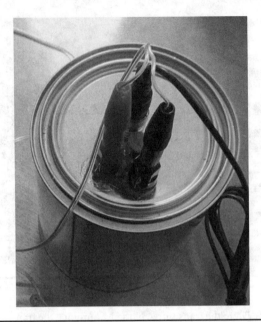

FIGURE 6-8   Attaching power leads to DL.

voltage reading via the terminal block on the panel meter shield to the software running on the Arduino. That voltage passes through the series resistor that we mentioned earlier (100K) and supplies the resulting voltage to analog pin A1 on the Arduino. From that point on, everything is done by the software.

Software

The code in Listing 6-1 makes use of an external reference voltage, as was mentioned in Chapter 5. The default behavior for the AREF pin is to use the internal reference voltage of 5 V. However, unless the analogReference(EXTERNAL) *function has been called and we apply an external voltage to the AREF pin, the two voltages are now connected together, essentially short-circuiting the two voltage sources. This may result in enough excessive current flow to permanently damage the Atmel processor. It is very important that you* NOT *attach the panel meter shield to the Arduino board until* AFTER *you have compiled and uploaded the code into the Arduino. Therefore, remove the panel meter shield before you compile and upload the code in Listing 6-1. Once the code is compiled and loaded into the Arduino, remove the power to the Arduino board and then attach the panel meter shield. Only then should you reapply power to the Arduino board.*

As you might expect, the software that works in conjunction with the DL is very similar to the code you examined in Chapter 5. Listing 6-1 presents the code used for the DL. You should connect the leads from the meter to the DL binding posts on the DL can, and then connect the BNC connector to the transmitter you wish to use as the signal source. Keying the transmitter applies the power to the DL and the watt meter reads the RF energy being supplied to the DL.

The code begins in the usual manner with a series of symbolic constant definitions and various working variables.

```c
/* Simple LCD Panel Meter ver1.0 23 December 2013 D.Kidder W6DQ
   Taken from Chapter 3. Modified for wattmeter, J. Purdum W8TEE
   11 March, 2014.
*/
#include <LiquidCrystal.h>
//                                     Function prototypes
void ShowTheOutput(int analogIn, float val);
void BuildTheOutput(int voltsIn);

#define DEFAULTPRECISION       3      // default decimal places
#define MAXANALOGVALUE      1000      // Set valid input to 0 - 1000
#define VALUEWIDTH             8      // Numeric field width, include '-', '.'
#define DIODEVOLTAGEDROP     1.2      // Two 1N4814...your mileage may vary
#define DISPLAYDELAY        2000      // Hold display for two seconds

#define UPDATEDELAY          100      // Delay for normal reading
#define ERRORDELAY           400      // Delay if error message displayed
#define STARTUPDELAY        3000      // Enough time to read splash screen

#define BARGRAPHROW            1      // Draw graph on LCD second row
#define BARGRAPHCOL            0      // Start graph in LCD column 0

const int LCDNUMCOLS =       16;      // number of columns on LCD
const int LCDNUMROWS =        2;      //      "    rows         "
const int DLSENSORPIN =       A1;     // define the analog input as A1

char buffer[LCDNUMCOLS + 1];          // Hold the output
float scaleFactor = 10.0;             // scaling factor for the meter 1000 = 100W

// Put your load resistance here
const float MESASUREDDUMMYLOADRESISTANCE = 50.4;

LiquidCrystal lcd(12, 11, 7, 6, 5, 4); // LCD hardware interface pins

void setup()                          // Step 1: Initialization
{
  analogReference(EXTERNAL);          // set the A/D to use the AREF input
  lcd.begin(LCDNUMCOLS, LCDNUMROWS);  // initialize the LCD

  lcd.print(" K&P Watt Meter");       //sappy stuff, name, version and date
  lcd.setCursor(0,1);
  lcd.print("ver 1.0 11Mar 14");
  delay(STARTUPDELAY);
  lcd.clear();

}

void loop()
{
  int analogInVolts;
```

LISTING 6-1 Software to calculate the DL wattage.

```
    float val;

    analogInVolts = analogRead(DLSENSORPIN);      // Step 2: Data Input
    if (analogInVolts) {
      BuildTheOutput(analogInVolts);              // Step 3: Process the input
      ShowTheOutput(analogInVolts, val);          // Step 4: Display the data
      delay(DISPLAYDELAY);
    }
            // There is no Step 5...just do everything again!
}

/*****
 * This function takes measured RF power and formats it for display on the LCD
 *
 * Parameter List:
 *    int analogIn      the voltage measured on DLSENSORPIN
 *    float watts       the calculated watts into the dummy load
 *
 * Return value:
 *    void
 *****/
void ShowTheOutput(int analogIn, float watts)
{
  lcd.setCursor(0,0);                     // Set up the display...
  lcd.print(buffer);

  if (analogIn <= MAXANALOGVALUE) {       // If analog value within scale...
    delay(UPDATEDELAY);
  } else {                                // ...otherwise complain about it.
    lcd.setCursor(0, 1);
    lcd.print("   Overrange!   ");
    lcd.noDisplay();                      // flash the display on and off
    delay(ERRORDELAY);
    lcd.display();
    delay(ERRORDELAY);
    lcd.clear();
  }
}
/*****
 * This function takes measured RF power and formats it for display on the LCD
 *
 * Parameter List:
 *    float watts       the calculated watts into the dummy load
 *
 * Return value:
 *    void
 *****/
void BuildTheOutput(int voltsIn)
{
  float scaledVolts;
```

LISTING 6-1 Software to calculate the DL wattage. (*continued*)

```
    float RMS;
    float watts;

    scaledVolts = (float (voltsIn)) + DIODEVOLTAGEDROP;     // Adjust for diode
                                                            // voltage drop
    scaledVolts = scaledVolts / scaleFactor;                // Scale it to meter
    RMS = scaledVolts / 1.414;                              // Get RMS of voltage
    watts = (RMS * RMS) / MESASUREDDUMMYLOADRESISTANCE;     // watts into DL
    dtostrf(watts, VALUEWIDTH, DEFAULTPRECISION, buffer);   // format value
    strcat(buffer, " watts");                               // Append units
}
```

LISTING 6-1 Software to calculate the DL wattage. *(continued)*

In *setup()*, the call to *analogReference(EXTERNAL)* allows us to modify the reference voltage used for the meter, as explained in Chapter 5. The rest of the code in *setup()* is used to display a splash screen telling the user what the program is. The code in the function corresponds to Step 1, the Initialization Step, of our Five Program Steps. The splash screen can be seen in Figure 6-9.

In the *loop()* function, the call to *analogRead()* takes the voltage reading from pin A1 and assigns it to *analogInVolts*. This is the program's Step 2, the Data Input Step, of our Five Program Steps. The statement:

```
    if (analogInVolts) {
```

FIGURE 6-9 DL wattmeter splash screen.

is used to avoid updating the display if there is no voltage on the input pin, A1. As you might guess, we noticed some flickering of the display on passes through the loop when there was a voltage present. This flicker is partially caused by small variations in the voltage coming into pin A1 from the DL. The power being applied to the DL may be constant, but the resistance may change slightly as the power being dissipated in the DL heats up the components. The call to *delay(DISPLAYDELAY)* in *loop()* holds the display measure for a period of two seconds and minimizes flickering. Obviously you can change this delay if you wish.

Assuming we do need to update the LCD display, function *BuildTheOutput()* is called. The function first adds back the voltage drop associated with the diode(s) you placed in your DL circuit. For our DL, we used two 1N4814 diodes, each of which has about a 0.6 V drop, or 1.2 V for both of them. We used a symbolic constant (*DIODEVOLTAGEDROP*) for the actual value because your diodes may be different from ours. Simply change the constant to whatever your value is.

Next we scale the reading from pin A1 according to the parameters used to build the meter. As you know from Chapter 5, the ADC in the Arduino uses 10 bits, or a maximum value of 1023. Because we want some margin for error, we treat a value of 1000 as full scale, or 100 W for our meter. Because 1000 corresponds to 100 W, our *scaleFactor* is 10.0 (10 = $^{1000}/_{100}$ W). Dividing the *scaledVolts* by *scaleFactor* gives us the actual *scaledVolts*. Because we want to use the *RMS* voltage in calculating our power measurement, the *scaleVolts* is divided by the square root of 2, or 1.414. Squaring *RMS* and dividing by the DL resistance, *MESASUREDDUMMYLOADRESISTANCE*, we get the power being dissipated in the DL. All that remains is to format the value of *watts* and place it into our output buffer via the call to *dtostrf()*. Note how this function equates to Step 3, the Processing Step, of our Five Program Steps.

Back in *loop()*, the final call is to *ShowTheOutput()*, which simply displays the power measurement that was just calculated on the LCD display. This function becomes Step 4, the Output Step, of our Five Program Steps. Once the value is displayed, control returns to *loop()* where we update the value of *oldAnalogVolts* and the process repeats itself with another pass through *loop()*.

There is no Step 5, Termination Step, because µC programs are designed to "run forever." In our case, the program continues to run until power is removed from the µC.

Conclusion

In this chapter you built a DL that also permits you to measure the output power of a transmitter. While we think you'll find the DL a very useful addition to your shack, other hams on the air will also appreciate it if you use the DL to tune your transmitter rather that doing it "live" on the air. Still, there are things that could be used to improve on our design.

First, there's no reason why you can't power the meter from a 9 V battery instead of using a wall wart and the onboard power connector. You would likely want to add a power switch if you convert the meter to battery power.

Second, there are LED displays that do not require 6 data pins to display numeric data. In Chapter 8 you will learn how to use the ATtiny85 chip as a replacement for a full-blown Arduino in certain uses. The "85" chip is very inexpensive (less than $1.50) but still has 8K of Flash memory. The primary limitation is that it has only six I/O pins. Still, with some LED displays now only using three I/O pins, you could build the panel meter without having to dedicate an Arduino to the task.

Whatever changes you do make to the DL design, make sure you adjust the code in Listing 6-1 for the specifics of your DL. When you're done with your modifications, let us know about what you've done so others might share your work.

CHAPTER 7

A CW Automatic Keyer

You may be thinking: Just what the world needs, another Arduino-based keyer. True, there are dozens of circuits for CW keyers based on several of the Arduino family of boards. However, using an Arduino board for something as simple as a CW keyer is an H-bomb-to-kill-an-ant approach to the problem. You could probably design your own keyer circuit using just three or so I/O pins. For that reason, we're going to depart slightly from our usual Arduino project and consider using one of the minimal Atmel µC chips: The ATtiny85.

Another wrinkle we're going to use to make our keyer a little different is that it uses capacitive reactance to activate the keyer circuit. This means you do not need to implement (or buy) a set of paddles to use the keyer. Instead, we provide two simple sensors that use the µC to detect which sensor (a dit or a dah) is activated and use your body's capacitive reactance to key the transmitter circuitry. The advantages of this approach are several: 1) no external paddle set required, 2) low cost, and 3) a more robust and rugged keyer. Because we can make the "paddle sensors" from simple aluminum brackets, the paddle sensors are rugged, cheap, and light weight. Because of the low current demands of the circuit, you should be able to get over 2000 hours of keyer use without having to change the battery. Indeed, you might decide to leave the on-off switch out of the project. Finally, as you will see shortly, the keyer is very small, which makes it easy to backpack into the field if you choose to do so.

Figure 7-1 shows a picture of both flavors of the keyer. The keyer on the left uses a bare ATtiny85 chip with a button battery and the keyer on the right uses a Digispark board based on the same ATtiny85 chip. As you can see in Figure 7-1, no expense was spared in making the capacitive touch paddles for the keyer input. (The paddles for the ATtiny85 keyer are corner braces from the local hardware store and the Digispark keyer uses wire connectors.)

The ATtiny85 is also produced by Atmel and has the feature set described in Table 7-1. As you can see, there is quite a bit of functionality crammed into an 8-pin DIP package. Given that our keyer design only requires three I/O lines, we can comfortably use an ATtiny85 for the keyer project.

Actually, we are going to present two versions of the keyer to you in this chapter. The first version uses the ATtiny85 chip in its stand-alone DIP package. The second version uses the Digispark board produced by Digistump (http://digistump.com). See Figure 7-2 for pictures of the two versions of the ATtiny85 chip.

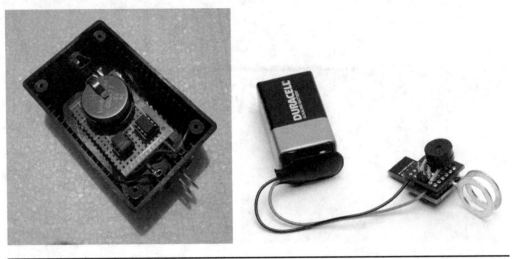

FIGURE 7-1   The ATtiny85 keyer (left) and the Digispark keyer (right).

The ATtiny85 chip can be purchased from domestic suppliers for about $2 each (see Appendix A). The Digispark board costs $9, but has a number of advantages. The biggest advantage is that funny-looking appendage sticking out of the left side of the Digispark board in Figure 7-2 is actually a USB connector. In other words, you can directly connect the Digispark board to your computer via a USB connection to program it. Not so with the barebones chip. To program the bare chip, we need to devise a simple circuit board and tie it to a 328-style Arduino board and use that board as the host programmer.

Description	Specification
Flash memory	8 kb
EEPROM	512 b
SRAM	512 b
Peripheral Features:	8-bit timer/counter
	10-bit ADC
	External and internal interrupt sources
Programmable I/O lines	6
Operating voltage	1.8–5.5 V[a]
Low Power Consumption	Active mode: 300 µA (1 MHz, 1.8 V)
Internal calibrated oscillator	0–10 MHz
Advanced RISC Architecture	120 instructions, most single cycle

[a]The ATtiny85 V is the low voltage version of the chip. The standard version requires at least 2.7 V.

TABLE 7-1   ATtiny85 Feature Set

FIGURE 7-2 The Digispark chip on the left and the 8-pin ATtiny85 chip on the right. *(Digispark courtesy of Digistump. ATtiny85 courtesy of cash)*

Required Software to Program an ATtiny85

You need to download and install certain libraries to be able to program the ATtiny85 chip. The "barebones" chip requires you to install the necessary files from the ATtiny-master.zip file, which can be found at https://github.com/damellis/attiny/. (The actual download button is near the lower-right side of the page.) The procedure you need to follow is a little different from what you would use for a "standard" library file, mainly because we need to alter the Arduino IDE in the process.

When you extract the zip file, you have a new directory named attiny-master created for you. Inside of that folder is another folder named attiny. That attiny folder needs to be copied to a new folder where your sketches are stored. You can use the File-Preferences menu option to see the location of your sketches folder, as shown in Figure 7-3.

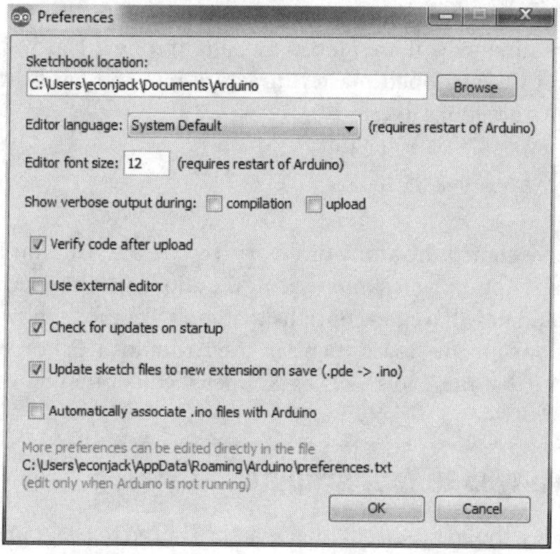

FIGURE 7-3 The preference menu option.

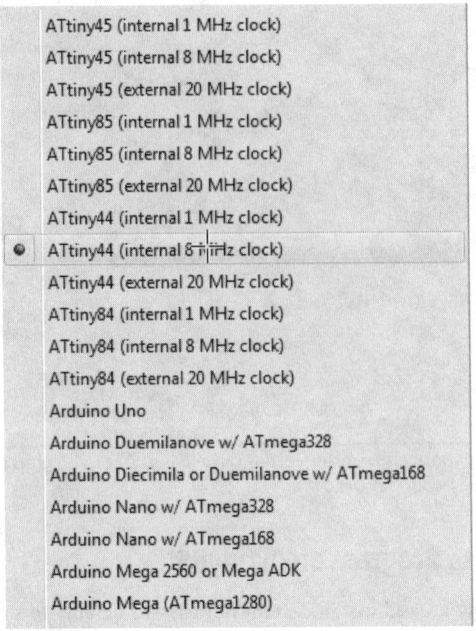

```
ATtiny45 (internal 1 MHz clock)
ATtiny45 (internal 8 MHz clock)
ATtiny45 (external 20 MHz clock)
ATtiny85 (internal 1 MHz clock)
ATtiny85 (internal 8 MHz clock)
ATtiny85 (external 20 MHz clock)
ATtiny44 (internal 1 MHz clock)
● ATtiny44 (internal 8 MHz clock)
ATtiny44 (external 20 MHz clock)
ATtiny84 (internal 1 MHz clock)
ATtiny84 (internal 8 MHz clock)
ATtiny84 (external 20 MHz clock)
Arduino Uno
Arduino Duemilanove w/ ATmega328
Arduino Diecimila or Duemilanove w/ ATmega168
Arduino Nano w/ ATmega328
Arduino Nano w/ ATmega168
Arduino Mega 2560 or Mega ADK
Arduino Mega (ATmega1280)
```

FIGURE 7-4   ATtinyXX board options.

As you can see at the top of Figure 7-2, we installed the Arduino IDE so the sketches are saved at:

`C:\Arduino1.0.5\Sketches.`

Inside your sketches directory, create a new subdirectory named Hardware. (You may already have a Hardware directory. If so, ignore creating the new Hardware directory.) Now copy the attiny folder from the unzipped master directory into the hardware directory. For our system, the directory structure looks like:

`C:\Arduino1.0.5\Sketches\Hardware\attiny`

After you have copied the attiny directory, restart the Arduino IDE. Now, when you click on the Tools → Board option, you should see new options that cover the ATtiny chips. That is, your Board menu should now have new options that look like those shown in Figure 7-4. If you do not see the new options for the Board menu in the Arduino IDE, you need to redo the steps above. (You will also note that the Tools → Programmer menu option now has a USBtiny option, too.)

Connecting the ATtiny85 to Your Arduino

You are now ready to build the circuit that is actually used to program the ATtiny85 chip. Figure 7-5 shows the pin assignments for the ATtiny85 chip. Notice that the labels for the pins are a little strange, in that what is labeled as circuit pin 0 is actually physical pin 5 on the chip. It's very important not to confuse the physical pin position with the circuit pin position or you risk the

Chapter 7: A CW Automatic Keyer   135

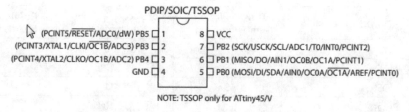

FIGURE 7-5   ATtiny85 pin assignments.

chance of causing your ATtiny85's untimely death. Figure 7-6 presents a breadboard view of the ATtiny85 programming shield.

To help you prevent confusing physical and logical pin numbers, Table 7-2 gives you the ATtiny85 labels as well as the chip numbers. The table is based on Figure 7-6. The third column of the table tells you where each pin from the ATtiny85 chip (often called just '85 in literature) is connected to your Arduino board. The fourth column relates to the ATmega1280 and ATmega2560 boards. So, for example, as you can see in Figure 7-6, the SCK line from the ATtiny85 runs from pin 7 of the '85 to pin 13 of the Arduino Uno. If you were using a Mega board, the SCK line would run to pin 52 on the Mega board. If you use Figure 7-6 in conjunction with Table 7-2, you should have no trouble wiring the shield.

Some Arduino boards aren't happy unless you connect a 10 µF electrolytic capacitor between the RESET and GND pins on the Arduino board. Because this capacitor is polarized, make sure

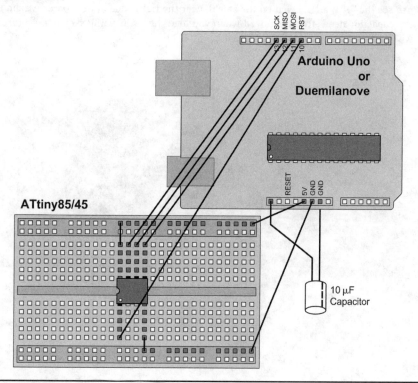

FIGURE 7-6   The layout of the ATtiny85 programming breadboard.

ATtiny85 Label name	ATtiny85 Pin Number on Chip	Connect to Arduino Pin Number	Arduino Mega Boards Pin Number
SCK	7	13	52
MISO	6	12	50
MOSI	5	11	51
Reset	1	10	53
VCC (+)	8	+5 V	+5 V
GND	4	GND	GND

TABLE 7-2 Pin Assignments for ATtiny and Arduino Connecting Pins

you connect the negative terminal to ground. Your connections should look similar to that shown in Figure 7-5.

Because we program ATtiny85 chips fairly often, we moved the breadboard circuit shown in Figure 7-6 to a "programming shield" that just piggybacks onto the Arduino board. (If you want more instructions on building an ATtiny85 programming shield, see: http://www.instructables.com/id/8-Pin-Programming-Shield/.) If this is the only ATtiny85 project you plan to build, the breadboard version is good enough. Figure 7-7 shows the shield we built for programming the ATtiny85.

The only thing about the shield is that we added two socket pins for the 10 μF capacitor. You can see the "+" mark labeled on the shield, near the right edge of the board. We did this because the capacitor needs to be removed when you load the ArduinoISP sketch, but must be replaced when you actually program the ATtiny85.

FIGURE 7-7 The programming shield for the ATtiny85 piggybacked to an Arduino.

The Proper Programming Sequence

When you wish to program an ATtiny85 chip, there's a specific sequence you must follow for everything to work properly. The following steps walk you through the sequence needed to program the ATtiny85:

1. Temporarily remove the 10 µF capacitor from the Arduino board.
2. If you have created a shield for the ATtiny85 programmer, plug it into the Uno or other 328-type Arduino board.
3. Load the ArduinoISP sketch found in the File → Examples directory.
4. Open "ArduinoISP" sketch from "Examples" folder.
5. Select your board from the Tools → Board menu. This should be the host Arduino board (e.g., Uno, Duemilanove).
6. Compile and upload the ArduinoISP sketch.
7. Replace the 10 µF capacitor and reinstall the programming shield (or the connections to the breadboard). Make sure you pay attention to its polarity.
8. From the Tools → Board menu, select the *ATtiny85 (internal 8 MHz clock)* option.
9. From the Tools → Programmer menu, select *Arduino as ISP*. (The ISP means the in-system programmer.)
10. At this point, you may want to upload the simple Blink program, change the program line that reads:

    ```
    int led = 13;
    ```

    to

    ```
    int led = 0;
    ```

    This is necessary because there is no pin 13 on the chip! Now connect a small LED between P0 (chip pin number 5) and the ground socket that holds the 10 µF capacitor. You should see the LED blink at a one second rate. This simply confirms that the program was successfully uploaded to the ATtiny85 chip.
11. You can now open the CW keyer sketch (presented later in this chapter) and compile it.
12. Upload the compiled sketch to the ATtiny85. For some host boards, you may see the messages:

    ```
    avrdude: please define PAGEL and BS2 signals in the configuration
    file for part ATtiny85
    avrdude: please define PAGEL and BS2 signals in the configuration
    file for part ATtiny85
    ```

    You can ignore the error messages that refer to PAGEL or BS2 signals if they appear. Chances are good that the upload worked.

You should now remove the chip from its programming socket and place it in the circuit that uses the chip. It should work as you designed it. However, if things are not working correctly, the next section presents some potential areas where things may have gone wrong.

Some Things to Check If Things Go South

There are some issues that you need to be aware of in case your setup doesn't correctly program the ATtiny85 chip. Consider the following checklist if you are having issues programming the chip using your Arduino:

1. Not all versions of the Atmel chip family are suitable for programming the ATtiny85. For example, the earlier ATmega168 won't work. Also, we have purchased some 328s online

that don't work with the 1.0.5 IDE, but do work with the pre-1.0 IDE. While we are not sure, it seems that such boards have issues with the bootloader software that's burned onto the chip. If you are unsure of your vendor's board and its compatibility, ask if their board has been tested with version 1.0.5 of the IDE before you order the board.
2. Check the pin assignments in your program's source code. It's pretty easy to forget that you're not working with a full Arduino board and to assume that you can still blink the LED on pin 13 when no pin 13 exists. While such flat-forehead mistakes do happen, usually they become obvious and no harm is done to the chip. Still, keep Figure 7-5 and Table 7-2 in mind when you move sketches from the Arduino to the ATtiny85.
3. Some versions of the Arduino IDE (e.g., version 1.0.2) simply don't work when trying to program the ATtiny85. If you have an earlier version than 1.0.5, we suggest that you upgrade the Arduino IDE to the latest version.
4. Sometimes, having the circuit shown in Figure 7-6 connected to the Arduino board appears to prevent the sketch from properly loading into the Arduino. If that is the case, completely disconnect the ATtiny85 breadboard circuit from the Arduino board, load the sketch, and then reconnect the breadboard to the Arduino board. This is one reason that we created an "ATtiny85 programming shield" based on Figure 7-6. It's a simple matter to remove the shield, load the sketch, and then plug the shield back into the Arduino board.
5. After programming the ATtiny85 chip, you may need to remove it from the programming circuit in order to make sure it is working properly. That is, testing the newly programmed chip in the circuit shown in Figure 7-6 may not work. It's better to remove the chip from the programming socket and put it into the circuit in which you plan to use the chip for testing purposes.
6. The 10 µF capacitor that spans the RESET and GND pins on the Arduino board should be removed while loading the ArduinoISP sketch. However, be sure to put the capacitor back onto the board before trying to program the ATtiny85.
7. Keep in mind which pins on the ATtiny85 may be used as analog inputs (i.e., 3, 4, and 7).

At this point, you are ready to program the ATtiny85 with the code for the keyer. However, before we get to that code, we also want to show how to use the Digispark for the keyer.

Using the Digispark

The first thing you need to do is download a modified version of the Arduino IDE that is designed for use with the Digispark. This is a free download and can be found at:

http://sourceforge.net/projects/digistump/files/DigisparkArduino-Win32-1.0.4-May19.zip/download

NOTE: *The Digispark IDE is a step or two behind the Arduino IDE in terms of version number, so make sure you keep the two IDEs in separate directories. Although the two IDEs look the same when they are run, they function differently under the hood, so don't try to use the Arduino IDE to program the Digispark.*

In Figure 7-8, you can see the +5 V and ground connections along the bottom of the board and the other pin connections (p) through P5, bottom to top along the right edge of the board. The USB connection is done via the connections on the left side of the board. If you are worried about the drain imposed on the battery by the LED, you can use a soldering iron to remove the LED (a little tricky because it is very small) or you can cut the traces using a sharp knife. Given the relatively small

Chapter 7: A CW Automatic Keyer

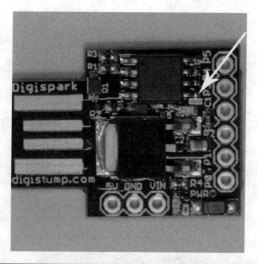

Figure 7-8 Close-up of the Digispark board with the onboard LED identified.

current drain for the LED, you could simply ignore the LED. Alternatively, you could add a switch to the circuit between the battery and the Digispark. We chose to ignore the LED's current drain.

Table 7-3 shows the pinouts for the Digispark. All pins (P0 through P5) can be used for digital I/O.

While it is true that the Digispark costs a little more than a naked ATtiny85 chip, there are a number of features that make the Digispark a worthy consideration. First, there is no need for the ISP setup used earlier in this chapter to program the ATtiny85. The Digispark has a USB port onboard that makes it work just like other Arduino boards. Second, it has an onboard voltage regulator that gives you more options in terms of powering the system. Although designed for 5 V operation (500 ma max), the regulator accepts up to 35 V at the ragged edge. (Digistump recommends not exceeding 12 V.) The regulator and USB chips do draw a little more current, but the board still has a decent battery life, especially if you use a 9 V battery. Third, the Digispark has an SMD LED similar to that on the Arduino Uno (see the arrow in Figure 7-8). Finally, the bootloader is already on the chip, which makes compiling and uploading programs very similar to the "normal" Arduino IDE.

Similar ... yes. Identical ... no.

Digispark Pin Number	Corresponding ATtiny85 Pin
P0	PB0 (PWM, MOSI, LED on Model B)
P1	PB1 (PWM, MISO, LED on Model A)
P2	PB2 (ADC1, SCK)
P3	PB3 (ADC3, USB+ when USB in use)
P4	PB4 (ADC2, USB − when USB in use)
P5	PB5 (ADC0, Reset)
Vcc	+5 V (8)
GND	GND (4)

Table 7-3 Digispark Pinouts Relative to ATtiny85

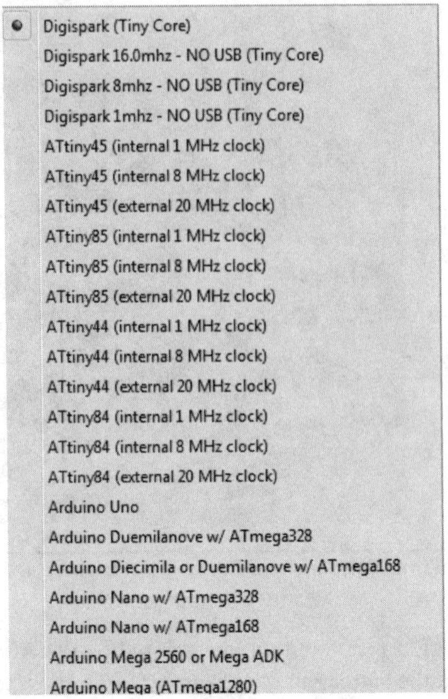

```
Digispark (Tiny Core)
Digispark 16.0mhz - NO USB (Tiny Core)
Digispark 8mhz - NO USB (Tiny Core)
Digispark 1mhz - NO USB (Tiny Core)
ATtiny45 (internal 1 MHz clock)
ATtiny45 (internal 8 MHz clock)
ATtiny45 (external 20 MHz clock)
ATtiny85 (internal 1 MHz clock)
ATtiny85 (internal 8 MHz clock)
ATtiny85 (external 20 MHz clock)
ATtiny44 (internal 1 MHz clock)
ATtiny44 (internal 8 MHz clock)
ATtiny44 (external 20 MHz clock)
ATtiny84 (internal 1 MHz clock)
ATtiny84 (internal 8 MHz clock)
ATtiny84 (external 20 MHz clock)
Arduino Uno
Arduino Duemilanove w/ ATmega328
Arduino Diecimila or Duemilanove w/ ATmega168
Arduino Nano w/ ATmega328
Arduino Nano w/ ATmega168
Arduino Mega 2560 or Mega ADK
Arduino Mega (ATmega1280)
```

FIGURE 7-9   Board options (partial list) for Digispark.

Compiling and Uploading Programs with Digispark

When you start the Digispark IDE for the first time, you need to set the Tools → Board menu option to reflect that you are using a Digispark. Figure 7-9 shows the selection you should make. The topmost menu option is the correct one to select. (We truncated the list of options, as it is fairly long.) Also make sure the Tools → Programmer menu option is set to Digispark.

You can now select a program to test the Digispark. Figure 7-10 shows the process for selecting the Digispark version of the Blink program found in the standard Arduino IDE. For the Digispark, the program is named Start. Note that you can also compile and run many of the standard Arduino examples, but you may have to make some minor changes. For example, the Blink program toggles the LED tied to pin 13 on the Arduino board. However, the Digispark doesn't have a pin 13. Its LED pin is pin 1, so the code needs to be changed to reflect that fact. (The Start program includes the necessary source code change.) Pins 3 and 4 are used for USB communication when you are uploading a sketch to the Digispark. However, once the program is uploaded, those pins can resume their normal I/O functions.

After the Start program is loaded, you can click the "check mark" icon to compile the Start code. You should *NOT* click the compile-upload "right-arrow" icon.

At this point, you should unplug the Digispark from the USB cable. While this may seem a bit weird, it's the way things work for the Digispark.

Once you see the *Binary sketch size* message as shown at the bottom of Figure 7-11, then you can click the "right-arrow" icon to upload the code. Note that the Digispark is still NOT connected to the USB cable at this time. (Interestingly, the Digispark compiles the Start program to 758 bytes, as can be seen in Figure 7-11. When you compile the Blink program using the standard Arduino 1.0.5 IDE, the program compiles to 1084 bytes. This suggests if you're thinking you would like to move an Arduino sketch to the Digispark but are afraid it won't fit, try compiling the Arduino code in the Digispark IDE and see what the final code size is. Also keep in mind that the Digispark library

Chapter 7: A CW Automatic Keyer 141

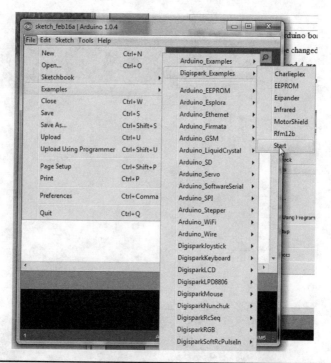

FIGURE 7-10 Selecting the Start (Blink) program example.

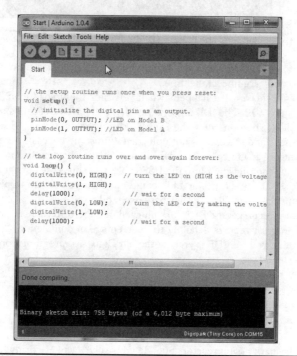

FIGURE 7-11 Compiling the Start program.

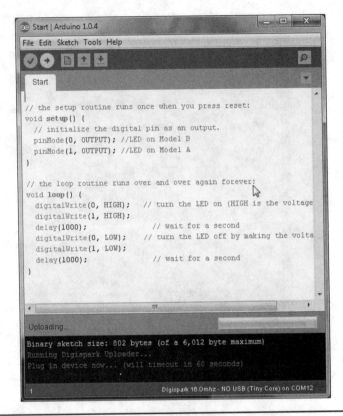

FIGURE 7-12   The upload message from the Digispark IDE.

environment is not as robust as the Arduino libraries. However, more and more new Digispark libraries are coming online all the time.)

Figure 7-12 shows what happens when you click the upload icon. Note the message at the bottom of Figure 7-12. The IDE informs you that you have 60 seconds to plug the Digispark into the USB cable. If you don't plug the Digispark into the USB cable within 60 seconds, you get a timeout message and the code is not uploaded. (On some boards, you may have to shut the IDE down and start over. On other boards, you just need to recompile and upload the source code.) When you plug the Digispark into the USB cable, you should see the green power LED come on. If you don't see the green power LED come on, chances are you've put the Digispark into the connector upside down. Flip the Digispark over and try again.

Assuming you do get the Digispark plugged into the USB cable in time, you will see messages giving you progress reports detailing the percentage of completion. You may also see progress messages telling you the IDE is erasing the old Digispark program along with upload messages. Finally, you should see:

```
>Starting the user app …
running: 100% complete
>> Micronucleus done. Thank you!
```

After the Digispark fission reactor shuts down (really?), you are ready to run the application that is now stored in the Digispark. You can disconnect the Digispark from the USB cable and place it

Chapter 7: A CW Automatic Keyer   143

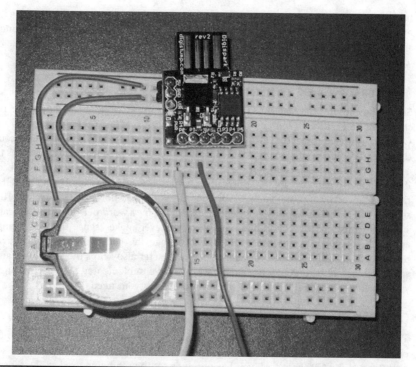

FIGURE 7-13   Prototype CW keyer using the Digispark.

in the circuit for which the software was written. For us, that is the CW keyer code, the subject of the next section.

Once you see the "Thank you" message, you know the program has been uploaded to the Digispark. It's then possible to add a power source and whatever components you need and run your program. Figure 7-13 shows the Digispark after uploading the keyer code and adding a battery plus two wires that serve to test the keying. As you can see in Figure 7-13, the battery (a CR2477) and its holder are about the same size as the Digispark. The CR2477 is rated at 1000 mAh, which, given the small current drain of the Digispark, should have the shelf life of granite even without a power switch. Another good battery choice is a standard 9 V battery, as shown in Figure 7-1. The two wires leading toward the bottom of the picture are attached to the keying paddles. You can actually test the circuit by just touching the wires.

The CW Keyer

Testing the prototype using the hardware shown in Figure 7-13 is pretty easy. There is very little to the circuit. The final circuit, however, may have an on-off switch. The reason we say "may have an on-off switch" is because it may not be worth it, especially if you are using the ATtiny85 by itself. With the bare chip drawing about 300 µA, the CR2477 should last over 2000 hours. The Digispark, however, has an onboard voltage regulator and power LED that increases the battery draw. Still, even assuming the worst-case assumptions about the additional current draw, the Digispark should provide at least 400 hours of operation on a single 9 V battery.

So, why all the fuss about a switch? After all, the switch only costs a buck or two, so why not include it? Actually, it's not the cost of the switch that is an issue to us, it's the "added appendage" that we object to. That is, we want a keyer with minimal moving parts, that is rugged, and doesn't have "stuff" sticking out that can break off in the field. Not using a switch is one less thingy sticking out of the case that can break in the field. Also, you can always remove the battery from its holder.

Adjusting Code Speed

Some of us are not the fastest fist in the West and appreciate it when an operator realizes we're struggling and QRS's for our benefit. Having the ability to adjust the keyer speed is a nice feature. Of course, you could have a small pot that is adjustable via a small screwdriver through a hole in the keyer case and use the chip's ADC capabilities to alter the speed. However, that's the beauty of a µC: we can set the speed in software. Simply stated, the program's source code recognizes that, if you hold the "dah key" for 10 consecutive strobes (i.e., greater than DAHTRIGGER), it lowers the keyer speed by approximately 1 word per minute. If you hold the "dit key" for 10 consecutive strobes (DITTRIGGER), it increases the keyer speed by 1 word per minute. (You can change the symbolic constants for the strobe counts to whatever makes sense to you.) Using a software approach to speed change makes it easy to change the code speed without having to use a potentiometer in the circuit. Not using a potentiometer also keeps the cost down and lowers the space requirements. If you don't like the way we have implemented the speed changing method, you can either change the circuit or the software to suit your needs.

Capacitance Sensors

The Arduino family of µCs are capable of making a touch-sensitive input using any of the Arduino pins. As you saw in Figure 7-13, two plain wires were used to make the keyer paddles for the prototype of the CW keyer. All you have to do is set the pin being used as the sensor to ground, turn on the chip's pull-up resistor, and measure how long it takes the pin to change state.

A number of programmers have refined a function, *readCapacitivePin()*, that is designed to return a value between 0 and 17 depending upon the level of capacitance on the pin. You can read the background information at http://playground.arduino.cc/Code/CapacitiveSensor. That reference points out that, although no hardware is required to make the touch sensor, a 1 nF capacitor in line with the pin being used helps reduce noise on the line. We chose to omit those caps because the keyer works fine without them. The authors also warn not to connect any voltage source to the sensor pin as it could damage the µC chip. Although not required for most modern rigs, we have added a 4N26 optoisolator to the keying circuit in much the same way it is used in Chapter 9 (see Figure 9-5).

The CW keyer source code is presented in Listing 7-1. The code begins with a series of *#define*s. The *#define DEBUG* preprocessor directive is used primarily to toggle the *digitalWrite()* method calls on the LED in and out of the program. This makes it easier to see what the code is doing and debug the program as needed. Obviously, you want the debug code removed when you are ready to program the chip for use in your circuit, so comment out the *#define DEBUG* directive at that time. The directives for DAHPADDLEPIN and DITPADDLEPIN are somewhat arbitrary since any digital pin can be used as a sensor. However, pin 1 on the Digispark is used for the onboard LED, which we do use in the debug mode, so we don't use it for one of the touch sensors. The OUTPUTPIN is the actual keying line and its output is fed into the 4N26 optoisolator. Figure 7-14 shows the complete schematic for the keyer using the ATtiny85 chip. The Digispark is essentially the same, but uses a 9 V battery.

The *#define MYTRIGGERVALUE* is one preprocessor directive value with which you may have to experiment. A whole host of factors can affect the actual behavior of the capacitance being placed on a sensor pin. Everything from the circuit ground to the size of the paddle sensor can affect the values being read. It is even possible to sense body capacitance without touching the sensor. We suggest you try the code starting with the value of 1 for this directive and see how

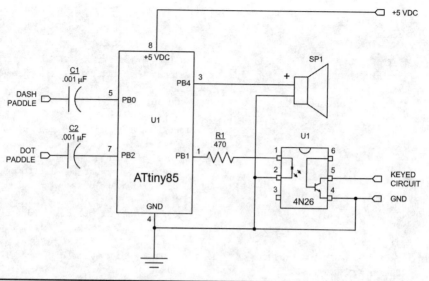

FIGURE 7-14   Schematic for ATtiny85 keyer.

the circuit behaves. In our initial tests, when the "paddles" were nothing more than wires connected to the pins, the value 1 worked fine. However, when we tried it with a large double-sided copper clad circuit board as the "paddle," we could actually trigger the circuit without touching it using the value of 1 for the threshold value.

```
/*****
  Capacitive touch sensor keyer

  Dr. Jack Purdum, Dec. 1, 2013

  This code is designed for either the ATtiny85 chip in a "stand-alone" mode
  or the Digispark board. It can also be used with all Arduino boards, too.

*****/

//#define DEBUG 1    // Used to watch debug values and LED. Comment out when happy

#define ATTINY85         1     // Defined when compiling for ATtiny85 chip.
                               // Comment out for Digispark

#define DAHPADDLEPIN     0     // Assign the paddle pins
#define DITPADDLEPIN     2
                               // For which board are we compiling?
#ifdef ATTINY85
  #define OUTPUTPIN      4     // Used to key the transmitter: Pin 4 for
                               // ATtiny85 (chip pin #3)
#else
```

LISTING 7-1   The CW keyer using touch sensors.

```
    #define OUTPUTPIN          5     // Use this one for Digispark
#endif

    #define LEDPIN             1     // LED pin for Digispark. No LED with ATtiny85.
    #define DAHTRIGGER         9     // If more dahs than this are sent consecutively,
                                     // slow down wpm
    #define DITTRIGGER         9     //     "   dits            "   speed up   "

    #define DITADJUST          7     // 92 = 1200/13 for 13 wpm. To speed up to 14 wpm,
                                     // the value becomes 85 = 1200/14. Therefore,
                                     // raising one wpm changes the dit time by
                                     // -7 milliseconds. Because we want to be able to
                                     // adjust up or down, this is an unsigned number.

    #define TOPSPEED          20     // 20 = 1200/60. The dit speed should not be
                                     // increased to a point where it exceeds 60wpm.

    #define SLOWSPEED       1200     // 1200 = 1200/1. This is the dit speed for 1 wpm.
    #define DITMULTIPLIER      1     // The dit is the basic unit of time
    #define DAHMULTIPLIER      3     // A dah is three times a dit

    #define MYTRIGGERVALUE     1     // The actual capacitance value will vary...
                                     // pick one that works


int dit;
int dah;
int wordsPerMinute;
int trackDits;
int trackDahs;

void setup()
{
#ifdef DEBUG
  pinMode(LEDPIN, OUTPUT);
#endif
  pinMode(OUTPUTPIN, OUTPUT);          // Use for keying circuit
  wordsPerMinute = 13;                 // Default
  dit = 92;                            // This value starts at 92 = 1200 / 13
  dah = dit * DAHMULTIPLIER;           // Three times a dit in length
}

void loop()
{
  // setWordsPerMinute();               // Not implemented yet...
  int i;

  uint8_t dahCycles = readCapacitivePin(DAHPADDLEPIN);
  if (dahCycles > MYTRIGGERVALUE) {
    sendDah();
```

LISTING 7-1   The CW keyer using touch sensors. (*continued*)

```
      trackDahs++;
      trackDits = 0;                            // Do this so we don't get a false speedup
      if (trackDahs > DAHTRIGGER) {
        setWordsPerMinute(DITADJUST);    // Slower speed, so raise dit value;
        trackDahs = 0;
      }
    }
    uint8_t ditCycles = readCapacitivePin(DITPADDLEPIN);
    if (ditCycles > MYTRIGGERVALUE) {
      sendDit();
      trackDits++;
      trackDahs = 0;                            // Do this so we don't get a false slow down
      if (trackDits > DITTRIGGER) {
        setWordsPerMinute(-DITADJUST);   // Slower speed, so lower dit value;
        trackDits = 0;
      }
    }

}

/*****
  This function is used to set the sending speed. This function also constrains
  the max keyer speed to 60wpm and the slowest speed to 1wpm.

  Parameter list:
    int newValue       Number of milliseconds to adjust dit speed. Can be + or -

  Return value:
    int                The new wpm value
*****/
int setWordsPerMinute(int newValue)
{
  dit += newValue;
  if (dit < TOPSPEED)      // If dit is reduced to 0 or negative, bad things happen.
    dit = TOPSPEED;
  if (dit >= SLOWSPEED)    // 1wpm is as slow as we want this to go.
    dit -= newValue;
  return dit;
}

/*****
  Function used to make a dit

  Parameter list:
    void

  Return value:
    void

*****/
```

LISTING 7-1 The CW keyer using touch sensors. (*continued*)

```
void sendDit()
{
#ifdef DEBUG
  digitalWrite(LEDPIN, HIGH);
#endif
   digitalWrite(OUTPUTPIN, HIGH);
   delay(dit);
   digitalWrite(OUTPUTPIN, LOW);
#ifdef DEBUG
  digitalWrite(LEDPIN, LOW);
#endif
  delay(dit);

}

/*****
 Function used to make a dash

 Parameter list:
   void

 Return value:
   void

 *****/
 void sendDah()
 {
#ifdef DEBUG
   digitalWrite(LEDPIN, HIGH);
#endif
   digitalWrite(OUTPUTPIN, HIGH);
   delay(DAHMULTIPLIER * dit);
   digitalWrite(OUTPUTPIN, LOW);
#ifdef DEBUG
   digitalWrite(LEDPIN, LOW);
#endif
   delay(dit);
 }


/*****
  This method is taken from the Arduino Playground and is used to read the
capacitance on a specific pin.
  See:  http://playground.arduino.cc/Code/CapacitiveSensor

  Parameter list:
    the pin being measured

  Return value:
```

LISTING 7-1 The CW keyer using touch sensors. (*continued*)

             uint8_t   an unsigned number from 0 to 17 used to indicate the
                          capacitance on the pin. Higher numbers indicate greater
                          capacitance.

*****/

uint8_t readCapacitivePin(int pinToMeasure) {
  // Variables used to translate from Arduino to AVR pin naming
  volatile uint8_t* port;
  volatile uint8_t* ddr;
  volatile uint8_t* pin;
  // Here we translate the input pin number from
  // Arduino pin number to the AVR PORT, PIN, DDR,
  // and which bit of those registers we care about.
  byte bitmask;
  port = portOutputRegister(digitalPinToPort(pinToMeasure));
  ddr = portModeRegister(digitalPinToPort(pinToMeasure));
  bitmask = digitalPinToBitMask(pinToMeasure);

  pin = portInputRegister(digitalPinToPort(pinToMeasure));
  // Discharge the pin first by setting it low and output
  *port &= ~(bitmask);
  *ddr |= bitmask;
  delay(1);
  // Prevent the timer IRQ from disturbing our measurement
  noInterrupts();
  // Make the pin an input with the internal pull-up on
  *ddr &= ~(bitmask);
  *port |= bitmask;

  // Now see how long the pin to get pulled up. This manual unrolling of the loop
  // decreases the number of hardware cycles between each read of the pin,
  // thus increasing sensitivity.
  uint8_t cycles = 17;
       if (*pin & bitmask) { cycles = 0;}
  else if (*pin & bitmask) { cycles = 1;}
  else if (*pin & bitmask) { cycles = 2;}
  else if (*pin & bitmask) { cycles = 3;}
  else if (*pin & bitmask) { cycles = 4;}
  else if (*pin & bitmask) { cycles = 5;}
  else if (*pin & bitmask) { cycles = 6;}
  else if (*pin & bitmask) { cycles = 7;}
  else if (*pin & bitmask) { cycles = 8;}
  else if (*pin & bitmask) { cycles = 9;}
  else if (*pin & bitmask) { cycles = 10;}
  else if (*pin & bitmask) { cycles = 11;}
  else if (*pin & bitmask) { cycles = 12;}
  else if (*pin & bitmask) { cycles = 13;}
  else if (*pin & bitmask) { cycles = 14;}
```

**LISTING 7-1**  The CW keyer using touch sensors. *(continued)*
```

```
      else if (*pin & bitmask) { cycles = 15;}
      else if (*pin & bitmask) { cycles = 16;}

      // End of timing-critical section
      interrupts();

      // Discharge the pin again by setting it low and output
      //   It's important to leave the pins low if you want to
      //   be able to touch more than 1 sensor at a time - if
      //   the sensor is left pulled high, when you touch
      //   two sensors, your body will transfer the charge between
      //   sensors.
      *port &= ~(bitmask);
      *ddr  |= bitmask;

      return cycles;
}
```

LISTING 7-1   The CW keyer using touch sensors. (*continued*)

In the *setup()* function, some debug code (which you've seen before) is run and then we set *OUTPUTPIN* to the output mode using a call to *pinMode()*. The default keyer speed is set to 13 words per minute (i.e., 65 characters per minute). Using the standard timing formula, a dit then corresponds to about 92 ms. A dah is then fixed to three times that time period. The inter-atom spacing is assumed to be one dit. That is, the letter "S" is 92 ms (dit), 92 ms (atom space), 92 ms (dit), 92 ms (atom space), 92 ms (dit), 92 ms (atom space). This means that every completed letter is automatically followed by one atom space. Obviously, final letter and word spacing is determined by the operator.

If you follow the comments for DITADJUST, you'll discover that changing the words per minute means that the dit speed in milliseconds rises (for a slower speed) or falls (for a faster speed) by 7 ms each time DITADJUST is changed. TOPSPEED and SLOWSPEED are used to set the maximum and minimum words per minute speeds.

The *volatile* Keyword

The *loop()* function first call the *readCapacitivePin()* for the *DAHPADDLEPIN*. Note the use of the *volatile* keyword for the first three variables used in the function. You use the *volatile* keyword to force the compiler to reload the rvalue of the variable each time it is referenced. Optimizing compilers often cache the rvalue of a variable, keeping it in a central processing unit (CPU) register to improve execution speed. However, if external resources can alter the value, it is possible to have the program be "out of sync" with the actual value for the variable. Using the *volatile* keyword forces the compiler to reload the most current value of the variable.

To get the most accurate reading from the pin as possible, the *readCapacitivePin()* code disables, interrupts, and then unrolls the polling of the pin using pointers to minimize the number of machine cycles required to read the pin. When the pin transitions, the value (0 to 17) is returned to the caller. If the return value stored in *dahCycles* is greater than *MYTRIGGERVALUE*, the *sendDah()* function is called. The same process is repeated for the dit sensor. Therefore, all the *loop()* function does is continually scan the dit and dah sensors looking for a change in capacitance. When that happens, the appropriate function is called to send a dit or a dah.

The *setWordsPerMinute()* function is used to change the sending speed. Two variables, *trackDits* and *trackDahs*, store the sequence of dits and dahs that have been sensed. The longest

Morse sequence using dits is the number 5, comprised of five dits. The longest sequence using dahs is the number 0, comprised of five dahs. The symbolic constants *DAHTRIGGER* and *DITTRIGGER* are defined as 9. This means that 10 consecutive dits raise the word per minute (*wpm*) by one. Likewise, 10 consecutive dahs lower the words per minute by one. These conventions make it fairly easy to adjust the code speed for the keyer.

Construction

The actual construction method you use is determined by which version of the keyer you plan to build. Let's build the ATtiny85 versions first. Figure 7-15 shows what our keyer looks like. You can use just about any case you wish as long as it's big enough to hold the components. (The case here was from Jameco Electronics, approximately 3 × 2 × 1.25 in. in size.) On the right side of the keyer in Figure 7-13, you can see the two small metal corner braces that we use for the dit and dah paddle levers. The ATtiny85 is the 8-pin chip on the top of the perf board and the 4N26 optoisolator is the 6-pin chip just below it. The battery is a CR2477, which provides 3 V for the keyer and is rated at 1 AHr. The ATtiny85 can operate with as little as 2.5 V, so the CR2477 supplies enough voltage for the chip. (The battery would not supply sufficient voltage for the Digispark, however, in part because of the onboard voltage regulator and LED.)

The output from the optoisolator is tied to the standard audio jack on the left edge of the case. You could leave this out and simply pass a wire through the case to a plug that is compatible with your rig. However, we have rigs that use both a ¼ in. and ⅛ in. jacks. Depending upon which rig we want to use, we have made a pair of connecting cables, one terminated with a ¼ in. plug and the other terminated with a ⅛ in. plug on one end and ⅛ in. plugs on the other ends. This approach means we don't have a loose plug wire dangling from the keyer when it's not in use. Pick whatever wiring best suits your needs.

We simply hot glued the perf board to the plastic case. While we can imagine that prying the battery from its holder may prove a little bit difficult, even with no power switch, we have yet to drain the battery. Some back-of-the-napkin calculations suggest that the battery should provide

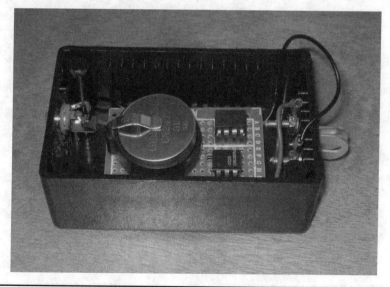

FIGURE 7-15 The ATtiny85 keyer. *(Case courtesy of Jameco)*

FIGURE 7-16 The Digispark version of the keyer.

power for several thousand hours even when left on continuously. Of course, you could simply remove the battery if you plan to leave it idle for any length of time.

Figure 7-16 presents another perspective of the Digispark version of the keyer that Dennis built. The glow emanating from the bottom of the keyer is from the Digispark onboard LED. Dennis also included a small buzzer, which you can see atop the small prototyping board available from Digistump. Dennis has his keyer tied to the Palm Paddle set of commercial paddles. He also has the Digispark version shown in Figure 7-1.

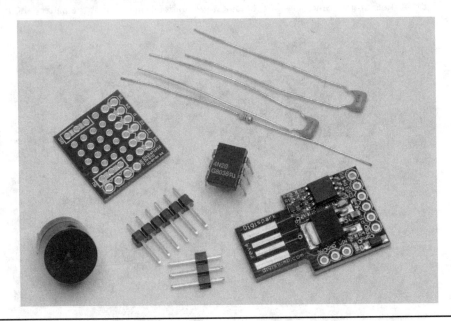

FIGURE 7-17 Digispark keyer parts.

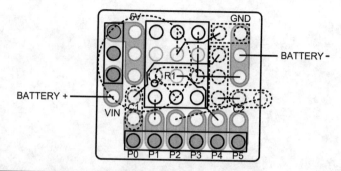

FIGURE 7-18 Parts placement for Digispark prototype board.

The parts Dennis used to construct the Digispark keyer shown in Figure 7-16 are presented in Figure 7-17. The parts show the two 1 nF capacitors that you might need to insert into the paddle leads under certain noisy conditions. The header pins are used to tie the prototype board to the Digispark.

Figure 7-18 presents a parts placement for the components shown in Figure 7-17. One advantage of Dennis's design is that the buzzer serves as a side tone for many low-cost QRP rigs that don't provide a sidetone.

Conclusion

This chapter has presented two designs for a simple keyer. Perhaps equally important, however, is that you learned how to build and program a μC circuit using the ATtiny85 chip. Such an approach is useful when the demands of your project don't warrant the expense of a full Arduino board. Knowing how to program the ATtiny85 gives you another tool to hang on your tool belt, and that's a good thing. After all, if the only tool you have is a hammer, all of your problems start to look like a nail.

Keep in mind that the ATtiny85 still has lots of program space left if you care to add new features to the keyer. You could, for example, hard code some messages (e.g., CQ, CQ, CQ DE...) into the chip and devise a way to activate those when needed. If you do add new features to the keyer, we hope you'll share your efforts with the rest of us via the web site.

CHAPTER 8

A Morse Code Decoder

The project discussed in Chapter 7 is a straight Morse code keyer that automatically generates dits and dahs for you according to the paddle lever you press. As interesting as those projects may be in and of themselves, they aren't terribly useful if you can't read Morse code.

The project for this chapter is a Morse code decoder. The function of a Morse code decoder is to read CW signals on some frequency, translate the dits and dahs into the appropriate ASCII characters, and display them on some output device (e.g., the LCD display from Chapter 3). Sounds pretty simple, right?

Wrong.

There are all sorts of difficulties in using electronics to decode a CW radio signal. First, it is very difficult to get a "clean" signal. Everything from adjacent signals, background noise, and QRM in general make it difficult to decode a given CW signal. While we can attempt to filter out the unwanted aspects surrounding a signal, even that approach is somewhat arbitrary. For example, you could construct a very narrow (e.g., 100 Hz) filter centered on a, say 700 Hz audio signal, in an attempt to reduce the unwanted elements of nearby signals. However, "reduced" is not the same as "eliminate." Second, and equally vexing, is that most humans don't send "perfect" CW code. In this context, by "perfect" we mean that the sender has a precise 3-to-1 timing ratio between dahs and dits, and that word spacing follows an exact 7-to-1 ratio in terms of dits. Try as we may, we each have our own "fist" and it's pretty likely it ain't perfect. Therefore, constructing a Morse code decoder that works perfectly on all CW signals simply isn't going to happen. We can, however, come close enough to make a decoder worthwhile.

Hardware Design Considerations

One of the first considerations is at what point in the receiver do we take the CW signal. While there are numerous choices, we opted to use the audio output from the receiver. There are several reasons for selecting this point for sampling the CW signal. First, every rig has a speaker or headphone jack that can be used to tap into the audio signal. Perhaps the most noninvasive way is to use an audio splitter jack, similar to that shown in Figure 8-1. The biggest advantage of this approach is that it does not modify the original state of the electronics in the receiver. Quite often, hams are reluctant to modify a piece of equipment for fear that they may diminish the retail value of the equipment. Another advantage of the splitter approach is that you can still continue to listen

Figure 8-1 Audio splitter jack.

to the audio output via a set of headphones while the decoder does its thing. Indeed, this is a great way to learn Morse code if you don't already know it. Many believe that listening to code "patterns" rather than individual letters is the best way to improve one's receiving speed. Because the implementation of an audio jack approach for attaching the CW decoder offers the advantage of not altering the original state of the equipment, that's the approach we employ.

While using the audio jack does allow us to implement the decoder without altering the equipment in a permanent manner, using the rig's audio output poses other issues. First, while many QRP systems can produce a signal with enough power to drive an Arduino analog pin, that signal contains components that we don't want passed into the circuit for processing. The biggest offenders are adjacent signals, strong local signals, and general background noise. An obvious solution is a filter to knock out these offending signals, but doing so with minimal compromise to the signal of interest.

Our design uses an LM324 op amp and an LM567 tone decoder to preprocess the signal before passing it along to the Arduino. Figure 8-2 shows the schematic that we used for our circuit

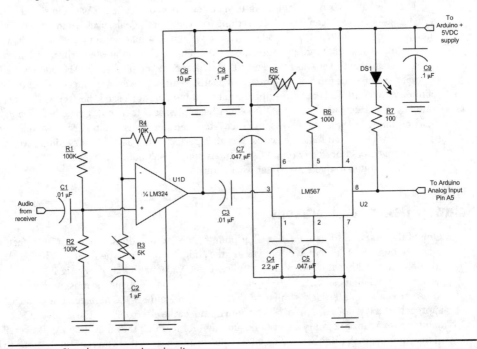

Figure 8-2 Signal preprocessing circuit.

Ref	Description	Source
C1, C3	0.01 µF monolithic, 15 V	Jameco, RadioShack, eBay, etc.
C2	1 µF monolithic, 15 V	
C4	2.2 µF monolithic, 15 V	
C5, C7	0.047 µF monolithic, 15 V	
C6	10 µF electrolytic	
C8, C9	0.1 µF monolithic, 15 V	
DS1	LED	
R1-2	100K Ω, ¼ W, 5%	
R3	5K Ω, potentiometer	eBay
R4	10K Ω, ¼ W, 5%	Jameco, RadioShack, eBay, etc.
R5	50K Ω, pot, multi-turn	eBay
R6	1K Ω, ¼ W, 5%	Jameco, RadioShack, eBay, etc.
R7	100 Ω, ¼ W, 5%	
U1	LM324 quad op amp	
U2	LM567 tone decoder	
Misc	Omega MCU Systems ProtoPro-B Prototyping shield	eBay, direct

TABLE 8-1   Morse Decoder Parts List

with the corresponding list of parts in Table 8-1. First, the audio signal is passed into the LM328 op amp to boost the signal. That signal is then passed into the tone decoder to filter out as much background noise as possible. The potentiometer/capacitor combine to form a filter network designed to set the bandpass near 700 Hz. An LED is used as an indicator that the signal of interest syncs with the filter. Once the signal is "tuned in," it is passed onto the analog input pin of the Arduino.

The story, however, is just beginning because the software takes over the task of delivering a useful output from the preprocessed signal.

Signal Preprocessing Circuit Description

The circuit for the Morse code decoder consists of an amplifier stage to boost the incoming signal followed by a tone decoder (see the schematic in Figure 8-2). In this case, we are using an LM324 op amp for signal amplification and the LM567 tone decoder to further process the amplified signal. We added the gain stage (LM324 op amp) just in case there is not enough audio signal to drive the LM567.

The parts placement for the circuit shown in Figure 8-2 is presented in Figure 8-3.

The amplifier uses one section of the LM324 quad op amp. It is designed as a non-inverting amplifier with adjustable gain. The LM324 is designed to operate off of a single supply; it is biased internally to provide "zero Volts out for zero Volts in." This biasing arrangement is great for a DC amplifier (such as was used in Chapter 5 for the general purpose panel meter), but it does not work well for an audio amplifier. We would much rather have an audio amplifier with the idling output halfway between the positive and negative supply voltages (in this case 5 V and ground), or about

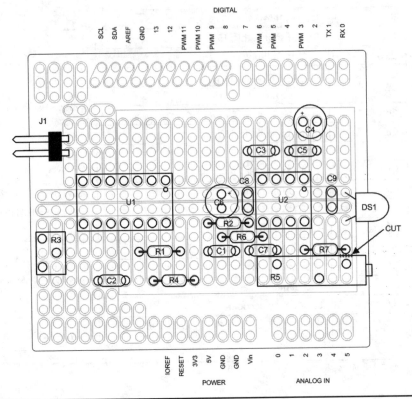

FIGURE 8-3 Decoder parts placement.

2.5 V. Resistors R201 and R202 provide the necessary input bias to set the idle output at approximately half the supply voltage. The actual output voltage varies depending on the tolerance of the resistors and the gain setting.

R203 and R204 set the gain for the op amp. C202 serves two purposes: 1) it provides a low frequency roll-off, and 2) prevents any DC offset on the inverting input from affecting the output. The formula for determining gain in a non-inverting configuration is:

$$V_{out} = V_{in} \times \left(1 + \left(\frac{R204}{R203}\right)\right)$$

You can use the formula to adjust the gain to a different value if you wish. C201 and C203 are coupling capacitors that prevent any stray DC from the previous stage from affecting the next stage in the circuit.

The LM567 is a simple phase-locked-loop that is used as a tone decoder. C207, R205, and R206 set the center frequency of the decoder. We used a 10-turn pot for R205. While you can use a different part for R205, the 10-turn pot makes it a lot easier to set the center frequency. R205 is used to adjust the LM567 center frequency to be the same as the centered audio tone using the narrowest CW filter on your receiver. The bandwidth of the decoder is set by C205, the loop filter. C204 is the "output filter" and it is used to smooth the output to remove noise in the decoded input. An LED (CR201) is used as a tuning indicator. When a signal is properly tuned in, CR201

FIGURE 8-4   Assembled decoder shield.

flashes in sync with the incoming Morse code audio. R207 is a current limiting resistor for CR201. The completed Morse decoder shield is shown in Figure 8-4.

Notes When Using the Decoder with the TEN-TEC Rebel

The Rebel presents a unique integration of a commercial product and "homebrew" add-ons. The Morse code decoder circuit can be built on shield to plug right into the Rebel. The Rebel utilizes a Digilent chipKIT Uno32 processor, which is an "Arduino-like" board. However, there are several things that are slightly different when using the Uno32 over any other Arduino.

While the Uno32 may look like an Arduino Uno or a Demilanove, you will notice that there are two rows of analog inputs and two rows of digital IO pins. Where the Arduino Uno and Demilanove have 6 analog inputs, the Uno32 has 12. Similarly, the Uno and Demilanove have 16 digital IO pins, the Uno32 has 30. When you build a shield for the Uno32, you must use the right kind of shield (one that supports two rows of pins for IO and analog inputs) and then the board must have some modifications made if you are to take advantage of the extra pins.

The Morse code decoder uses analog 6 (A6) from the Rebel for the input audio. The Rebel calls this pin the "Code Read" pin. It is a very low level audio signal that is tapped off before the final audio PA and volume control but after the AGC. This signal is about 40 mV, hence the need for a gain stage before the LM567 tone decoder. The decoded audio is sent back to the processor on pin analog 11 (A11). A11 is an unassigned analog input on the Rebel.

The typical Uno32 shield with two rows of pins has the adjacent pins connected together. Therefore it will be necessary to separate those pins. We prefer using a Dremel tool with a cutting disk, but an Xacto knife would do the trick as well. Unless you are planning on using other analog and digital pins on the Rebel, it should only be necessary to cut apart A0 and A6 as well as A5 and A11. You can then use the breakaway headers to cut off two single pins for A6 and A11. Use the Rebel as a guide to hold the pins as you solder them to the shield. Of course, we don't need to remind you to make sure the power is disconnected from the Rebel before you use it to hold the header pins for soldering to the shield.

Of course, you can also use a prototyping shield that is made specifically for the chipKit Uno32. No cutting required! If you are building the decoder for different equipment, a regular Arduino shield works fine.

Decoder Software

There are probably dozens of workable algorithms that can be used to process the CW signal that comes into the Arduino. Some of these algorithms are like sledgehammers being used to craft a fine piece of jewelry. Indeed, some are sufficiently crude that they surely must degrade the performance of the decoder, both in terms of the code speed they can read and the reaction time for break-in keying. While we could spend a bucket load of time crafting our own solution, we chose to stand on the shoulders of others to cut down on the coding cycle. We drew upon the work of a number of other people.

One very interesting approach was developed by a Norwegian individual named Ragnar O. Aronsen. His web site http://raronoff.wordpress.com/2010/12/16/morse-endecoder/ provides an excellent description of how his software works. If you wish to experiment with his code, you can download his code at https://code.google.com/p/morse-endecoder/.

A key element in his approach to decoding the incoming CW is the use of a binary tree. While you may not be familiar with the term "binary tree," you have probably used it yourself in various guessing games. Binary trees work for data that is arranged in an order that permits a binary search on the data. For example, suppose you are charged with guessing a number between 1 and 100 and, after each guess, you are told that your guess is too high or too low. You could just start with 1, increment by 1 after each guess until you reach the correct number. On a large group of random numbers, your average number of guesses should approach 50. Not good.

A more efficient way is to divide the list in half and make your first guess 50. If they tell you your guess is too low, you can immediately throw away half of the list (i.e., numbers 1 through 50) and concentrate on the numbers that remain. Because your first guess was too low, your next guess should be halfway up the numbers that remain, or 75. (If the guess of 50 was too high, you would halve the difference and guess 25 and you would discard all numbers from 50 through 100.) Note that after just two guesses, you have eliminated three-quarters of the numbers. You repeat this process until you zeroed in on the number. Using this approach you will know the number within six guesses which, on average, is about eight times faster than linear guessing. This is what is called a binary search: You divide the range in half each time, throwing out half of the remaining list on each guess.

Search a Binary Tree of ASCII Characters

In the Aronsen code, he arranges the Morse code characters in a manner similar to that seen in Figure 8-5.

Just as you started the guessing game at the mid-range point, so does the search of a binary tree start at the midpoint. Let's call each aspect of an incoming Morse code character an atom. In other words, an atom can be a dot or a dash, depending upon its duration. If the table has 128 characters in it, the pointer starts at the midpoint, or position 64, labeled Start in Figure 8-5. If the first atom is a dit, the pointer moves from Start halfway to the left, or position 32. If the first atom is a dah, the pointer moves halfway to the right, position 96. Therefore, the rule is dits move halfway left, dahs move halfway right.

If the first atom of a character read is a dit, the pointer moves midway to the left (i.e., position 32 = Start / 2 = 64 / 2) and reads the letter E. If the next atom equals the spacing for a letter, we know the character is complete and the Morse character is an E. However, if the next atom is a dah, then we move halfway to the right (48 = (64 − 32) / 2 + 32) and find the letter A. If, instead of a dah

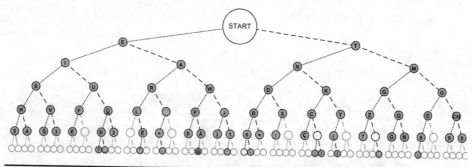

FIGURE 8-5   Morse code arranged as a binary tree. (*Source*: http://commons.wikimedia.org/wiki/File: Morse_code_tree3.png)

the next atom is a dit, we move halfway to the left again and find the letter I (16 = 32 / 2). Once again, if the next atom is a letter space, we either have an A or an I, depending upon which atom was actually read. You can see that six-level tree as shown in Figure 8-3 means that we can find any given letter in six comparisons or less. Binary trees using a binary search are a very efficient way of locating an individual element in a list of organized data.

Upon seeing this binary tree approach for character lookup, we immediately were interested in implementing Ragnar's code. We figured it would be more than fast enough to keep up with any human code that might come across the airwaves. In fact, Ragnar has a video that shows two Arduinos; one receiving code and the other sending it, decoding successfully at over 300 words per minute (wpm)! Of course, the advantage is that the signal was not going out "over the airwaves" and the code being sent was perfect (i.e., 1:3 ratio) code. Still, that's a pretty impressive speed. However, after some experimentation, we discovered that almost any table lookup is fast enough for human-generated code, even badly programmed examples.

Another algorithm we examined closely was by Budd Churchward, WB7FHC. We liked his approach because it is similar to the encoding used by Mark VandeWettering, K6HX, which we adapted to the PS2 keyer presented in Chapter 9. The actual coding scheme used by Budd is explained very clearly in a tutorial you can read at http://www.honorlevel.com/data/arduino/readcode/readCode.01.html.

In essence, each code character is encoded as a byte value where a start bit is binary 1 and the actual code sequence follows. Each dit is encoded as a 1 bit, and each dah as a 0 bit. Therefore, if you were going to send the letter A, which is dit-dah, the binary representation Budd uses becomes 00000110. Reading the byte from left to right, the first bit 1 digit is the start bit and is ignored, which leaves 10 as the remaining bit pattern. Because a dit is 1 and a dah is 0, the bit pattern is dit-dah, or the letter A.

Budd stores the alphabet in character form in the following array:

```
char mySet[] = "##TEMNAIOGKDWRUS##QZYCXBJP#L#FVH09#8###7#####/-'
                61######2###3#45";
```

You binary aficionados know that binary 00000110 is decimal 6. Now, look at *mySet[6]* above and what do you find? Because C arrays start with element 0, *mySet[6]* is the letter A. What about the letter N, which is dah-dit, or using Budd's encoding, 00000101 in binary is 5 in decimal. *mySet[5]* is the letter N. If you pick various elements in the array and work out the binary value for its index, you'll find that the array contains the Morse alphabet using Budd's encoding scheme. The '#' characters in the array correspond to array indexes that do not contain a valid Morse character. Pretty slick! We also like it because any bit shifting we may need to do to extract letters is an extremely efficient operation on a μC.

Morse Decode Program

As we mentioned earlier, most of the Morse decode program is based on Budd Churchward's code. Listing 8-1 presents the code for the Morse decoder.

```
/* Barnacle Budd's Morse Code Decoder v. 1.4
   (c) 2013, Budd Churchward - WB7FHC

   This project makes use a custom built tone decoder module using
   the LM567C microchip. Details of this module will eventually be posted
   on line. This module allows you to tune to the frequency of a specific
   tone while ignoring noice and other tones of different frequencies

   The program will automatically adjust to the speed of code that
   is being sent. The first few characters may come out wrong while it
   homes in on the speed. If you are not seeing solid copy, press the
   restart button on your Arduino. You can try adjusting the tone decoder.
   Lowering the volume of the incoming CW can also help. If the tone decoder
   is not centered on the frequency of the incomming signal, you may have
   to fine tune the module as you lower the volume.

   The software tracks the speed of the sender's dahs to make
   its adjustments. The more dahs you send at the beginning
   the sooner it locks into solid copy.

   After a reset, the following text is very difficult to lock in on:
   'SHE IS HIS SISTER' because there are only two dahs in the whole
   phrase and they come near the end. However, if you reset and then
   send 'CALL ME WOODY' it will match your speed quite quickly.

   This project is built around the 20x4 LCD display. The sketch includes
   funtions for word wrap and scrolling. If a word extends beyond the 20
   column line, it will drop down to the next line. When the bottom line
   is filled, all lines will scroll up one row and new text will continue
   to appear at the bottom.

   This version makes use of the 4 digit parallel method of driving the
   display. A two line, I2C version will be forthcoming.

   Hook up your LCD panel to the Arduino using these pins:
      LCD pin  1 to GND
      LCD pin  2 to +5V
      LCD pin  4 to D7
      LCD pin  6 to D6
      LCD pin 11 to D5
      LCD pin 12 to D4
      LCD pin 13 to D3
      LCD pin 14 to D2
```

LISTING 8-1 The Morse decode code.

```
          LCD pin 15 to +5V
          LCD pin 16 to GND

    Data from pin 8 of the LM567C will be fed to D8 on the Arduino
    When this pin is HIGH there is no tone detected.
    When this pin is LOW a tone of the set frequency has been detected.

*/

#include <LiquidCrystal.h>
#include <stdlib.h>

#define LEDPIN      13
#define AUDIOPIN    A5

#define FALSEREAD   10      // Caused when an audio burst, like QRN, generates
                            // a short false signal.

#define LCDROWS     2
#define LCDCOLUMNS  16

// initialize the library with the numbers of the interface pins
// Could also use: (7, 6, 5, 4, 3, 2) without interrupts
LiquidCrystal lcd(12, 11, 5, 4, 3, 2);   // For our LCD shield

int audioPin = AUDIOPIN; // Jack's decoder pin
int audio = 1;           // will store the value we read on this pin

// Array index matches myNum that we parsed out of the code.
// #'s are miscopied characters
char mySet[] = "##TEMNAIOGKDWRUS##QZYCXBJP#L#FVH09#8###7#####/-61#######2###3#45";
char lcdGuy = ' ';              // We will store the actual character decoded here
char lcdBuffer[LCDCOLUMNS + 1];

boolean justDid = true;       // Makes sure we only print one space during long gaps
boolean characterDone = true; // A full character has been sent
boolean ditOrDah = true;      // We have either a full dit or a full dah

// The following values will auto adjust to the sender's speed
int averageDah = 240;     // A dah should be 3 times as long as a dit
int dit = 80;             // We start by defining a dit as 80 milliseconds
int realDit = dit;
int myBounce = 2;         // Used as a short delay between key up and down
int myNum = 0;            // Turn dits - dahs into a binary number stored here

long downTime = 0;        // How long the tone was on in milliseconds
long upTime = 0;          // How long the tone was off in milliseconds
long startDownTime = 0;   // Arduino's internal timer when tone first comes on
long startUpTime = 0;     // Arduino's internal timer when tone first goes off
```

LISTING 8-1   The Morse decode code. (*continued*)

```
///////////////////////////////////////////////////////////////////
void setup() {
  pinMode(AUDIOPIN, INPUT);
  pinMode(LEDPIN, OUTPUT);      // We're going to blink Arduino's onboard LED
  lcd.begin(LCDCOLUMNS, LCDROWS);      // Cuz we have a 16x2 display
  lcd.clear();                  // Clear display
  lcd.print(" K&P CW Decoder");
  lcd.setCursor(0, 1);
  memset(lcdBuffer, '\0', LCDCOLUMNS + 1);  // Clear buffer
}

void loop() {
  audio = digitalRead(audioPin); // What is the tone decoder doing?

  if (!audio) {
    keyIsDown();        // LOW, or 0, means tone is being decoded
  }
  if (audio) {
    keyIsUp();          // HIGH, or 1, means no tone is there
  }
}

/*****
  This function is called when the µC senses the start of a Morse character.
  There are a number of global variables set here.

  Parameter list:
    void

  Return value:
    void

*****/

void keyIsDown() {
  // The decoder is detecting our tone
  // The LEDs on the decoder and Arduino will blink on in unison
  digitalWrite(LEDPIN,1);              // turn on Arduino's LED

  if (startUpTime > 0){
    // We only need to do once, when the key first goes down
    startUpTime = 0;    // clear the 'Key Up' timer
  }
  if (startDownTime == 0) {
    startDownTime = millis();  // get Arduino's current clock time
  }
  characterDone = false; // we're still building a character
  ditOrDah = false;      // key still down we're not done with the tone
```

LISTING 8-1 The Morse decode code. *(continued)*

```
      delay(myBounce);        // Take a short breath here

   if (myNum == 0) {          // myNum = 0 at the beginning of a character
      myNum = 1;              // Start bit - it only does this once per letter
   }
}

/*****
   This function is called when the µC senses the end of a Morse character by
   the absence of an audio tone. Again, there are a number of global variables
   set here.

   Parameter list:
      void

   Return value:
      void

*****/
void keyIsUp() {
  int farnsworthCutoff;

  // If we haven't already started our timer, do it now
  if (startUpTime == 0){
    startUpTime = millis();
  }

  // Find out how long we've gone with no tone
  upTime = millis() - startUpTime;

  if (upTime < FALSEREAD)              // Static?
    return;

    // If it is twice as long as a dah print a space
  if (realDit < 90) {
    farnsworthCutoff = averageDah * 2;
  } else {
    if (realDit > 90 && realDit < 105) {
      farnsworthCutoff = averageDah + realDit;
    } else {
      farnsworthCutoff = averageDah * 3;
    }
  }

  if (upTime > (realDit * 3)) {        // Space??
    printSpace();
  }
```

LISTING 8-1   The Morse decode code. (*continued*)

```
    // Only do this once after the key goes up
    if (startDownTime > 0){
      downTime = millis() - startDownTime;  // how long was the tone on?
      startDownTime = 0;                    // clear the 'Key Down' timer
    }

    if (!ditOrDah) {   // We don't know if it was a dit or a dah yet
      shiftBits();     // let's go find out! And do our Magic with the bits
    }

    // If we are still building a character ...
    if (!characterDone) {      // Are we done yet?
      if (upTime > realDit) {  // BINGO! we're done with this one
        printCharacter();      // Figure out character and print it
        characterDone = true;  // We got him, we're done here
        myNum = 0;             // Setup for getting the next start bit
      }
      downTime = 0;            // Reset our keyDown counter
    }

  }
}

void shiftBits() {
  // we know we've got a dit or a dah, let's find out which
  // then we will shift the bits in myNum and then add 1 or not add 1

  if (downTime < dit / 3)
    return;                 // ignore QRN

  myNum = myNum << 1;       // shift bits left
  ditOrDah = true;          // we will know which one in two lines

  // If it is a dit we add 1. If it is a dah we do nothing!
  if (downTime < dit) {
    myNum++;                // add one because it is a dit
    realDit = (downTime + realDit) / 2;
  } else {
    // The next four lines handle the automatic speed adjustment:
    averageDah = (downTime + averageDah) / 2;  // running average of dahs
    dit = averageDah / 3;                      // normal dit would be this
    realDit = dit;                             // Track this for timing
    dit = dit * 2;          // double for threshold between dits and dahs
  }
}
void printCharacter() {
  justDid = false;          // OK to print a space again after this

  // Punctuation marks will make a BIG myNum
  if (myNum > 63) {
```

LISTING 8-1   The Morse decode code. *(continued)*

```
    printPunctuation();  // Value parsed is bigger than our character array
                         // Probably a punctuation mark so go figure it out.
    return;              // Go back to the main loop(), we're done here.
  }
  lcdGuy = mySet[myNum]; // Find the letter in the character set

  DoMyLCD();             // Go figure out where to put in on the display
}

void printSpace() {
  if (justDid) {
    return;  // only one space, no matter how long the gap
  }

  justDid = true;        // so we don't do this twice
  lcdGuy = ' ';          // this is going to go to the LCD
  DoMyLCD();             // go figure out where to put it on the display
}

/*****
  Punctuation marks are made up of more dits and dahs than letters and
  numbers. Rather than extend the character array out to reach these higher
  numbers we will simply check for them here. This funtion only gets called
  when myNum is greater than 63.

  Parameter List:
    void

  Return value:
    void
*****/
void printPunctuation() {
  switch (myNum) {
    case 71:
      lcdGuy = ':';
      break;
    case 76:
      lcdGuy = ',';
      break;
    case 84:
      lcdGuy = '!';
      break;
    case 94:
      lcdGuy = '-';
      break;
    case 97:
      lcdGuy = 39;      // Apostrophe
      break;
```

LISTING 8-1 The Morse decode code. *(continued)*

```
      case 101:
        lcdGuy = '@';
        break;
      case 106:
        lcdGuy = '.';
        break;
      case 115:
        lcdGuy = '?';
        break;
      case 246:
        lcdGuy = '$';
        break;
      case 122:
        lcdGuy = 's';
        DoMyLCD();
        lcdGuy = 'k';
        break;
      default:
        lcdGuy = '#';     // Should not get here
        break;
  }
  DoMyLCD();      // go figure out where to put it on the display
}

/*****
  This function moves the current character read to the LCD display. It
  assumes that lcdGuy has been set prior to calling this function.

  Parameter list:
    void

  Return value:
    void

*****/
void DoMyLCD() {
  static int passCounter = 0;
  int wpm;
  char numBuff[5];
  char outBuff[17];

  if (passCounter % 100 == 0) {
    wpm = 1200 / realDit + (realDit * .06);
    itoa(wpm, numBuff, 10);
    strcpy(outBuff, "Approx WPM = ");
    strcat(outBuff, numBuff);
    strcat(outBuff, " ");
    lcd.setCursor(0,0);
    lcd.print(outBuff);
```

LISTING 8-1   The Morse decode code. (*continued*)

```
        passCounter = 0;
    }
    passCounter++;
    memcpy(lcdBuffer, &lcdBuffer[1], LCDCOLUMNS - 1);
    lcdBuffer[LCDCOLUMNS - 1] = lcdGuy;
    lcd.setCursor(0, 1);
    lcd.print(lcdBuffer);
}
```

LISTING 8-1   The Morse decode code. (*continued*)

The code begins with the usual definitions for symbolic constants and working variables. The code that appears in the *setup()* function establishes the parameters for the LCD display and set the pin modes for the audio input and LED. We use the LED to verify that the code is in sync with the code being received.

In essence, the decoder is a state machine that is controlled by the audio input to the system. In *loop()*, the first thing that is done is to sample the audio pin for the presence or absence of an audio signal via a call to *digitalRead()*, assigning the value to *audio*. If there is no audio input, either there is no Morse signal or the signal is being decoded. This state is such that the function *keyIsDown()* is called. If there is a tone, we assume that a Morse character is being read, *keyIsUp()* is called. This seems like the logic is reversed from what it should be, but keep in mind that determining a dit or a dah depends upon the start and end times for a given atom. (Recall that an atom can be either a dit or a dah.) Either way, the duration of the audio signal is important in decoding the incoming character.

Consider a key down event. Because a single dit atom can be read multiple times during *loop()*, we need a way to know whether this is the continuation of an atom that has previously been sensed or if we are starting a new atom. We do this by reading the variable named *startDownTime*. If *startDownTime* is zero, we know we are at the start of an atom so we assign the millisecond count from a call to *millis()* into *startDownTime*. Because this is the start of a new character, we set *myNum* to 1. *myNum* is the variable we use to calculate the binary bit pattern used to index into the *mySet[]* character array mentioned earlier. Some working variables (*characterDone* and *ditOrDah*) are also set at this point. Program control now returns back to *loop()*.

Once again, *digitalRead()* is called and its value assigned into *audio*. If no signal is detected, either we have the end of an atom or we are in a space of some sort (e.g., between letters, words, or just a plain pause). If *startUpTime* is zero, we assign the current millisecond count into it. We then calculate the amount of time that has transpired since *startUpTime* was read and assign it into *upTime*. A very short value for *upTime* would take place if we just read *millis()* or if there was a short signal burst, like static. In either case, we simply return to *loop()*. However, if the value of *upTime* is fairly long, we remain in the *keyIsUp()* function for further processing.

Farnsworth Timing

A good portion of the remainder of the *keyIsUp()* function is an attempt to cope with Farnsworth timing. Simply stated, Farnsworth timing is recognition of the fact that people who can copy Morse code well do so by listening to the rhythm of the characters rather than individual dits and dahs. Because a rhythm is more difficult to detect at slower speeds, Farnsworth timing speeds up the characters being sent, but increases the spacing between characters. For example, if the code speed is to be 13 wpm, Farnsworth timing might send the Morse characters at 18 wpm, but adds additional time between characters so that the average speed is 13 wpm. (You can find an ARRL paper on the topic at http://www.arrl.org/files/file/Technology/x9004008.pdf.)

Clearly, this messes up the traditional timing characteristics between letters and words, which means our attempt to decode a signal using Farnsworth timing is going to be difficult at best. The ARRL code practice sessions, for example, use Farnsworth timing at speeds less than 18 wpm, but uses conventional timing on speeds of 18 wpm or greater. The ARRL feels that speeds at and above 18 wpm have sufficient speeds to sense the rhythm of the characters being sent. However, trying to write code that copes with this variance isn't easy. We used the ARRL Morse audio MP3 files to try to process Farnsworth timing as well as regular timing. Quite honestly, we don't feel we have a good solution yet. Our efforts are a starting point, not an end point. If you come up with a solution, we sure hope you'll share it with the rest of us.

If *upTime* is greater than three times a dit spacing, we assume that we have just finished "reading" a space and *printSpace()* is called. If *startDownTime* is nonzero, we know that an atom (at least) has just been finished. *downTime* then becomes the difference between the current value of *millis()* and *startDownTime*. If we don't yet know whether we are processing a dit or a dah, a call to *shiftBits()* is done. Based on the value of *downTime*, we can determine whether we need to increment *myNum* (which becomes the index value into the *mySet[]* character array) or not. Recall that a dit increments *myNum* while a dah does not. The *shiftBits()* function also tracks the average time for dahs in an attempt to detect a change in sending speed.

If *upTime* is greater than what a dah should be, we assume that we are between characters and we can display the character via a call to *printCharacter()*. In *printCharacter()*, we check to see if *myNum* is greater than 63, which is the largest index permitted in the *mySet[]* character array. If it is, special case processing is performed by a call to *printPunctuation()*. Otherwise, the character is displayed on the LCD display by a call to *DoMyLCD()*.

The *DoMyLCD()* also attempts to display an approximation of the current sending speed. Again, this is at best an approximation and Farnsworth speeds make it even more so. Still, perhaps a hint of the current speed is better than nothing. Also, note the statement:

```
memcpy(lcdBuffer, &lcdBuffer[1], LCDCOLUMNS - 1);
```

Often you will see code where someone is copying a sequence of characters from one array to another using a *for* loop. The *memcpy()* function is a considerably faster way to do that. The first parameter is the destination array, the second parameter is the source for the copy, and the third parameter is the number of bytes to copy. Note what our *memcpy()* does. Our source is everything from the second element of the array onward, but it copies it back onto itself starting at the first element. If you wrap your head around this, the result is that we are "scrolling" the contents of the array on the LCD display because we update the last character on the display with the character that we just parsed.

Another *mem*()* function worth knowing is *memset()*, which can be used to set a section of memory, like an array, to a single value. For example, the statement:

```
memset(buffer, 32, sizeof(buffer));
```

sets all of the valid bytes in *buffer* to a space character. (The ASCII value for a space is 32.) The first parameter is the area of memory to use, the second is the value to use for each byte in that memory space, and the final parameter is the number of bytes to process. Note how using the *sizeof* operator makes the idiom for the third parameter portable for any data array on any system. Using the idiom would work, for example, whether your host system uses 2- or 4-byte *int*s.

Once the call to *DoMyLCD()* finishes, control ultimately returns to *loop()* and the process begins again.

Conclusion

This chapter presents a CW decoder that preprocesses the signal before decoding it. Using W1AW code practice MP3 files for testing, we were able to copy speeds of around 35–40 wpm. The software is not terribly complex, but there are so many variations in sending CW that it is difficult to get a 100% decode message. Farnsworth timing complicates things, but even that could be handled if everyone had a "perfect fist." Alas, that is not the case. Some hams have a rhythm that is almost lyrical, why others make you feel like you're listening to a buzz saw. Still, given its relatively low cost, it is easy to build the shield and experiment with the software. This is one project where some innovative change to the software could bring about significant improvement in the results. Give it a try and let us all know how you fared.

CHAPTER 9

A PS2 Keyboard CW Encoder

In 2006, the Federal Communications Commission (FCC) dropped the Morse code requirements for amateur radio licenses. People have different views about dropping the code requirement as part of becoming a licensed radio amateur. It seems reasonable to assume that the disappearance of the requirement probably did bring more people into amateur radio. However, not knowing Morse code has its downside, too.

First, you can buy a low power continuous wave (CW) transceiver for less than $50, making the entry cost into the hobby fairly inexpensive. Second, CW affords more "miles per watt" than other transmission modes. With a good antenna system and favorable band conditions, QRP rigs that output a couple of watts are capable of worldwide communications. Third, CW rigs are well suited to emergency operation since they offer reliable communication with relatively low power requirements. Battery operation is routinely done with many QRP rigs. (Chapter 17 features a project that uses solar power for emergency communications.) Also, for certain classes of licenses, only CW can be used in certain segments of the bands. Finally, we feel that amateurs who haven't experienced CW operation have really missed out on something. We can't really express the feeling clearly, but there is an inner satisfaction of being able to communicate in Morse code.

We actually started wondering about CW and keying in general when a friend of ours with a wrist injury said that he had backed off of using CW because of the wrist pain. (He preferred a straight key.) Like many other hams, however, he could type quite well and he could do so without much pain. Hence, the project of this chapter. While typing on an old PS2 keyboard may not be your exact cup of tea, the project does show how easy it can be to interface an external device to an Arduino plus how to isolate your rig from the Arduino. Finally, there are some interesting software techniques used in this project's software that are worth knowing.

The PS2 Keyboard

Most keyboards for today's computers are USB devices. Unfortunately, the USB specification and interfacing to a USB port is a fairly complex process. PS2 keyboards, on the other hand, are easy to interface. Another advantage is that you can buy PS2 keyboards fairly inexpensively. The keyboard we use in this project was picked up at a church donation center for $2.

The pin out for a PS2 connector is shown in Figure 9-1.

Pin 1 +DATA
Pin 2 Not Connected
Pin 3 GND
Pin 4 Vcc
Pin 5 +CLK
Pin 6 Not Connected

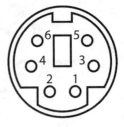

FIGURE 9-1 PS2 connector (facing female socket).

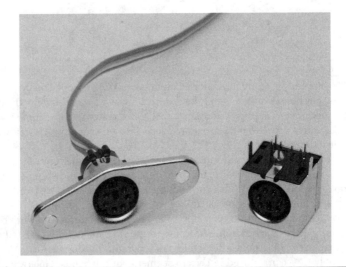

FIGURE 9-2 PS2 connector sockets.

Only four of the six pins are needed to interface the PS2 keyboard to the Arduino. Figure 9-2 shows two of the female PS2 sockets that are available at a fairly reasonable cost.

We chose to use the connector on the right in Figure 9-2 because we were able to buy 10 of them for less than $0.30 each and they are easily fitted to a perf board. (We called the sockets "PS2 boxes" but technically, it is called a "6-pin Mini-DIN" connector.) The connector on the left is more expensive but would be better suited for mounting the electronics in some kind of enclosure rather than on perf board.

The PS2 box connector we are using has the leads coming from the bottom of the connector in the pattern shown in Figure 9-3. The pin out interpretation is that shown in Figure 9-1. Note that the pins shown in Figure 9-3 are as viewed from the bottom of the connector. If you are unsure of the pin outs, just stick a piece of wire in each socket hole and use a multimeter to check the pins. The PS2 box connectors are a standard connector so there shouldn't be any surprises. We tied the four pins from the PS2 box to a 4-pin header to make it easier to connect it to the Arduino during testing. The small perf board is shown in Figure 9-4.

Reading from left to right in Figure 9-4, the first pin connects to the PS2 ground connection (pin 3). The second header pin connects to the PS2 clock pin (pin 5). The third pin connects to the

```
              4        2    1       3
              +  -------------------  +
              | *      *    *     *  |
              | *                 *  |
              +  -------------------  +
                 6                5
```

FIGURE 9-3 PS2 pin out for PS2 boxes (bottom view).

data pin (pin 1), and the final pin connects to the V_{CC} positive voltage pin (pin 4). Pins 2 and 6 are not connected.

Testing the PS2 Connector

Once you have connected the PS2 connector leads to the header pins, you can connect the keyboard's PS2 male connector to the PS2 box female connector. Connect the ground header pin to the Arduino GND pin and the V_{CC} positive voltage header pin to the Arduino 5 V pin. When you attach the USB connector to the Arduino board, you will likely see one or more LEDs on the keyboard briefly flash once. That's a good sign, because it not only means you've got the power connections wired correctly, but also that your keyboard is alive and well. (For $2, there were some thoughts that our used keyboard might be DOA.)

Since the power test was passed, go ahead and connect the clock pin (pin 5 on the PS2 connector and the "C" pin in Figure 9-4) to pin 3 on the Arduino. Also connect the data pin (pin 1 on the PS2 connector and the "D" pin in Figure 9-4) to pin 4 on the Arduino. You are now ready for a more robust test of the keyboard. That test, however, involves using the software that supports the keyboard CW encoder. We explain how to perform that test later in the chapter.

FIGURE 9-4 Connecting the PS2 pins to the header.

The PS2 Keyboard Encoder Software

The software used to convert the keystrokes sent by the keyboard to their Morse code equivalent is shown in Listing 9-1. The code is a slightly modified Open Source version written by Mark VandeWettering, K6HX. He has provided some nice features with the keyboard, including a "type-ahead" buffer feature that lets you type in characters faster than the selected code speed outputs them to the transmitter. Before we get into the details about the code shown in Listing 9-1, you need an additional Open Source library specifically written for the PS2 keyboard.

Adding the PS2 Library Code to Your IDE

The first thing you need to do before you attempt to compile the code in Listing 9-1 is download the keyboard library files used in the program. You can download these files without charge from:

http://www.pjrc.com/teensy/td_libs_PS2Keyboard.html

When the page loads in your browser, you can see a download link for the PS2Keyboard.zip file. The browser does the usual inquiry about where you want to save the downloaded file. We usually create a directory that reflects the nature of whatever it is we are downloading. In this case, you might create a new directory name PS2KeyboardLibrary. When the download finishes, use whatever program you have that can extract the compressed files from the ZIP file. Under Windows, simply double-click on the file and the files are extracted.

```
/*
  Based on code written by Mark VandeWettering, K6HX, as found on his web site:

  http://brainwagon.org/2012/01/21/an-arduino-powered-ibm-ps2-morse-keyboard/

  You need four connections from the PS2 keyboard:

  5V        Connects to pin 4 of the PS2 connector (and to 5v on the Arduino board)
  ground           "        3         "            (and to GND on the Arduino board)
  clock            "        5         "            (to pin 3 on the Arduino board)
  data             "        1         "            (    "    4        "           )

  Pins 2 and 6 of the PS2 connector are not used.

*/

// Version for the PS2 Keyboard
// using the library from http://www.pjrc.com/teensy/td_libs_PS2Keyboard.html
//

#include <PS2Keyboard.h>
```

LISTING 9-1 The PS2 encoder program.

```
#define DEBUG             1         // For debugging. Comment out when not debugging

#define PS2CLOCKPIN       3
#define PS2DATAPIN        4


#define SIDETONEFREQ      700
#define ESCAPEKEY         27
#define PERCENTKEY        37
#define NEWLINE           '\n'

#define LEDPIN            13
#define TONEPIN           12        // Used if you want audio via a small buzzer

#define QUEUESIZE         128
#define QUEUEMASK  (QUEUESIZE-1)

#define DEFAULTWPM        15        // MIN_WPM <= DEFAULT_WPM <= MAX_WPM
#define MINWPM            5
#define MAXWPM            50
#define NOCHARAVAILABLE  -1
// ========================= Global data definitions =====================
char buffer[QUEUESIZE];

#ifdef DEBUG
int aborted = 0;
#endif

int bufferHead = 0;
int bufferTail = 0;

int wordsPerMinute = DEFAULTWPM;       // Default speed
boolean sideTone = false;              // Default is no sidetone
boolean speedChange = false;           // 'true' indicates speed change requested

int ditlen = 1200 / wordsPerMinute;

PS2Keyboard kbd;

void setup()
{
  pinMode(LEDPIN, OUTPUT);
  pinMode(TONEPIN, OUTPUT);
  kbd.begin(PS2DATAPIN, PS2CLOCKPIN);

#ifdef DEBUG
  Serial.begin(9600);
  Serial.println("PS2 keyboard ready:");
#endif

}
```

LISTING 9-1   The PS2 encoder program. (*continued*)

```
void loop()
{
  ps2poll();              // Loop for a keystroke

  if (bufferHead != bufferTail)    // If there's a keystroke present in the
                                   // buffer...
     send(BufferPopCharacter());   // ...send it along.
}

/*****
 * This method adds a character to the buffer.
 *
 * Parameters:
 * char ch        the character to add
 *
 * Return value:
 * void
 *****/
void BufferAdd(char ch)
{
  buffer[bufferTail++] = ch;
  bufferTail &= QUEUEMASK;
}

/*****
 * This method adds a character to the buffer. See text as to how this method
shares the same method name as
 * the one above without creating a duplicate definition error
 *
 * Parameters:
 * char *s        the character(s) to add
 *
 * Return value:
 * void
 *
 *  CAUTION: If any part of an existing message is in the buffer when this
function is called, it is lost.
 *****/
void BufferAdd(char *s)
{

  int len;
  len = strlen(s);
  if (len > QUEUEMASK) {  // If the message is too long, kill it.
     *s = '\0';
     return;
  }
  BufferReset();          // Override anything in buffer
  while (*s)
```

LISTING 9-1 The PS2 encoder program. *(continued)*

```
      BufferAdd(*s++);
}

/*****
 * This method removes a character from the buffer.
 *
 * Parameters:
 * void
 *
 * Return value:
 * char         the character that is copied from the buffer
 *****/
char BufferPopCharacter()
{
  char ch;
  ch = buffer[bufferHead++];
  bufferHead &= QUEUEMASK;
  return ch;
}


/*****
 * This method reset the buffer.
 *
 * Parameters:
 * void
 *
 * Return value:
 * void
 *****/
void BufferReset()
{
  bufferHead = 0;                    // Reset to first character in buffer
  bufferTail = 0;
  memset(buffer, 0, QUEUESIZE);  // Clear out previous contents
}

/*****
 * This method polls the keyboard looking for a keystroke. If a character is
available, it is added to the
 * keyboard input buffer. The inline modifier is a hint to the compiler to
generate inline code if possible,
 * to improve the performance of the method.
 *
 * Parameters:
 * void
 *
 * Return value:
 * void
 *****/
```

LISTING 9-1   The PS2 encoder program. (*continued*)

```
inline void ps2poll()
{
  char ch;
  while (kbd.available()) {
    if (((bufferTail+1)%QUEUESIZE) == bufferHead) {        // is buffer full ?
#ifdef DEBUG
      Serial.println("== Buffer Full ==");
#endif
      break;
    } else {
      switch (ch=kbd.read()) {
        case ESCAPEKEY:                                    // case '\033':
          BufferReset();
  #ifdef DEBUG
          Serial.flush();
          Serial.println("== Buffer reset ==");
            aborted = 1;
  #endif
          aborted = 1;
          break;
        case PERCENTKEY:
          BufferAdd("CQ CQ CQ DE W8TEE W8TEE W8TEE K");
          break;
        case '(':              // Start buffering without transmitting
          DelayedTransmit();
          break;
        case '#':              // Change keying speed
          ChangeSendingSpeed();
          speedChange = true;
          break;
        case '~':              // Change sidetone. Default is no tone (0)
          sideTone = !sideTone;
          break;
        /*
        case YOURCHARACTER:
          BufferAdd("Your special message here");
          break;
        */
        default:
          BufferAdd(ch);
          break;
      }
    }
  }
}

/*****
 * This method generates a delay based on the millis() method call. This is
 * the preferred way to perform a delay because it doesn't put the board to sleep
 * during the delay.
```

LISTING 9-1 The PS2 encoder program. (*continued*)

```
 *
 * Parameters:
 * unsigned long ms    the number of milliseconds to delay
 *
 * Return value:
 * void
 *****/
void mydelay(unsigned long ms)
{
  unsigned long t = millis();
  while (millis()-t < ms) {
    ps2poll();
  }
}

/*****
 * This method generates a tone if speakers are hooked to the system. The
 TONEPIN value determines which pin is used to generate the tone. Not
 implemented.
 *
 * Parameters:
 * void
 *
 * Return value:
 * void
 *****/
void scale()
{
  long f = 220L;
  int i;

  for (i=0; i<=12; i++) {
    tone(TONEPIN, (int)f);
    f *= 1059L;
    f /= 1000L;
#ifdef DEBUG
    Serial.println(f);
#endif
    delay(300);
  }
  noTone(TONEPIN);
}

/*****
 * This method generates a single dit in the sequence necessary to form a
 character in Morse Code.
 *
 * Parameters:
 * void
 *
```

LISTING 9-1 The PS2 encoder program. (*continued*)

```
 * Return value:
 * void
 *****/
void dit()
{
  digitalWrite(LEDPIN, HIGH);
  if (sideTone)
    tone(TONEPIN, SIDETONEFREQ);
  mydelay(ditlen);
  digitalWrite(LEDPIN, LOW);
  if (sideTone)
    noTone(TONEPIN);
  mydelay(ditlen);

#ifdef DEBUG
  Serial.print(".");
#endif
}

/*****
 * This method generates a single dah in the sequence necessary to form a
 character in Morse Code.
 *
 * Parameters:
 * void
 *
 * Return value:
 * void
 *****/
void dah()
{
  digitalWrite(LEDPIN, HIGH);
  if (sideTone)
      tone(TONEPIN, SIDETONEFREQ);
  mydelay(3*ditlen);
  digitalWrite(LEDPIN, LOW);
  if (sideTone)
    noTone(TONEPIN);
  mydelay(ditlen);

#ifdef DEBUG
  Serial.print("_");
#endif
}


// The coded byte values for the letters of the alphabet. See text for explanation
char ltab[] = {
```

LISTING 9-1   The PS2 encoder program. (*continued*)

```
    0b101,              // A
    0b11000,            // B
    0b11010,            // C
    0b1100,             // D
    0b10,               // E
    0b10010,            // F
    0b1110,             // G
    0b10000,            // H
    0b100,              // I
    0b10111,            // J
    0b1101,             // K
    0b10100,            // L
    0b111,              // M
    0b110,              // N
    0b1111,             // O
    0b10110,            // P
    0b11101,            // Q
    0b1010,             // R
    0b1000,             // S
    0b11,               // T
    0b1001,             // U
    0b10001,            // V
    0b1011,             // W
    0b11001,            // X
    0b11011,            // Y
    0b11100             // Z
};

// The coded byte values for numbers. See text for explanation

char ntab[] = {
    0b111111,           // 0
    0b101111,           // 1
    0b100111,           // 2
    0b100011,           // 3
    0b100001,           // 4
    0b100000,           // 5
    0b110000,           // 6
    0b111000,           // 7
    0b111100,           // 8
    0b111110            // 9
};

/*****
 * This method generates the necessary dits and dahs for a particular code.
 *
 * Parameters:
 * char code    the byte code for the letter or number to be sent as take from
the ltab[] or ntab[] arrays.
 *
```

```
 * Return value:
 * void
 *****/
void sendcode(char code)
{
  int i;

  for (i=7; i>= 0; i--) {   // Look for start bit
    if (code & (1 << i))
      break;
  }
  for (i--; i>= 0; i--) {   // Remaining bits are the actual Morse code
    if (code & (1 << i))
      dah();
    else
      dit();
  }
  mydelay(2*ditlen);     // space between letters
#ifdef DEBUG
  Serial.println("");
#endif
}

/*****
 * This method translates and sends the character.
 *
 * Parameters:
 * char ch       the character to be translated and sent
 *
 * Return value:
 * void
 *****/
void send(char ch)
{
  int index;

  if (speedChange) {                  // use new speed
    ditlen = 1200 / wordsPerMinute;
    speedChange = false;
  }

  if (isalpha(ch)) {
    index = toupper(ch) - 'A';       // Calculate an index into the letter array
                                     // if a letter...
    sendcode(ltab[index]);
  } else if (isdigit(ch)) {
    sendcode(ntab[ch-'0']);          // Calculate an index into the numbers
                                     // table if a number...
  } else if (ch == ' ' || ch == '\r' || ch == '\n') {
```

LISTING 9-1 The PS2 encoder program. (*continued*)

```
      mydelay(4*ditlen);
#ifdef DEBUG
      Serial.print("  ");
#endif
   } else {
      switch (ch) {                    // Punctuation and special characters. NOTE;
                                       // Tree depth is 6, so
         case '.':                     // characters max out at 7 dit/dah combinations
            sendcode(0b1010101);
            break;
         case ',':
            sendcode(0b1110011);
            break;
         case '!':
            sendcode(0b1101011);
            break;
         case '?':
            sendcode(0b1001100);
            break;
         case '/':
            sendcode(0b110010);
            break;
         case '+':
            sendcode(0b101010);
            break;
         case '-':
            sendcode(0b1100001);
            break;
         case '=':
            sendcode(0b110001);
            break;
         case '@':
            sendcode(0b1011010);
            break;
         case '\'':                    // ' apostrophe
            sendcode(0b1011110);
            break;
         case '(':
            sendcode(0b110110);
            break;
         case ')':
            sendcode(0b1101101);
            break;
         case ':':
            sendcode(0b1111000);
            break;
         case ';':
            sendcode(0b1101010);
            break;
```

LISTING 9-1 The PS2 encoder program. *(continued)*

```
        case '"':
          sendcode(0b1010010);
          break;

        default:
          break;
      }
  }

#ifdef DEBUG
  if (!aborted) {
    Serial.println(ch);
    if (ch == 13)
      Serial.print((char) 10);
  }
  aborted = 0;
#endif

}


/*****
 * This method flashes the LED on pin 13 if the input buffer is close to being
full.
 *
 * Parameters:
 *    void
 *
 * Return value:
 *    void
 *****/

void FlashBufferFullWarning()
{
  int i;
#ifdef DEBUG
  Serial.print("************* Approaching buffer full ==============");
#endif

  for (i = 0; i < 10; i++) {
    digitalWrite(LEDPIN, HIGH);    // Visual
    delay(100);
    digitalWrite(LEDPIN, LOW);
    delay(100);
    tone(TONEPIN, SIDETONEFREQ);              // Audio...if available
    delay(100);
    noTone(TONEPIN);
```

LISTING 9-1   The PS2 encoder program. (*continued*)

```
    delay(100);
  }
}

/*****
 * This method buffer all keystrokes after reading the leading '(' until it
reads a ')'. At that
 * time, the buffer is sent to the transmitter.
 *
 * Parameters:
 *    void
 *
 * Return value:
 *    int         the number of characters buffered.
 *****/
int DelayedTransmit()
{
  char ch;
  int charCount = 0;
  int i;

  memset(buffer, '\0', sizeof(char));

  bufferTail = 0;
  BufferReset();              // Clear the buffer and start over...

  while (true) {
    if (kbd.available()) {
      ch = kbd.read();
      if (ch == ESCAPEKEY) {  // They want to terminate message
        BufferReset();        // Clear all and start over
        return 0;
      }
      charCount++;
                              // Is long message finished or terminated?
      if (ch == ')' || charCount == QUEUEMASK || ch == NEWLINE) {
        for (i = 0; i < charCount; i++) {
          send(buffer[i]);
        }
        BufferReset();        // Clear the buffer and start over...
        break;
      } else if ((ch == '(') || (ch == ESCAPEKEY) || (ch == NOCHARAVAILABLE)) {
        ; // ignore character
      } else {
        buffer[charCount++] = ch;
        if (charCount > (QUEUEMASK - 20)) {   // Approaching full buffer
#ifdef DEBUG
```

LISTING 9-1   The PS2 encoder program. (*continued*)

```
      Serial.print("  charCount = ");
      Serial.println(charCount);
#endif
            FlashBufferFullWarning();
          }
        }
      }
    }
    return charCount;
}


void ChangeSendingSpeed()
{
  char ch;

  int wereDone = 0;

#ifdef DEBUG
  Serial.print(" wordsPerMinute at start =");
  Serial.println(wordsPerMinute);
#endif
  while (true) {
    if (kbd.available()) {
      ch=kbd.read();
#ifdef DEBUG
      Serial.print("Speed change ch =");
      Serial.println(ch);
#endif
      switch (ch) {
        case '>':
          if (wordsPerMinute < MAXWPM)
            wordsPerMinute++;
          break;
        case '<':
          if (wordsPerMinute > MINWPM)
            wordsPerMinute--;
          break;
        case '#':
          wereDone = 1;
          break;
        default:
          break;
      }
    }
    if (wereDone)
      break;
  }
  ditlen = 1200 / wordsPerMinute;
}
```

LISTING 9-1   The PS2 encoder program. (*continued*)

If you are using Windows, load Windows Explorer if you haven't already done so. (You can run Windows Explorer by right-clicking on the Windows Start button in the lower-left corner of your display and select Open Windows Explorer.) Now navigate to the directory where you extracted the PS2Keyoard library files. You should see a subdirectory named PS2Keyboard. Inside the PS2Keyboard directory is yet another directory with the same name (PS2Keyboard). For example, if you copied the ZIP file to a directory named C:\Temp, when you are finished extracting the library files from the ZIP file, you should be able to see:

```
C:\Temp\PS2Keyboard\PS2Keyboard
```

Highlight the *second* PS2Keyboard directory and copy it (e.g., select Organize → Copy from the Windows Explorer menu). Now go to the directory where you have installed your Arduino IDE. On our systems, we have the IDE stored on drive C in a directory named Arduino105. If we double-click on the IDE directory, you will see a subdirectory named *libraries*. For us, the *libraries* directory is found at:

```
C:\Arduino105\libraries
```

Once you are in the *libraries* subdirectory, Paste the new PS2Keyboard library files into the *libraries* subdirectory (e.g., Organize → Paste). When you are finished, you should see the new PS2Keyboard subdirectory along with the many other standard library files that are distributed with the IDE (e.g., LiquidCrystal, Stepper, etc.). If you have the IDE open at the moment, you must close it before you can use the new library. Once you restart the Arduino IDE, the new PS2Keyboard library is available for use in your programs.

Code Walk-Through on Listing 9-1

While we don't pretend that this is a programming book, there are a few things in the listing that help you better understand what the code is doing. The program begins with a *#include*, which makes the PS2Keyboard library methods available for use in the program. Following that is a series of *#define*s for numeric constants used in the program. Following that is the definition of a number of global variables, including the *buffer[]* array for storing the characters typed by the user. The size of the type-ahead buffer is 128 characters. Unless you are a *really* fast typist and transmitting at a *really* slow speed, this should be more than large enough.

We left a number of *DEBUG* directives in the code. These lines enable you to view the code being produced on the Serial monitor. Just comment out the *#define DEBUG 1* preprocessor directive when you are finished debugging/viewing the code.

Next is the statement:

```
int ditlen = 1200 / WORDSPERMINUTE;
```

This is one of the few times we leave a magic number in the source code; 1200 in this case. The details for this magic number can be found at http://en.wikipedia.org/wiki/Morse_code. Basically, it assumes a 50 dit duration period for a standard word and can be stated in units of time. (The word PARIS is often used as the standard word.) Once the time for a dit is calculated, all other spacings are based on it. Note that we have arbitrarily set the words per minute (*wordsPerMinute*) to 15 in the listing. You can change this to whatever reasonable speed you wish as the default speed at program startup.

Next are the standard *setup()* and *loop()* methods. We use pin 13 to blink the onboard LED in sync with the code being generated. If you tie a speaker to pin 10, make sure it is small and that you use a current limiting resistor in the line. Within *loop()*, the code constantly calls *ps2poll()* to look for a character from the keyboard, placing any character read into the *buffer[]* array. If there is a character in the buffer, the *send()* method is called to ultimately send the character to the transmitter.

Overloaded Methods

If you look at lines 92 and 110 in the listing (place the cursor on a source code line in the IDE and the line number is displayed in the lower left corner of the IDE), you can see:

```
void BufferAdd(char ch)
void BufferAdd(char *s)
```

Whoa! How can this be? Both methods have the same name, so why doesn't the compiler issue a duplicate definition error? The reason is because C++ allows methods to share the same name as long as their parameter lists are different. In the first method, the parameter is a character, like 'A' or 'B'. In the second method, the parameter is a pointer to a character. Because a valid pointer variable can only have one of two values, *null* or a valid memory address (and *not* a character), the compiler is smart enough to know which one of the two methods to use when the program needs to call one of these methods. (The method name and its parameter list are often referred to as the *signature* of the method. The process of using two methods with the same name, but different parameter lists, is called method overloading. You'd be surprised how often that little factoid comes up in cocktail conversation. Like Jack's four-year-old grandson says . . . more "qwap" for your brain! Jack has no clue where he learned such terms.)

The compiler figures out which method to call by looking at the type of parameter(s) being passed to it and comparing signatures. Of course, the programmer can screw things up by using the wrong parameter type; the compiler is only smart enough to do what the programmer tells it to do. You can see the difference between the calls by looking near lines 189 and 192 in Listing 9-1. (Recall that the small number in the lower-left corner of the IDE is the current line number of the cursor in the source code window.) The first use is a long string of characters that is resolved to a pointer (an lvalue). The second method call simply passes in a single character.

The *sendcode()* Method

Perhaps the hardest method to understand is the *sendcode()* method. We'll use a single letter to help you understand what the method (and the program) is doing. First, suppose you type in the letter 'a' at the PS2 keyboard. The *ps2poll()* method reads that letter and stuffs it into the *buffer[]* array by calling *BufferAdd()*. Because the buffer is not empty now, *BufferPopCharacter()* is called. *BufferPopCharacter()* simply gets whatever character is being pointed to by *bufferHead* (and adjusts its value to prepare for the next character) and returns the character back to *loop()*.

However, the character returned from the call to *BufferPopCharacter()* becomes the argument passed to the *send()* method. In essence, line 80 now looks like:

```
send('a');
```

The first statement in *send()* is:

```
if (isalpha(ch)) {
```

which becomes

```
if (isalpha('a')) {
```

Because the character 'a' is an alpha character, the *if* expression is logic True. Therefore, the next statement executed is:

```
index = toupper(ch) - 'A';
```

which may be viewed as:

```
index = toupper('a') - 'A';
```

Because the *toupper()* method (actually, it's a macro, but we need not get into that) converts the lowercase 'a' to an uppercase 'A', the last statement becomes:

```
index = 'A' - 'A';
```

At first blush, it seems weird to subtract letters from one another. However, recall that the keyboard is actually sending ASCII (American Standard Characters for Information Interchange) values for every character that is pressed on the keyboard. The ASCII value for 'A' is 65; therefore, the last statement above becomes:

```
index = 65 - 65;
```

which resolves to 0, and that value is assigned into *index*. (You should ask yourself what would *index* be if the character was a 'b' or an 'n'. If you understand those values, you're right on track! An easy-to-read ASCII table can be found at http://www.asciitable.com/.) Because the ASCII values are in alphabetical order, it's pretty easy to figure out what happens to *index* using different letters.

Now look at the next statement:

```
sendcode(ltab[index]);
```

Because we now know that *index* for 'a' becomes 0, the call is actually:

```
sendcode(ltab[0]);
```

If you look at the first element in the *ltab[]* array (i.e., the letter table array), its value is 0b101. The leading "0b" tells the compiler that what follows is a *byte* value expressed in binary terms (the "0b" component in the statement stands for "binary"). You can see by the comment in the listing that this is the letter 'A'. If you know Morse code, you know that the letter 'A' is di-dah. If you look at the second entry in the table, you see the letter 'B' is coded as 0b11000. The letter 'B' in Morse code is dah-di-di-dit. Hmmm. If we strip away the first binary bit (which is always a 1 in the *ltab[]* array), and treat a 1 as a dah and a 0 as a dit, it seems that the letter array is actually the Morse code equivalents for all of the letters of the alphabet. Shazam . . . an Ah-Ha moment!

Some Bit-Fiddling

You can see the following code snippet in the *sendcode()* method:

```
for (i=7; i>= 0; i--)
    if (code & (1 << i))
        break;
```

If you're not familiar with bit shifting, the second statement above may look a little strange because of the bit shift left operator (<<). Recall that *code* at this point in the program equals 0b101. However, expressed as a complete binary byte value, *code* is actually 00000101. Note that the *for* loop starts with 7 and decreases the value of *i* on each pass through the loop. In other words, the loop starts looking at the bits in *code* starting with the last (7th or high) bit. Because the first five bits are all 0s, the *if* test spins past these five bits because the bitwise AND operator, &, and the bit shift left expression ($1 << i$) evaluates to logic False for each 0 bit read. This means the *break* statement to terminate the *for* loop is not executed until the first 1 is read from the *code* variable. Table 9-1 shows what is happening to the *if* expression on each pass through the *for* loop.

i	(1 << i)	Code & (1 << i)	Logical Result
7	10000000	00000101 10000000 ------------ 00000000	False
6	01000000	00000101 01000000 ------------ 00000000	False
5	00100000	00000101 00100000 ------------ 00000000	False
4	00010000	00000101 00010000 ------------ 00000000	False
3	00001000	00000101 00001000 ------------ 00000000	False
2	00000100	00000101 00000100 ------------ 00000100	True

TABLE 9-1 Evaluation of the Statement: *if (code & (1 << i))*

Notice how the bit mask in column two in Table 9-1 changes with each pass through the *for* loop because of the bit shift left operator. Also note that, when $i = 2$ the *if* expression evaluates to logic True, which causes the *break* statement to be executed. This is because the bit mask is ANDed with *code* and produces a nonzero result.

The *break* statement causes execution of the first *for* loop to end and execution proceeds to the first statement that is not part of the controlling *for* loop. The next statement that is executed, however, is actually a second *for* loop as seen in the next code snippet:

```
for (i--; i>= 0; i--) {
    if (code & (1 << i))
        dah();
    else
        dit();
}
```

The same *if* test is applied to the remaining part of the variable named *code*. Note that variable *i* is not initialized in the second *for* loop; variable *i* remains equal to the value of *i* from the previous *for* loop. The *for* loop decrements *i* before it is used, so *i* is now equal to 1 when the *if* test in the second *for* loop is executed. Table 9-2 shows what is happening on each pass through the second *for* loop.

Notice that when the *if* expression evaluates to logic False a dit is sent via the call to method *dit()*. Likewise, when the *if* expression evaluates to logic True, a dah is sent via a call to method *dah()*.

You should now be able to figure out why the *ltab[]* array has elements whose binary values all begin with a 1. The algorithm uses the leading 1 to indicate that the bits that follow it decode into the appropriate dits and dahs for the characters in the array. While you could also implement this algorithm as a more simple rat's nest of cascading *if* statements, this algorithm is much more efficient.

It's worth the effort to spend enough time with this section to understand what the code is really doing. Bit shifting instructions execute extremely fast in the processor and are often a good solution to a variety of programming problems.

i	(1 << i)	Code & (1 << i)	Outcome of Logical Result
1	00000010	00000101 00000010 ------------ 00000000	dit()
0	00000001	00000101 00000001 ------------ 00000001	dah()

TABLE 9-2 Second for Loop Evaluation of the Statement: *if (code & (1 << i))*

Isolating the Arduino from the Transmitter

With the software under your belt, we can finish up the circuit that ends up actually keying the transmitter. Because it's just a good idea to isolate the Arduino circuitry from the transmitter, we decided to use an optoisolator IC to segregate the two circuits.

We chose to use the 4N26 chip simply because we had some lying around. You can purchase these for under $0.50 each either using some of the suppliers mentioned in Appendix A or online. The selection of this chip is not etched in stone and you can use others provided they can handle the I/O voltages and current.

The 4N26 is a 6-pin IC that is configured as shown in Figure 9-5. Note the current limiting resistor (470 Ω) on pin 1 of the 4N26, which is then tied to pin 13 of the Arduino. Pin 2 of the 4N26 is tied to ground and pins 3 and 6 are not connected. Pin 5 goes to the positive lead in your transmitter's keying circuit and pin 4 is tied to its ground. Usually, this just means that pins 4 and 5 are terminated with a jack that plugs into your CW rig. This simple circuit can key most modern rigs that have a positive line going to ground when keyed.

Because we're not big fans of soldering ICs directly into a circuit, we modified an 8-pin IC socket for use with the 4N26. In our case, we simply took a pair of needle-nosed pliers, grasped what is pin 4 on the 8-pin IC socket, and pushed the pin up and out through the top of the socket. We then repeated the process for pin 5 on the socket. The results can be seen in Figure 9-6.

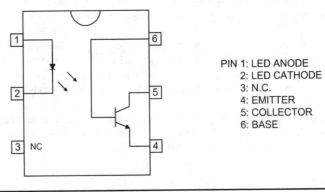

FIGURE 9-5 Optoisolator circuit for 4N26.

FIGURE 9-6 An 8-pin IC socket with pins 4 and 5 removed.

Chapter 9: A PS2 Keyboard CW Encoder

FIGURE 9-7   The finished circuit.

After butchering the socket as we did, we moved it to our small perf board and wired it according to Figure 9-5. The final results can be seen in Figure 9-7. The connection to pin 13 of the Arduino is partially blocked in the photo, but you can see the current limiting resistor at the edge of the perf board. The two leads labeled Key Out are to the transmitter's keyed circuit. The PS2 leads were discussed earlier in this chapter. The two leads heading north go to a super-small 8 Ω speaker that has a 1K Ω resistor in the lead that ties into pin 12 (see Listing 9-1 and look for TONEPIN near the top of the listing). The other speaker lead is attached to a GND pin on the Arduino board. (You could also use a small buzzer instead of the speaker.)

The schematic for the complete circuit is shown in Figure 9-8. (The circuit does not show the power source for the Arduino, although the USB supplies the power while testing the circuit and

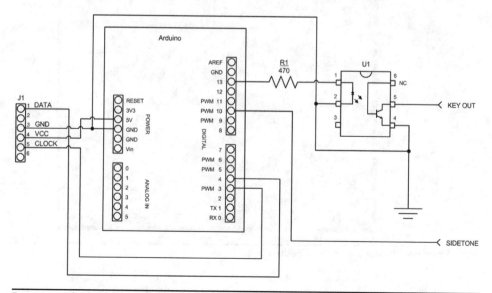

FIGURE 9-8   The PS2 keyboard encoder circuit.

the software. Most people will power the final circuit with an inexpensive 6 V wall wart power adapter using the barrel power connector on the Arduino.)

Testing

Testing both the software and hardware should be done before you commit the circuit to its final state (e.g., moving it to a "real" shield). If you look at Listing 9-1 closely, the first statement after the *#include* is:

```
#define DEBUG    1     // For debugging statements. Comment out when not debugging
```

(which you have seen before). The DEBUG symbolic constant is used to toggle debugging code into and out of a program. In this program, the DEBUG symbolic constant changes the output so it can be viewed on the Serial output tied to your PC. When you first see the string "PS2 keyboard ready:" appear on the monitor, you can start typing on the PS2 keyboard. (Did any of you type on your PC keyboard instead of the PS2 keyboard and wonder why nothing happened? We did . . . briefly.) If all goes well, you should see the letters you typed appear on the Serial monitor. For example, we typed in "This is a test" and the Serial monitor showed the results as shown in Figure 9-9.

Of course, you should to comment out the *#define DEBUG 1* statement line when you actually want to use the program so time isn't spent sending data to the Serial output.

You can compare the cobbled test version of the PS2 keyer in Figure 9-7 with the shield version shown in Figure 9-10. In Figure 9-10 you can see a small buzzer onboard to act as a side tone using pin 10 as shown in the code listing.

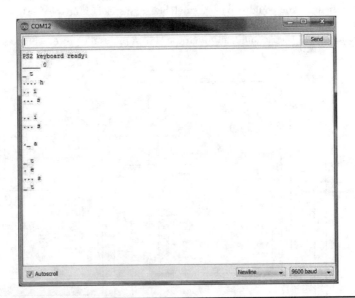

FIGURE 9-9   Program output in DEBUG mode.

Figure 9-10   PS2 keyer shield.

Other Features

The following sections detail several features that may not be obvious when the program is executing. (Of course, you already know about these features because you poured over Listing 9-1 with a fine-toothed comb, right?)

Change Code Speed

The default code speed is 15 words per minute (wpm). Obviously, this doesn't mean a whole lot because words have differing lengths. However, way back when, the FCC's rule was that a word consisted of five letters, so 15 wpm is about 75 characters a minute, or a little more than a character a second. For experienced CW enthusiasts, 15 wpm is a glacial pace. For us mere mortals, it's a pretty comfortable speed. However, no matter how fast you can send and receive, you may wish to make a contact with someone who is sending faster or slower than you are currently set up to transmit.

Rather than change the speed setting in the program and make you recompile and upload it, the code supports on-the-fly speed modifications. Around line 200 in Listing 9-1 you can see a *case* statement based on the '#' sign. When the code reads a '#' sign from the PS2 keyboard, method *ChangeSendingSpeed()* is called. If the next character read from the keyboard is a greater than sign ('>'), the wpm is bumped up by one. So if you want to increase the speed from 15 to 20 wpm, you would type in the following characters:

```
#           // Says we're going to change the keying speed
>  // Speed is now 16
>  // Speed is now 17
>  // Speed is now 18
>  // Speed is now 19
>  // Speed is now 20
#  // Done adjusting the speed
```

As you can see, the '#' character triggers the call, but the same character is also used to terminate the speed adjustment. The program then resumes its normal operation, but at the newly assigned speed. If, after pressing the first # key you had pressed the less than key ('<') five times, you would have decreased the keying speed to 10 wpm. These key combinations allow you to change the speed without recompiling the code.

Sidetone

The default mode is for the program not to generate a sidetone. Usually, most rigs have their own sidetone and have the program code default to generating a second sidetone would just confuse things. However, during testing or perhaps trying to improve your code speed, you may want to have a sidetone.

At about the same place in the code listing, you'll see that the tilde character ('~', located to the left of the '1' key on the keyboard) is used in another *case* statement. When the tilde is read, it causes the variable named *sideTone* to be toggled. Since *sideTone* is a *boolean* data type, it starts with the value False so there is no sidetone. If you press the tilde key, *sideTone* is toggled, changing its value to True. If you look at the *dit()* and *dah()* methods, you'll see how the *sideTone* works. Pressing the tilde key a second time again toggles the *sideTone* variable, thus setting it back to False and turning the sidetone off.

The *sideTone* feature assumes that you have a small speaker or buzzer attached to pin 12 on the Arduino and the other side tied to ground. We used a 1000 Ω current limiting resistor on the lead going to pin 12, which produces a very weak background sidetone. Almost any value between 400 and 1000 Ω will work. Forgetting a current limiting resistor could fry your Arduino, so make sure you add it to your circuit.

Long Messages

Sometimes when you're rag chewing it's nice to be able to make notes so that when your turn comes, you remember what it was you wanted to say. (We're old, remember?) Again, looking near line 200 in Listing 9-1 you'll see a *case* statement that executes when it sees an opening parenthesis ('(') character. After reading a '(', the *DelayedTransmit()* method is called. All this does is buffer whatever keystrokes follow the '(' until a closing parenthesis (')') character is read. At that time, the contents of the buffer are sent in normal fashion.

In the code, if you start to fill up the buffer to the point where it might overflow, the Arduino LED starts to flash at a rate of 10 times per second. As the code currently stands, the flashing starts when the buffer has QUEUEMASK – 20, which resolves to 107, characters of empty buffer space remaining. Of course you can change this, but it's probably a good idea to have some indicator for pending buffer overflow.

So, how would you use this feature? Probably, you would type in the opening parenthesis followed by the normal QSO resume pattern (e.g., W8XYZ DE W8TEE R) and then start typing the message that you wanted to send. If you're using a break-in type of QSO where you go back and forth without sending call signs, the long message feature doesn't makes a lot of sense. If it's a feature you don't think you'll use, don't send an opening parenthesis!

Conclusion

The program could benefit from some additions. For example, another addition would be to add the LCD display from Chapter 3 to the circuitry so you could see what you are typing as you type it. There are a lot of keys on the keyboard that remain uncoded. You can use these for additional commands you may think useful. If you do make improvements to either the hardware or software, don't forget to share it with the rest of us by posting it on the web site, http://arduinoforhamradio.com.

CHAPTER 10

Project Integration

In this chapter, our goal is to take three previous projects and combine them into one package, both in terms of hardware and software. The three projects are: 1) the PS2 keyboard keyer, 2) the CW decoder, and 3) the station clock/ID timer. You could easily integrate the capacitive keyer you built in Chapter 8, but we wanted to find a middle ground between the complexities of integration and narrative clarity. Three projects are enough complexity to illustrate the issues of integration, but simple enough to make the discussion clear. Also, many rigs already have a keyer integrated into them so there is less need to add that feature in this project.

One of the primary purposes of this chapter is to discuss some of the issues you need to think about when doing a more complex project. On the surface, the idea of project integration seems simple enough: take the code from the three projects, cut-and-paste the code together, and stack the three shields onto a board in a huge "Arduino Sandwich" and you're done! Right?

Well, not really. There are a number of complications that make it a little more difficult than it might seem at first blush. First, the three independent projects discussed in earlier chapters now have to "play nice" with each other and share resources. Second, there's almost too much hardware and software for a simple ATmega328 µC. While there's more than enough Flash memory to shoehorn all of the source code into a 328 board, it would be running on the ragged edge of the board's other resources (especially SRAM) and that's almost never a good idea. Third, this is a good opportunity to introduce you to a board based on the ATmega2560 (simply "2560" from now on) for the first time. The 2560 has 256 kb of Flash memory (minus 8 kb for the bootloader), 8 kb of SRAM, and 4 kb of EEPROM. Cap it off with 54 digital I/O pins plus some additional nice features (e.g., more interrupt pins) and you've got a substantially deeper resources pool at your disposal. Despite the resource depth of the 2560, the board can be purchased online for less than $15.

For this project, we have also elected to use an Arduino Expansion shield from DFRobot, as shown in Figure 10-1. The shield allows you to add up to four independent shields on a single board without the need to stack the shields. Not only does the expansion board do away with most heat issues that might arise from using the Arduino Sandwich approach of handling multiple shields, it also gives you a few more nano-acres of board real estate to work with. If you look closely at Figure 10-1, you can see how four shields can be fit onto the board. Each station on the board has its own 5 V and GND point, which makes it easier to power each shield on the board. The 2560 plugs into the board from below (you can see the outline of the 2560's pins), and the tie-ins for the 2560 are shared across the board. That is, for example, not all of the analog pins are dedicated to a single shield but are distributed across all four stations. The same is true for the interrupt pins.

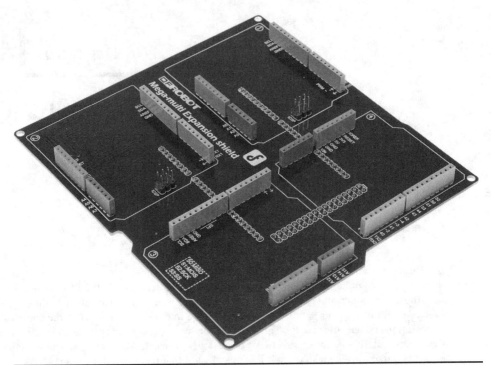

FIGURE 10-1   The DFRobot Mega-multi Expansion shield. *(Shield courtesy of DFRobot)*

It's a perfect solution for our integration project. You can still use the stacked shield approach if you wish; it's just that the DFRobot shield makes it easier to build.

Because we've already discussed these projects in previous chapters and because hardware and software elements are much the same as before, we concentrate our focus in this chapter on the issues that arise when multiple projects with different feature sets must share a limited resource pool. Both the hardware and software are affected when combining the three projects that must now share a single processor.

As usual, we discuss the hardware issues first and then the software. Unlike our earlier projects, however, this project requires multiple source code files ... something we haven't done before. The Arduino IDE has some fairly strict rules we must follow to allow a single project to have multiple source code files and we discuss those rules as well. Because the Arduino IDE is happiest with *.cpp (i.e., C++) files when there are multiple source files in a single project, we also show you the conventional form used for such source code and header files. At the close of the chapter, we bring it all together in an integrated project.

Integration Issues

First, from an operational approach, we want to view the hardware aspects of each project as being etched in stone. That is, we assume that you have already built the three projects as a set of stand-alone shields. From our perspective, this means that we want to use the shields as they already exist, making as few changes to the hardware as possible. We know that we must make some changes, but we want to minimize those changes as much as we can.

Second, because we have software with multiple tasks to solve, and yet must function within a single memory space, those functions need to be handled a little differently. That is, when each project stood as a solution to a single task, we didn't have to worry about interaction effects with other elements of the software. We had little concern about sharing the limited board resources. Not so when you wish to combine multiple projects onto a single processor. Also, initially we wrote those software solutions with little regard to *scalability* (the ability to alter the scope of a project); we didn't write the code with the intent of adding disparate functionality at a later date. As a result, we do need to make some software changes in the way a given shield interacts with the other shields.

Because different projects often use the same pins, pin conflicts are often an issue when you try to integrate multiple projects into one larger project. Appendix C presents a list of the Arduino pins that are used for each of the various projects. A quick look at the table should help you understand where a pin conflict may arise. Creating your own table is not a bad idea as you start to develop projects of your own.

Finally, it is the software that's going to have a large impact on how well each element of the hardware plays with the other elements. The word "integration" in the chapter has substance. If we didn't care about the interplay between projects in a software realm, we could have titled the chapter "Project Stacking." If we just wanted to put three projects into one box, that's a much simpler task. We could, for example, have two dedicated LCD displays; one for the RTC and one for the decoder. However, if we want to share resources between tasks, the software needs some refinement. Sharing resource takes a little more "software thought" and that's what we need to address first in this chapter.

The Real Time Clock (RTC) Shield

In order to communicate with the DS1307 RTC module discussed in Chapter 4, we used the I2C interface. Using the I2C interface also means we needed to use the Wire library for the Arduino. Table 10-1 shows the connections the Wire library expects when using different Arduino boards.

Because the 2560 uses a different set of I/O pins for the I2C interface than the Uno or Duemilanove, we need to modify the pin assignments used by the I2C clock (SCL) and data (SDA) pins.

Another complication is that we used a dedicated LCD display for both the RTC and the CW decoder. While we could keep both displays, that seems redundant. To that end, and not wanting to move the display lines for the LCD display, we now have a decision to make. The SDA and SCL lines are available at every station on the expansion board. By doing this, DFRobot makes the I2C interface available to all shields that might be placed on the expansion board. This makes sense because the I2C interface is a shared bus and can use the 2560 as the Master and each of the support shields could have its own I2C Slave device on it. In that sense, it doesn't matter where we place the RTC shield. As far as the I2C wiring goes, all we need to do is tie the SDA and SCL lines on the expansion board (i.e., pins 20 and 21) to the corresponding lines on our RTC shield (pins A4 and A5). Rather than constructing a new RTC shield, we simply attached Dupont jumpers from pins 20 and 21 on the expansion board to pins A4 and A5 on the existing RTC shield.

While the I2C interface demands the use of certain pins, it matters little which I/O pins are used for the LCD display. In earlier projects, we chose to use the pin arrangement shown in row one (i.e., the 328 row) of Table 10-2. Because the examples provided with the Arduino IDE

Board	SDA Pin	SCL Pin
Uno, Duemilanove	A4	A5
ATmega2560	20	21
Leonardo	2	3
Due	20, SDA1	21, SCL1

TABLE 10-1 The I2C Interface and Arduino Boards

Board	RS	EN	D0	D1	D2	D3
328	12	11	7	6	5	4
2560	29	13	5	4	3	2

TABLE 10-2 The LCD Display Pin Map

LiquidCrystal library tend to use pins 12 and 11 for the Register Select (RS) and Enable (EN) pins, we used those same pins when we constructed the LCD display in Chapter 3. We did retain the data pins 5-2 for the previous projects. We did this because most Arduino examples use these pins.

If you view the expansion board with its labeling correctly oriented for reading, we ended up with the RTC shield in the North-West corner (position 1) of the expansion board. As long as you are aware of the pin placements, you could place the RTC shield in any one of the four positions (see Table 10-3). Keep in mind, however, that if you place the expansion board in a case and keep the LCD display "on board" the RTC shield, its position on the expansion board dictates where the display appears on the case. If that positioning poses a problem for your design, you can always "cable" the connections to the LCD display to the LCD shield and move the display to whatever position makes sense for your design. (We used this cabling approach for the LCD display for the Dummy Load project in Chapter 6.)

The pushbutton switch used to reset the station ID timer can use any I/O pin you wish as long as you remember to change the program source code accordingly. We elected to use pin 28 simply because it's available at the N-W position on the expansion board.

CW Decoder Shield

The decoder shield uses analog pin A5 as its input from the speakers. The A5 analog pin on the expansion board is found at the South-West position (2) on the board. For that reason, we placed the decoder shield at the S-W position of the expansion board. However, because pin A5 on the decoder shield does not line up with A5 on the expansion board, we soldered a small jumper between the two pins on the decoder shield.

Because we intend to share the LCD display on the RTC board with the decoder shield, we simply unplugged the decoder LCD display from its socket header. Whatever output might be generated by the decoder shield is routed to the RTC shield using the same LCD object (i.e., *lcd*) we create in software. From the perspective of the decoder shield, it appears that the LCD shield is the same as before. The multiple use of the LCD display is our first example of a shared resource in this project.

PS2 Keyboard Keyer

The shield used for the PS2 keyer is almost unchanged. We did move the PS2 socket for the keyboard from the shield itself to a case-mounted socket. Because there are no special pin requirements you are free to place the shield at either of the two remaining positions. Figure 10-2 shows that we opted for the North-East position (4) on the expansion board. You can see the off-board PS2 socket on the right side of Figure 10-2. The two wires leading away from the same shield go to the jack for the keyed circuit.

Chapter 10: Project Integration   203

Figure 10-2 The three project shields on the expansion board.

The Expansion Board

The RTC shield is in the upper-left corner of Figure 10-2 and holds the LCD display for both the RTC and the CW decoder. You can also see the pushbutton switch that is used to reset the station ID timer connected to the RTC shield. In the same figure, the bottom-left corner is occupied by the CW decoder shield with its LCD display removed. The two wires from the shield are ultimately connected to the speaker of the transceiver. The 2560 µC board is "under" the expansion board. As you can see, the S-E position of the expansion board is empty. If you have some other project that you would like to add to the integrated project, you could use that position.

You should keep in mind that using three shields produces an increased power load on the supporting 2560 board. For that reason, we encourage you to use an external power source (e.g., a wall wart) plugged into the external power connector on the 2560 board. Because the 2560 board has its own voltage regulator, you can use any source capable of supplying 7 to 12 V. (The regulator can handle higher voltages, but it's not a good idea to stray too far from the 7-12 V range. The maximum current on the regulator is about 1 amp.)

Table 10-3 shows how the DFRobot expansion board maps to the 2560 pins. Table 10-3 allows you to determine which pins are available for each of the four positions on the expansion board.

| Arduino R3 Pins | DFRobot Mega-Multi Expansion Shield |||||
| | Mega Pins |||| SPI |
	Position 1	Position 2	Position 3	Position 4	
D0	RX0/D0	RX1/D19	RX3/D15	RX2/D17	
D1	TX1/D1	TX1/D18	TX3/D14	TX2/D16	
D2	D22	D31	D45	D40	
D3	D23	D32	D5	D2	
D4	D24	D33	D46	D41	
D5	D25	D34	D6	D3	
D6	D26	D35	D7	D4	
D7	D27	D36	D47	D42	
D8	D28	D37	D48	D43	
D9	D11	D8	D49	D44	
D10	D12	D9	D53	D53	SS
D11	D13	D10	D51	D51	MOSI
D12	D29	D38	D50	D50	MISO
D13	D30	D39	D52	D52	SCK
			SPI	SPI	
SCL	SCL	SCL	SCL	SCL	
SDA	SDA	SDA	SDA	SDA	
AREF	AREF	AREF	AREF	AREF	
GND	GND	GND	GND	GND	
A0	A0	A4	A12	A8	
A1	A1	A5	A13	A9	
A2	A2	A6	A14	A10	
A3	A3	A7	A15	A11	
A4	NC				
A5	NC				

ICSP1		ICSP2	
1	D50	1	D50
2	5V	2	5V
3	D52	3	D52
4	D51	4	D51
5	RST	5	RST
6	GND	6	GND

TABLE 10-3 DFRobot Mega-Multi Expansion Shield Pin Mapping

Software Project Preparation

We don't know how you have organized the available source code files for the projects in this book. The way we have them organized is with the main directory named P-KHamRadioBook as the main directory and then a chapter number for each chapter in the book. Within each chapter we have additional directories for topics of interest in that chapter. For example, we use the E: drive as our primary storage drive and we named this project "IntegrationCode." Therefore, our directory for the code in this project is:

E:/P-KHamRadioBook/Chapter10/IntegrationCode

Into that directory, we copied each of the source code files from the other related chapters. There is a file for each chapter project: 1) the decoder chapter (e.g., Decoder.cpp and Decoder.h), 2) the PS2 keyer (PS2Keyer.cpp and PS2Keyer.h), and 3) the RTC (RTCTimer.cpp and RTCTimer.h). In addition, we have the main project file, IntegrationCode.ino, and its associated header file IntegrationCode.h. You can copy these files from the McGraw-Hill web site for this book or make empty files using a text editor. Eventually, of course, you're going to have to add in the source code, so you may as well download the source code files now if you haven't already done so.

There are three rules you must follow when using multiple source files in a single project. First, all of the source files used in the program sketch must appear in the directory that uses the same directory name as the name of the sketch. In other words, because we have named this program sketch IntegrationCode, the Arduino IDE wants all of the files used in the program to reside in a directory that shares the same name as the project file. For us, the result of our directory looks like:

```
E:\P-KHamRadioBook\Chapter01
                   Chapter02
                   ...
                   Chapter10\IntegrationCode\Decoder.cpp
                                             Decoder.h
                                             IntegrationCode.ino
                                             IntegrationCode.h
                                             PS2Keyer.cpp
                                             PS2Keyer.h
                                             RTCTimer.cpp
                                             RTCTimer.h
```

If you use this structure for your source files, when you load the IDE and then load the IntegrationCode program sketch, all of the files in the Chapter 10 directory appear as tabs in your IDE. If you've done things correctly, your IDE should look similar to Figure 10-3. Notice the multiple tabs just above the source code window. You should see one tab for each of the files in the IntegrationCode directory.

The second rule is that there can only be one *.ino file for each sketch, regardless of how many source code files there are in the program sketch. As you already know, the name of the *.ino file must be the same as the directory that holds the *.ino file. In our case, the directory holding the source files is IntegrationCode, so the *.ino file must be named IntegrationCode.ino. Using this *.ino naming convention also means that the IDE expects to find the *setup()* and *loop()* functions in the file bearing the ino secondary file name extension. (Actually, the C++ compiler

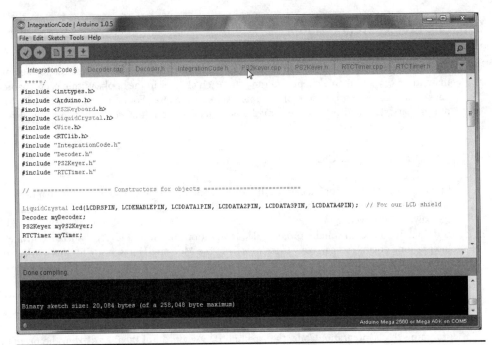

FIGURE 10-3   The Arduino IDE with IntegrationCode sketch loaded.

that lurks below the Arduino IDE surface uses the ino information to log the file associated with the *main()* function, which is the primary entry point for all C and C++ programs.)

The third rule is that the IDE wants any file other than the *.ino file to be either a C or C++ source code file (i.e., uses the *.c or *.cpp secondary file name) or a C or C++ header file (i.e., uses the *.h secondary file name). Any other files in the directory are simply ignored.

Okay, because the IDE can use either C or C++ files, which should you use? The easy answer is to use the language you're comfortable with. We've been using C for almost 40 years now, so we are quite comfortable with it. However, C++ and Object-Oriented Programming (OOP) in general bring so much to the table, we have elected to build our integration project using C++. What follows is a brief explanation of some of the software idioms that C++ uses. If software just isn't your thing, you can skip the next section if you wish. However, we think understanding the OOP at some level helps you understand the ins-and-outs of this program and may help you to create your own projects down the road.

C++, OOP, and Some Software Conventions

First, we should admit that you could build this project using plain C and it would probably work just fine. Given that, why mess around with C++? The first reason is because the compiler behind the Arduino IDE is based on a C++ compiler. Knowing something about C++ lets you take advantage of some of the features that C++ offers. Second, C++ is an OOP language. We touched on some of the advantages OOP has in Chapter 2, but we actually use some of them in this chapter. Third, most of the libraries are written in C++. Reading and understanding their source code should better equip you to use those libraries. After all, being able to stand on the shoulders of others makes seeing things a whole lot easier. Finally, once mastered, practicing OOP principles and techniques in your own programming efforts is going to produce better code that is more easily maintained. To learn OOP in general and C++ specifically takes some time, but is well worth the effort. With that in mind, let's dive in and get our feet wet ...

C++ Header Files

A header file contains information that is required for its associated program file to work properly in the environment for which it is defined. Many beginners tend to place program code in their header files, and that's not what they are intended for. Instead, they are used to provide overhead information that the compiler needs to do its job effectively. Listing 10-1 presents the header file (PS2Keyer.h) associated with the PS2 keyer source code file (PS2Keyer.cpp).

The first line in the header file is a preprocessor directive. All preprocessor directives start with a sharp (or pound) sign (#). Recall that preprocessor directives do *not* use the semicolon to terminate the directive. The directive ends at the end of the line holding the directive. That's one reason that we call them a directive rather than a statement. Most C++ and C programmers write the directive in a way that makes the directive fit on a single line, as you see in the first line of Listing 10-1. You can continue a long directive to a second line if the first line ends with a single backslash character (\), but most programmers tend to avoid multiline directives.

#ifndef ... If Not Defined

The *#ifndef* directive is used like a normal *if* statement in that it can be used to toggle directives and statements into the program. The general form is:

```
#ifndef expression1
    // statements or directives controlled by the directives
#endif
```

If expression1 is *not* defined at this point in the program, everything that follows the directive up to the *#endif* is compiled into the program. If you look at Listing 10-1, you might be wondering where the *#ifndef*'s matching *#endif* is. Look at the last line in Listing 10-1. What this means is that everything that appears between these two preprocessor directives is compiled into the program, provided that PS2KEYER is not defined at this point in the program. In this file, the first preprocessor directive controls all of the statements in the header file!

Wait a minute! The very next preprocessor directive *#define*'s PS2KEYER. What's the purpose of that? Think about it. Suppose you included this header file two times in the same program by mistake. You would end up defining everything in the file twice, which would cause the compiler to generate a bunch of duplicate definition error messages. Using this technique makes it impossible to "double include" the contents of the header file. You will see this idiom used in most library files.

```
#ifndef PS2KEYER
#define PS2KEYER

#include "IntegrationCode.h"
#include <PS2Keyboard.h>

class PS2Keyer
{
    public:
//  =========================  public Method Prototypes  =====================
```

LISTING 10-1   The PS2Keyer.h header file.

```
        PS2Keyer();
        void PS2Startup();
        void PS2Poll();
        void BufferAdd(char ch);
        void BufferAdd(char *s);
        char BufferPopCharacter();
        void ps2poll();
        void mydelay(unsigned long ms);
        void scale();
        void ditPS2();
        void dahPS2();
        void sendcode(char code);
        void send(char ch);
        void FlashBufferFullWarning();
        void ChangeSendingSpeed();
        int DelayedTransmit();
// =========================== public Members ==================================

        void BufferReset();

        boolean speedChange;        // 'true' indicates speed change requested
        boolean sideTone;           // Default is no sidetone
        char buffer[QUEUESIZE];

        int aborted;
        int wordsPerMinute;         // Default speed
        int ditlen;
        int bufferTail;
        int bufferHead;

        PS2Keyboard kbd;

 private:

// =========================== private Members ==================================

// =========================== private Method Prototypes ====================
};
#endif
```

LISTING 10-1 The PS2Keyer.h header file. *(continued)*

#include *Preprocessor Directive ... Read a File into the Program*

The *#include* preprocessor directive causes the compiler to read the specified file into the program at that point. In Listing 10-1 we see:

```
#include "IntegrationCode.h"
#include <PS2Keyboard.h>
```

This tells the compiler to open and read both the integrationCode.h and the PS2Keyboard.h header files into the program. The compiler opens those files and behaves as though the contents of those two files appear at this exact point in the program as part of the source code.

Why is the first file name surrounded by double quotation marks ("") but the second file name is surrounded by angle brackets (< >)? The *double quotation marks* tell the compiler to look in the current working directory to find the IntegrationCode.h file. For our program, that file is indeed in the current working directory, because that's where the *.ino file is located. By default, the location of the *.ino sketch file defines the location of the current working directory.

When *angle brackets* surround a file name, that tells the compiler to look in the default include directory for the file. For the Arduino IDE, such files are typically found in the Arduino\libraries directory.

Class Declaration

The next few lines in the header file are:

```
class PS2Keyer
{
    public:
//  =========================== public Method Prototypes =====================
        PS2Keyer();
```

The first line says that we are about to declare a formal description of an C++ class named PS2Keyer. Everything after the opening brace ({) through to the closing brace (}, near the end of the file) tells what we can expect to find in the PS2Keyer class.

public and *private* Members of a Class

The keyword *public* followed by a colon character (:) states that the following items are public members of the class. The term *member* means that a specific variable or method "belongs to" the class. The *public* keyword means that those things that have the *public* attribute are accessible at all points in the program. You can also think of public meaning that any *public* variable or method as having global scope. If you look a little farther down in Listing 10-1, you find the keyword *private* followed by a colon. The keyword *private* means that only program elements (e.g., methods) declared between the opening brace and closing braces of the class declaration have access to the items designated as *private*. The *PS2Keyer* class does not have any *private* members or methods.

Function Prototypes

If you look a few lines farther down in the public section of the class declaration, you find the statement:

```
void BufferAdd(char ch);
```

This is called a function prototype. A *function prototype* consists of the data type the function returns (*void* in this case), the name of the function (*BufferAdd*), and the parameters (if any) that are used when calling the function (*char ch*). (Sometimes you may hear everything from the function name to the end of the statement referred to as the *function signature*.)

Why have function prototypes? The reason is that function prototypes can help you from shooting yourself in the foot. How so? Well, after reading the function prototype, the compiler knows that the function named *BufferAdd* cannot return a value to the caller because its return type is *void*. The compiler also now knows that this function requires precisely one parameter and it better be a *char* data type or it's going to let you know about it. Essentially, prototypes enable the compiler to verify that you are using the class functions in the way in which they were intended to be used. This process of checking the function signature against how you are actually using the function is called *type checking*.

Some OOP jargon is in order. To make a distinction between C functions and C++ functions defined within a class, most programmers refer to a "function" declared within the class as a *method*. As a memory-jogger, just remember that "methods have class."

As mentioned earlier, another OOP piece of jargon is to refer to the data defined within the class as *members* of the class. Therefore, the lines:

```
boolean speedChange;        // 'true' indicates speed change requested
boolean sideTone;           // Default is no sidetone
char buffer[QUEUESIZE];
```

define specific members of the PS2Keyer class because they are defined within the braces of the class. Because they are defined within the *public* access specifier's purview, they are more specifically referred to as *public* members of the PS2Keyer class. Whereas plain C likely refers to the data definitions as simply variables, data defined within a class are members of the class.

cpp Files

You saw earlier in this chapter that the program sketch ended with the ino secondary file name and that each source code file ended with cpp. The C Plus-Plus (cpp) files contain the actual code for a particular class. While you could define code in the header file, that practice is frowned upon. Instead, the actual class code normally is found in the cpp files. Listing 10-2 contains part of the PS2Keyer.cpp file.

```
// Some preprocessor directives and some global data definitions
PS2Keyer::PS2Keyer()
{
  bufferTail = 0;
  bufferHead = 0;

  speedChange = false;       // 'true' indicates speed change requested
  sideTone = false;          // Default is no sidetone
}

void PS2Keyer::PS2Startup()
{
  pinMode(PS2OUTPUTPIN, OUTPUT);
  pinMode(TONEPIN, OUTPUT);
  kbd.begin(PS2DATAPIN, PS2CLOCKPIN);

  wordsPerMinute = DEFAULTWPM;
  int ditlen = 1200 / wordsPerMinute;
}
```

LISTING 10-2   PS2Keyer class source code (partial listing).

Class Constructor Method

Every C++ class has a class constructor. You can identify the class constructor by two of its characteristics: 1) it always has the same name as the class, and 2) it is never allowed to have a type specifier. That is, the constructor can never return a value and it can't even use the keyword *void*. The interesting thing is that, if we had not written the constructor you see in Listing 10-2, the compiler automatically would have written one for us. Equally interesting is the fact that, if you look for this "automatically generated" constructor, you'll never find it. That's why we sometimes call it the *ghost constructor*.

Really? If the compiler can automatically write a constructor for us, why do we need to write one? Well, in many cases, you don't. The primary responsibility of the ghost constructor is a default initialization of all data members of the class. That is, all data types are initialized either to 0, *false*, or *null*, whichever is appropriate for the data type in question. However, there will be times when you're not happy with those default initialization values. That's when you should write your own constructor. You would write your own constructor when you want the class object to come to life with a specific, known, state that you want to control.

The first line in Listing 10-2 is:

```
PS2Keyer::PS2Keyer()
```

Verbalizing this line actually makes more sense if you read it from right to left. Doing so, we might verbalize this line as: "The *PS2Keyer()* method belongs to the *PS2Keyer* class." You know that it is a method because of the two parentheses following the method's name. You also know that it is the class constructor method because it has the same name as the class itself. The two colon characters (::) are formally called the "scope resolution operator." The purpose of the scope resolution operator is to tie the method named *PS2Keyer()* to the class named *PS2Keyer*. For us, it makes more sense to verbalize the scope resolution operator as the "belongs to" operator when reading from right to left. Verbalize it in whatever way makes the most sense to you.

As we just pointed out, the *PS2Keyer()* method shares the same name as the class and, hence, must be the class constructor. Notice that we initialized four members of the class in the constructor. Technically, we didn't need to write a constructor at all. The reason is because the default initialization values that would have been used by the ghost constructor are the same values we used. We wrote our own constructor simply to show you how you can write a constructor and use it to assign whatever values you wish to class members.

So, what does the following line tell you?

```
void PS2Keyer::PS2Startup()
```

This line tells you that: "*PS2Startup()* is a method that takes no parameters, returns a *void* data type, and is a method that belongs to the *PS2Keyer* class." The statements that appear between the opening brace ({) and closing brace (}) of the method hold the statements that are designed to fulfill the task that this method performs.

If you look through the PS2Keyer.cpp file, you can see the task each method is designed to perform. The other *.cpp files follow the same C++ conventions. The actual code in the class methods is pretty much "pure" C code. We encourage you to download and review the code in each of the source files so you have an understanding of how C++ files are constructed and their purpose. It will be time well spent.

IntegrationCode.ino

The sketch file for the integration program is IntegrationCode.ino. Like all other Arduino IDE projects, the primary purpose of the *.ino file is to hold the code that initiates and controls the program. The *.ino file is always the file that contains the *setup()* and *loop()* functions that are common to all Arduino programs. Using multiple source code files doesn't change the purpose of the *.ino file. Listing 10-3 presents the code for IntegrationCode.ino.

Header Files

The code begins with a number of *#include* preprocessor directives. Most of the calls are to include special library header files used in the program. The header files surrounded by double quotation marks are the header files we wrote and appear in the working directory. Note that some of the header files are "nonstandard." That is, they are associated with special libraries that are not included as part of the standard Arduino library files. If you haven't downloaded those libraries yet, please refer to the chapter that discusses the related project to find where you can download the necessary library file.

```
/*****
   This code is the main entry point into the Integration Project
   discussed in Chapter 10. It combines the following projects:
     1. PS2 keyboard keyer
     2. CW decoder
     3. Station ID timer
   Each of these projects is separated into its own class. This
   source file coordinates the action contained within those
   classes

   Jack Purdum, W8TEE
   Jan 10, 2014
*****/
#include <inttypes.h>
#include <Arduino.h>
#include <PS2Keyboard.h>
#include <LiquidCrystal.h>
#include <Wire.h>
#include <RTClib.h>
#include "IntegrationCode.h"
#include "Decoder.h"
#include "PS2Keyer.h"
#include "RTCTimer.h"

//#define DEBUG 1     // Uncomment if you want to add debug print statements

// ====================== Constructors for objects ============================

LiquidCrystal lcd(LCDRSPIN, LCDENABLEPIN, LCDDATA1PIN, LCDDATA2PIN, LCDDATA3PIN,
             LCDDATA4PIN);  // For our LCD shield
Decoder myDecoder;
PS2Keyer myPS2Keyer;
RTCTimer myTimer;
```

LISTING 10-3   Source code for IntegrationCode.ino.

```
int flag = 0;

void setup() {
#ifdef DEBUG                            // Use when debugging
Serial.begin(115200);
#endif

  pinMode(LEDPIN, OUTPUT);       // The Arduino LED pin
  pinMode(AUDIOPIN, INPUT);      // Audio input pin for decoder
  digitalWrite(AUDIOPIN, HIGH);  // Set internal pull-up resistor
  lcd.begin(LCDCOLUMNS, LCDROWS);

  myDecoder.DecoderStartup(AUDIOPIN);  // Decoder setup() start
  myPS2Keyer.PS2Startup();       // PS2 Keyer setup() start
  myTimer.RTCStartup();          // RTC setup() start
  lcd.clear();
}

void loop() {
  int i;
  int sentBack;
  int audio;

  audio = digitalRead(AUDIOPIN); // What is the tone decoder doing?

  if (audio == 0) {
    lcd.clear();
    myDecoder.DecoderPoll();     // Pulled low when a signal appears
  } else {
    myTimer.RTCPoll();           // Otherwise, just run the RTC
  }

  myPS2Keyer.PS2Poll();          // Are they using the PS2 keyer?

#ifdef DEBUG                     // freeRam() gives approximate SRAM left
if (flag == 0) {
  Serial.print("Free SRAM = ");
  Serial.println(freeRam());
  flag++;
}
#endif
}

#ifdef DEBUG
/*****
  Function that gives an estimate of the amount of SRAM available. It is only
  an approximation because SRAM ebbs and flows as the program executes, so the
  amount of SRAM is only taken at a point in the program, not for the entire
  execution sequence.
```

LISTING 10-3 Source code for IntegrationCode.ino. (*continued*)

```
    Parameter list:
      void

    Return value:
      int           the amount of unused SRAM at this point in the program

*****/
int freeRam()
{
  extern int __heap_start, *__brkval;
  int v;
  return (int) &v - (__brkval == 0 ? (int) &__heap_start : (int) __brkval);
}
#endif
```

LISTING 10-3   Source code for IntegrationCode.ino. (*continued*)

The next four statements in Listing 10-3 are:

```
LiquidCrystal lcd(LCDRSPIN, LCDENABLEPIN, LCDDATA1PIN, LCDDATA2PIN, LCDDATA3PIN,
                  LCDDATA4PIN);   // For our LCD shield
Decoder myDecoder;
PS2Keyer myPS2Keyer;
RTCTimer myTimer;
```

Constructors

You've used the *LiquidCrystal* statement before, in your projects that used the LCD display shield described in Chapter 3. Actually, each of the four statements above are calls to the class constructor for each class used in this project. The *LiquidCrystal* constructor passes in six parameters that are used to initialize specific members of the *LiquidCrystal* class. The other three constructor calls do not use any arguments when their constructor is called.

When these four statements are finished executing, chunks of memory will have been reserved for *lcd*, *myDecoder*, *myPS2Keyer*, and *myTimer*. In OOP parlance, these four statements *instantiate* an object for each class. The process of instantiating an object is the duty of the class constructor. The result of instantiation is an object of that class.

How the Terms *Class*, *Instantiation*, and *Object* Relate to One Another

When readers first start using OOP practices in their coding, many are confused about the terms *class*, *instantiation*, and *object*. Actually, understanding each item is pretty easy. First, a class is a description of what an object of that class is and can do. For example, suppose you wanted to create a cookie cutter class. You would likely use a header file (e.g., cookiecutter.h) to describe the members and methods of the class. Your file would follow the same general form shown in Listing 10-1. Next, you would then proceed to specify in detail how each member and method works in the class using the code in the class code file (e.g., cookiecutter.cpp). Again, you would probably follow something similar to the style shown in the partial listing shown in Listing 10-2. When you have finished writing the *.h and its associated *.cpp file, you have a "blueprint" for the class. The *members* of the class are variables defined within the class that are capable of holding specific values, like the thickness of the cookie cutter metal, the number of bends in the metal, the angle of each bend, and so forth. The *methods* defined within the class describe the actions you want the cookie cutter class to be able to perform (e.g., *pressCutterIntoDough()*, *ejectCookieFromCutter()*, *frostTheCookie()*). Some people like to think of class members as nouns and class methods as verbs.

The important thing to understand is that a class is simply a description, or a template, of an object. Just as blueprints describe what a house will look like, blueprints themselves are not a house object. Likewise, a class description provides the details about the class in the *.h and *.cpp files, but the class description is not an object of the class. Think of a class as a set of blueprints for an object of the class.

The term *instantiation* refers to the process of actually creating an object of the class. Object instantiation occurs when the class constructor is called. When the class constructor finishes its job, there is a chunk of memory set aside for an object of that class. In other words, that chunk of memory is the object of the class. When the four statements above finish executing, the four class objects, *lcd*, *myDecoder*, *myPS2Keyer*, and *myTimer*, are available for use in your program.

Using our cookie cutter class analogy, the cookie cutter is described by the class *.h and *.cpp files. Your code calls the cookie cutter class constructor, which is the same as you picking up the cookie cutter and pushing it into some dough. After you've squiggled the cookie cutter around in the dough, you pull the cookie cutter away, which is the process of instantiation. You then shake the cookie cutter lightly and a raw cookie falls onto the cookie sheet. That cookie is the object of the class. Therefore, the class describes the exact cookie cutter you want to use, and in front of you is a huge sheet of dough (i.e., SRAM); you push the cookie cutter into the dough (instantiation) and extract an object called a cookie (e.g., *lcd*, *myTimer*).

The Arduino IDE provides a number of predefined classes for you to use, many of which you can see in the libraries subdirectory of the Arduino main directory. (If you want to find information on other class libraries, just Google "Arduino class libraries." We got over a million and a half hits!) The other *.h and *.cpp files in the IntegrationCode project are descriptions of how we want each object to perform in the program. Because most of the code is simply a repeat of the code from earlier chapters, we don't repeat it here. However, you should download the code and spend a little time looking at it. If you keep the discussion presented here in your mind as you examine the source code, you will find there is nothing mysterious about using OOP in your projects.

The Dot Operator (.)

Below are several statements from the *setup()* function presented in Listing 10-3.

```
lcd.begin(LCDCOLUMNS, LCDROWS);

myDecoder.DecoderStartup(AUDIOPIN);   // Decoder setup() start
myPS2Keyer.PS2Startup();              // PS2 Keyer setup() start
myTimer.RTCStartup();                 // RTC setup() start
lcd.clear();
```

All five statement lines begin with the name of a class object that was defined earlier in the program. Let's look at the statement:

```
myPS2Keyer.PS2Startup();      // PS2 Keyer setup() start
```

as an example of what the five lines are doing. First, note that the class object is named *myPS2Keyer*. As we said earlier, there is a chunk of memory associated with that object named *myPS2Keyer* caused by the instantiation process that takes place when the class constructor is called.

The *dot operator* is the period that you see between the class object's name (*myPS2Keyer*) and the class method named *PS2Startup()*. We like to think of the dot operator as being verbalized with the words "fetch the memory address of." With that in mind, you can verbalize the entire statement as: "Go to the memory address allocated for *myPS2Keyer* and fetch the memory address of *PS2Startup()*." Because *PS2Startup()* is a method defined within the class, the program takes the memory address it just fetched, jumps to that memory address, and starts executing whatever code it finds at that address. If you look at Listing 10-2, you can see the statements associated with the *PS2Startup()* method.

Some people find it useful to think of the dot operator as a "table of contents" for a class. That is, if *myPS2Keyer* corresponds to a book, the dot operator serves as a table of contents where you can find each member and method that belongs in the book (i.e., the class). Each entry in the dot operator's table of contents is a memory address of where to find a specific member or method. Use whatever imagery for the dot operator that works for you.

If you look at Listing 10-1, the PS2Keyer header file contains the statement:

```
int wordsPerMinute;
```

Now suppose you wanted to change the words per minute being sent by the PS2 keyer to 20 words per minute. How would you do it? To answer that question, look at it from the compiler's point of view: What does the compiler need to know to change the current value of the class member variable? The compiler needs to know three things: 1) where does the *wordsPerMinute* member variable of the *myPS2Keyer* object live in memory? 2) how many bytes of memory are associated with that member variable? and 3) what is the new value to assign to the member variable?

You already know that the expression:

```
myPS2Keyer.
```

says: "Go to the memory address of *myPS2Keyer* and fetch the memory address of ..." So, part 1 of the things the compiler needs to know is now resolved. Part 2 asks how big the variable is. Well, the type checking the compiler performed when it read (and memorized!) the header file tells it that *wordsPerMinute* is an *int*, so the compiler already knows that two bytes are associated with the class member variable named *wordsPerMinute*. Part 3 is the value we wish to assign into the variable, which is 20. Therefore, we can complete the statement to assign the *wordsPerMinute* class member a new value of 20 by writing:

```
myPS2Keyer.wordsPerMinute = 20;
```

You can verbalize the statement as: "Go to the memory address assigned to the *myPS2Keyer* object and fetch the memory address of the two bytes associated with *wordsPerMinute* and assign the value 20 into those two bytes of memory."

Suppose you want to read the new value for the *wordsPerMinute* class member. To retrieve the class member value for *wordsPerMinute*, you would use:

```
int val = myPS2Keyer.wordsPerMinute;
```

As you can see, there is nothing mysterious about using classes or OOP in general. The dot operator always appears between the class object name and the member or method name you are trying to access. It just takes a little practice and a slightly different way of thinking about how to write code. Before reading on, spend some time reviewing the code in the header and source code files for this project. With a little practice, we think you will find that OOP code is pretty easy to understand.

The *loop()* Function

Listing 10-4 repeats the *loop()* function code from Listing 10-3 so you don't have to flip back and forth while reading the code description.

```
void loop() {
  int i;
  int sentBack;
  int audio;

  audio = digitalRead(AUDIOPIN);  // What is the tone decoder doing?

  if (audio == 0) {
    lcd.clear();
    myDecoder.DecoderPoll();      // Pulled low when a signal appears
  } else {
    myTimer.RTCPoll();            // Otherwise, just run the RTC
  }

  myPS2Keyer.PS2Poll();           // Are they using the PS2 keyer?

#ifdef DEBUG                      // freeRam() gives approximate SRAM left
  if (flag == 0) {
    Serial.print("Free SRAM = ");
    Serial.println(freeRam());
    flag++;
  }
#endif
}
```

LISTING 10-4 The *loop()* function source code.

The *loop()* function begins by defining a few working variables and then calls *digitalRead()* to see if there is a signal on *AUDIOPIN*, which is the pin that is tied to the receiver's speaker. Because of the way the hardware on the decoder shield and software work, variable *audio* holds the value 1 when there is no signal present on the speakers. In that case, the statement:

```
myTimer.RTCPoll();                // Otherwise, just run the RTC
```

is executed. You should be able to verbalize what this statement does. If you look at the class code for the *myTimer* object, you can see that the method *RTCPoll()* updates the RTC and moves the appropriate data to the LCD display. If you compare the code found in the *RTCPoll()* method, you will discover that it is almost identical to the code found in the *loop()* function in Listing 4-1 in Chapter 4.

If *audio* equals 0, the *myDecoder.DecoderPoll()* method is called and the display switches away from the date and time and begins displaying the Morse code associated with the signal being received. Because of the way we wrote the *DecoderPoll()* method, as long as the decoder shield is reading Morse code, the program continues to process and display the translated Morse code. When the code stops, control returns to *loop()*.

When control returns to *loop()*, the program checks to see if the PS2 keyer is active. If so, control remains in *PS2Poll()* until the PS2 keyer stops sending code, at which time control returns to *loop()* and another iteration of the code is begun.

Listing all of the source code here would be repetitive for several reasons. First, the code is little changed from the chapters that presented the original projects. Second, you should be able to read and understand what the code is doing now that you know how OOP works. Again, we urge you to spend a little time with the code so you can see how multiple source files can be used in a single program sketch.

Conclusion

As you experiment more with µCs, you will want to develop larger and more complex projects. While you could keep all of the code in one large, monolithic source file, you will discover that using multiple source files and OOP techniques makes developing such projects much easier. Breaking down a large task into a series of smaller tasks always makes that task seem more manageable. You will also discover that OOP techniques make debugging a project a lot easier, too. A final benefit is that, done with a little thought, you might be able to reuse the code in other projects.

There is no way that we can give C++ and OOP in general its due in a few pages. Our goal here was to give you enough OOP understanding to be able to read the code associated with OOP and C++ source code files. There are a number of C++ tutorials available on the Internet if you'd like to learn more about C++. Once the penny drops relative to all that OOP brings to the table, you'll never code any other way.

CHAPTER 11

Universal Relay Shield

One of Dennis's many interests in ham radio involves "boatanchor" radios ... you know, the ones that cause the house lights to dim when you turn them on and then glow in the dark if you turn the lights out. Every now and then he runs across an application for an Arduino that would be ideal for controlling one of these old sets, but the digital IO pins just can't handle the higher voltages used in these old radios. The solution is to use a relay.

There are many relays that can be controlled by a 5 VDC signal and with sufficiently low current as to not stress the IO pins on the Arduino. These relays can switch several hundred volts and a respectable amount of current and come in a very compact size. Figure 11-1 shows some examples of miniature relays. The relays shown all have 5 VDC coils and can switch from

FIGURE 11-1   Typical PCB mounted miniature relays.

219

FIGURE 11-2 An array of high current, high voltage relays.

250 mA (the reed relay in the lower right) up to 30 A (the brute in the upper left). These relays are designed to be mounted on a circuit board, such as our Arduino prototyping shields.

If you need to control even higher current or voltage than the miniature relays we have chosen, you can always use an external relay with contacts rated for your application. The relays and sockets shown in Figure 11-2 are some examples that can handle high current and high voltage, such as would be used to switch mains power.

This chapter presents a project that constructs a relay shield. Figure 11-3 shows a version of the relay shield. The shield contains four double pole double throw (DPDT) relays that can switch up to 250 VAC at 2 A per contact! The relays we chose only draw 25 mA at 5 VDC, which is probably within the acceptable range for one or two relays being driven (the Arduino can source or sink up to 33 mA on an IO pin), but driving four relays simultaneously can potentially damage the µC. We chose to use the digital IO pins with a current driver to operate the relays. By using the current driver, we do not exceed the current limits of the Arduino.

Each relay includes an LED to indicate the operational status of the relay. Another nice feature: a removable jumper that allows you to test your programming without actually engaging the relays, but using the LEDs as an indicator of the relay operation. This can be very helpful during debugging when you might not want to actually operate the piece of equipment the relays are going to control. In addition, the relay contacts are brought out to heavy-duty screw terminals, making connection to the external controlled device much easier.

This project is the basis of three projects later in the book; an antenna rotator controller and a sequencer for controlling low noise preamplifiers and power amplifiers. A third project is based on the relay shield circuit, but is assembled using a different style prototyping board to build a sequencer.

Chapter 11: Universal Relay Shield

FIGURE 11-3   A Universal relay shield.

Construction

The relay shield is assembled on a Mega prototyping shield. This shield provides sufficient "real estate" for all four relays and the associated circuitry. The good news is that the Mega shield does fit an Arduino Uno, Demilenova, and so forth, without any problem. We just don't use all of the headers; only those headers that are common to an R3 shield. As always, there is no rule that says you must build the circuit in this manner. In fact, you might not need four relays or even DPDT relays for your application; remember that the circuit is quite scalable both up and down in size. One possible option would be to use an R3 shield and a header to a ribbon cable to connect to a larger prototyping board with more than four relays. The possibilities are limited only by your needs and your creativity.

Circuit Description

The schematic for the relay shield is shown in Figure 11-4. For the sake of clarity, only one relay circuit is shown. The remaining three are wired identically. Each relay is driven by one section of a DS75492 hex driver chip. The DS75492 is rated at 250 mA current per output pin, not to exceed 600 mA total for the device. The DS75492 is an "open-collector" output device, meaning that the relay is connected to the positive supply while the DS75492 provides switching to ground. A "HIGH" output from the Arduino actuates the relay. The driver output is connected directly to the LED indicator for that section, and the relay is connected through a two-pin jumper. By using the jumper (JP1 in Figure 11-4), the relay can be disabled for testing by removing the jumper while the LED provides an indication that the output is being driven. This prevents "accidents" from happening when driving the external circuits connected to the relay.

The parts list for the relay shield is shown in Table 11-1. Most of the components are sourced from eBay; however, the DS75492 is easily purchased from Jameco (http://www.jameco.com).

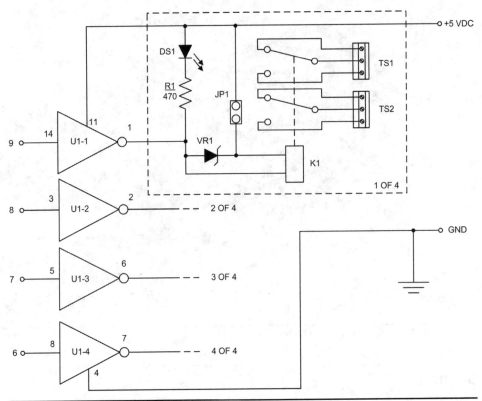

FIGURE 11-4 Universal relay shield schematic diagram.

Ref Designator	Description	Part No	Mfg	Source
DS1,DS2,DS3,DS4	LED, Red		various	eBay, etc.
K1,K2,K3,K4	Relay, 5VDC, DPDT or equiv	DS2YE-S-DC5V	Aromat	Jameco
JP1-JP4	Header, 1x3 pin, 2.56 mm (0.1 in.)		various	eBay, etc.
R1,R2,R3,R4	Resistor, 470 Ω, ¼ W, 5%		various	eBay, etc.
VR1,VR2,VR3,VR4	Diode, Zener, 1N4735A, 6.2 V, 1 W	1N4735A		Jameco
U1	DS75492, Hex driver	DS75492		Jameco
	Mega Prototyping board			eBay, etc.

TABLE 11-1 Universal Relay Shield Parts List

Construction of the Relay Shield

Our example relay shield, as shown in Figure 11-3, is assembled on a Mega prototyping shield. This shield functions on multiple Arduino platforms as they share a common layout for the power, six analog in and 14 digital IO pins. We didn't use the other IO pins and only installed the four headers needed for those pins. Figure 11-5 shows the layout we have used.

The reverse side of the shield is shown in Figure 11-6. Wiring is done as described in Chapter 3, using bare wire from components leads, and Teflon tubing for insulation. Because of limited

Chapter 11: Universal Relay Shield 223

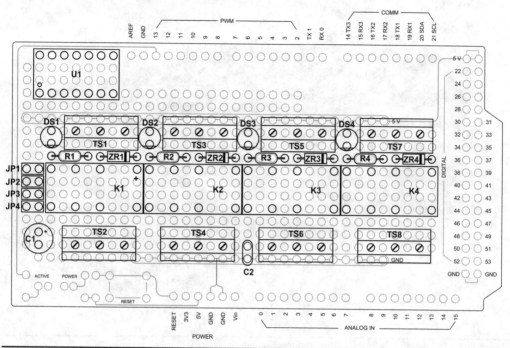

FIGURE 11-5 One possible layout of the relay shield.

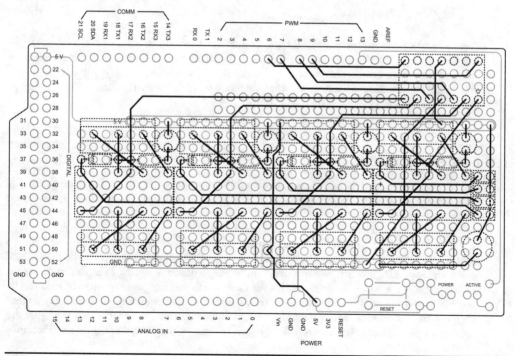

FIGURE 11-6 The "Backside" of the relay shield showing the wiring.

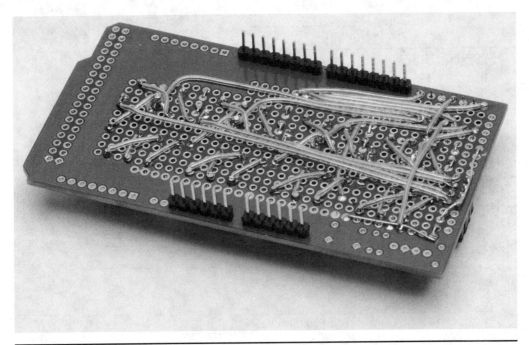

FIGURE 11-7   The wiring side of the completed relay shield.

"real estate" on the prototyping shield and clearance issues with the screw terminals near the current driver IC, we chose to connect directly to the leads of the IC rather than using a socket. A socket would have obstructed the wiring to the adjacent screw terminals. If you decide on this method of construction, use extra care when soldering wires to the current driver. The wiring side of the completed relay shield is shown in Figure 11-7.

Testing the Relay Shield

We have provided a sketch that you can use to test that the relays are functioning as expected. The program energizes each relay in sequence. The corresponding LED should also light up when that relay is energized.

After having checked all of your wiring (you have checked your wiring, yes?), you can attach the relay shield to an Arduino and apply power. Assuming there is no program in the Arduino that would set any of the output pins we are using to HIGH, none of the LEDs should be lit. If there are any LEDs lit, then either a program in the Arduino is setting that particular output HIGH or there is a wiring error. If none of the LEDs are lit you can proceed with the next test by compiling and loading the sketch shown in Listing 11-1. This sketch turns each relay on and then off, in sequence, on about a one-half second interval. You can observe that the LEDs are cycling in sequence. Next, insert the jumpers to enable the relays and observe that they are cycling as well.

```
/*
 *
 * This is a test sketch for the Relay Shield.  It is designed to
 * sequence through each relay, first turning each one on and then
 * turning each off.  If the shield is wired correctly, the LEDs and
 * relays will all come on in sequence and then go off in the same
 * order
 *
 */

int relay0Pin = 11;           // Relay 1 connected to digital pin 11
int relay1Pin = 10;           // Relay 2 connected to digital pin 10
int relay2Pin = 9;            // Relay 3 connected to digital pin 9
int relay3Pin = 8;            // Relay 4 connected to digital pin 8

void setup()
{
  pinMode(relay0Pin, OUTPUT);  // these lines set the digital pins as outputs
  pinMode(relay1Pin, OUTPUT);
  pinMode(relay2Pin, OUTPUT);
  pinMode(relay3Pin, OUTPUT);
}

void loop()
{
  digitalWrite(relay0Pin, HIGH);   // sets relay 0 on
  delay(100);                      // waits for a 1/10 second
  digitalWrite(relay1Pin, HIGH);   // sets relay 1 on
  delay(100);                      // waits for a 1/10 second
  digitalWrite(relay2Pin, HIGH);   // sets relay 2 on
  delay(100);                      // waits for a 1/10 second
  digitalWrite(relay3Pin, HIGH);   // sets relay 3 on
  delay(100);                      // waits for a 1/10 second
  digitalWrite(relay0Pin, LOW);    // sets relay 0 off
  delay(100);                      // waits for a 1/10 second
  digitalWrite(relay1Pin, LOW);    // sets relay 1 off
  delay(100);                      // waits for a 1/10 second
  digitalWrite(relay2Pin, LOW);    // sets relay 2 off
  delay(100);                      // waits for a 1/10 second
  digitalWrite(relay3Pin, LOW);    // sets relay 3 off
  delay(100);                      // waits for a 1/10 second
}
```

LISTING 11-1   A test sketch for the relay shield.

Test Sketch "Walk-Through"

The code used to test the relay shield is pretty straightforward. We define the Arduino digital IO pins that drive the relays and set their mode to *OUTPUT*. Next, we set each pin HIGH in sequence on $\frac{1}{10}$ second intervals and then set each pin LOW in sequence on $\frac{1}{10}$ second intervals. It should take $\frac{8}{10}$ second to complete one cycle for all four relays.

Conclusion

This shield along with the test sketch does not do a great deal on its own other than verify the functionality of the relay shield. The relay shield, however, becomes an integral part of three upcoming projects; a rotator controller and two versions of a sequencer. In any case, you may find other applications for the relay shield. In Dennis's case, he is using one to control a repurposed AM broadcast transmitter. The transmitter uses four momentary contact pushbuttons and two latching relays to control the AC power and high voltage, respectively. Dennis wanted to be able to control the transmitter remotely using a pushbutton to turn the power on and off and a lever switch to turn on the high voltage for transmit. The program Dennis uses applies a 100 ms pulse to the appropriate relay to simulate the pushbuttons on the transmitter. Latching relays do not like to have continuous power applied to the coil. As for the lever switch, he just determines if the switch has been moved and to what position it moved and triggers the appropriate relay with a 100 ms pulse.

Maybe you have similar applications that can benefit from the relay shield. There's nothing etched in stone that requires this shield to be used only with your amateur radio equipment. A little thought "outside the box" will likely reveal other uses. While it takes a little more hardware and software than we want to get into here, you can activate an Arduino via a cell phone and Internet connection. Using that type of setup, the relay shield could be used for everything from remote warm-up of the rig to activating a burglar alarm system. Think about it ...

CHAPTER **12**

A Flexible Sequencer

Have you ever had a transmit-receive relay fail and witnessed the ensuing carnage that happens when you pump 100 W or more of RF into the front end of your multithousand-dollar receiver? Not a pretty sight. Dennis has experienced something similar to this with predictable results. In his case, it was a relay on the output of a very expensive 2 meter low noise amplifier (LNA) that was in the wrong position when the transmitter keyed up. You guessed it. The result was an ex-LNA.

If you spend much time around VHF and UHF "weak signal" work, referring to operating on CW or sideband, you eventually want to improve the performance of your station. The means to improving performance are simple: more power, more antenna gain, and more receiver gain. More power is easy, we add a high-power amplifier. Antennas? Well, bigger and higher Yagis come to mind. Receiver gain increases? We add an LNA. When we add amplifiers and LNAs we also need the means of controlling them, in other words, ways to switch them in and out of the circuit. By placing these items under control, we prevent things like what Dennis experienced when placing several hundred Watts of RF into the *output* of his very expensive LNA. What we need is a sequencer. Figure 12-1 shows the completed sequencer that is the subject of this chapter.

FIGURE 12-1   The flexible sequencer.

Just What Is a Sequencer?

Just like the name implies, a *sequencer* is a device that can turn other devices on and off in a specified order. A typical sequencer has four outputs that can be energized in order with an ever-so-slight delay between each output turning on, and then can de-energize the four outputs in the reverse order, with the same ever-so-slight delay between each output turning off. Why four outputs? The answer is pretty simple. You generally have four items that you want to control, namely, a low-noise amplifier, a transmit-receive relay, a high-power amplifier, and of course, the transmitter and receiver or transceiver, or possibly a transverter. Most sequencers have fixed time delays between outputs, say 100 ms. For most operations, 100 ms are more than enough time for things like relays to settle out after switching (they do take time to stop bouncing around after switching) and we do want them settled before applying the high-power RF. So, a sequencer is a pretty essential item in a high-power, weak-signal VHF/UHF station.

However, VHF/UHF stations are not the only place sequencers are used. Even on HF, if you are using a high-power linear amplifier, keying the amplifier in sequence with an antenna changeover relay is not a bad idea. Sequencers do have multiple applications in an amateur radio station.

The Sequencer Design

The sequencer shield described here is based on the relay shield constructed in Chapter 11. You can use the relay shield as it was built, with a few minor additions, connecting the external devices using the screw terminals, or you can use an alternative approach, which we describe later in this chapter. In the latter case, we provide a design that is "purpose built." That is to say, it is a complete system, down to an enclosure to house it.

We chose not to use an Arduino board and shield; rather, the Digispark (as was used in Chapter 7 for the Keyer) seemed to be the perfect fit for this project. With six digital IO ports available on the Digispark, we can use one port as an input, four ports to drive the four relays, and the remaining port to drive an LED indicating there is an input signal present. (You could also use an ATtiny85 and program it as we did in Chapter 7. The Digispark, however, is a little easier to use because of its built-in USB port. The Arduino Nano 3.0 also offers a small footprint and attached USB port.) All four relays have LEDs to show their state.

The sequencer is shown in schematic form in Figure 12-2. For the sake of simplicity, the diagram depicts only one relay section, as shown within the dashed outline. There are three additional relay sections that are not shown, but are identical to the relay section shown in the diagram. Module A1 in Figure 12-2 is the Digispark (or an ATtiny85) and IC U1 is the hex driver (DS75492).

We have tried to create a flexible design, one that allows use of different voltage relays (we have a mix of 5, 12, and 24 VDC relays in use), different keying schemes (some pieces of equipment may require a grounded input to operate while others require a circuit to open). However, unlike the relay shield, we have added jumpers JP1 through JP12 to allow reconfiguration of the shield for normally open, normally closed, contacts to ground, to 12 VDC or to an auxiliary input that might use 24 VDC, a common voltage used for coaxial relays. There is no reason why the jumpers couldn't be replaced with hardwired connections for your specific needs, simplifying construction.

Timing

The software is designed to allow the timing to be easily changed for each relay as well as the order in which the relays are energized and de-energized. This adds another layer of flexibility not found in the typical commercially-available sequencers. To help visualize what the output of the sequencer looks like, a representative timing diagram is shown in Figure 12-3. As the input goes to TRUE, each output is enabled in sequence. When the input goes to FALSE, each output is disabled in sequence. The sequence depicted is the default ordering of the outputs. The software is designed to allow the sequence to be changed easily.

Chapter 12: A Flexible Sequencer   229

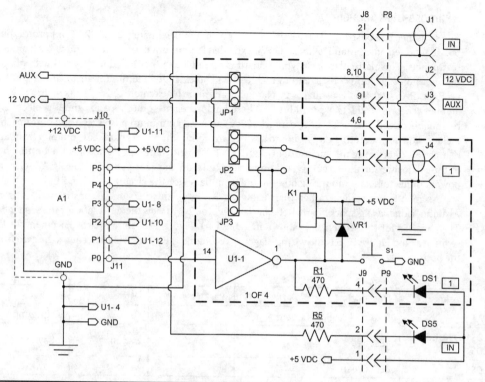

FIGURE 12-2   Sequencer schematic diagram using Digispark.

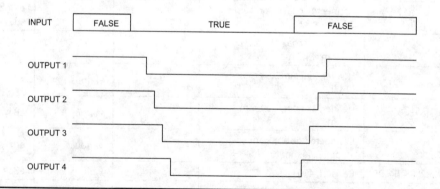

FIGURE 12-3   Default timing for the sequencer.

Constructing the Sequencer

As we mentioned earlier, we provide two methods of construction. First, we present a purpose-built design that is not assembled as a shield and is mounted in a nice enclosure. As an alternative, the relay shield from Chapter 11 can be used with several minor modifications and can be used with a second sketch described later in this chapter and shown in Listing 12-2.

A Purpose-Built Sequencer

The sequencer is constructed on a small 5 cm by 7 cm (approximately 2 in. × 2.75 in.) prototyping board obtained on eBay and is housed in an extruded Aluminum chassis with removable front and rear panels. This particular prototyping board is quite nice in that it is double sided with plated-through holes, making construction much easier. The extruded aluminum enclosure is an LMB product, EAS-100, and is available in plain aluminum or black anodized finish (see: http://www.lmbheeger.com/products.asp?catid=68). These enclosures are quite nice for small projects and reasonably priced. It consists of two interlocking extruded aluminum pieces (top and bottom) that include slots for mounting circuit boards, with two aluminum plates that screw on for a front and rear panel. You can see the enclosure in Figure 12-1. We drilled mounting holes in the bottom half of the enclosure for the circuit board; mounting holes in the rear panel for the RCA jacks and power connectors; and mounting holes in the front panel for the circuit board and LEDs.

As mentioned earlier, rather than using an Arduino, this version is constructed using an Arduino "clone," a Digispark board. Refer back to Figure 12-2 to view the schematic for the sequencer. The parts list for the sequencer is shown in Table 12-1. Figure 12-4 shows parts placement for the sequencer and Figure 12-5 is the wiring diagram. As before, the wiring diagram shows the card from the bottom side.

Ref Designator	Description	Part No	Mfg	Source
A1	Digispark USB Dev Board			Digistump
DS1, DS2, DS3, DS4	LED, Red		various	eBay, etc.
DS5	LED, Green		various	eBay, etc.
K1, K2, K3, K4	Relay, 5 VDC, DPDT or equiv.	DS2YE-S-DC5V	Aromat	Jameco
J1, J4, J5, J6, J7	RCA connector, female chassis mount		various	eBay
J2	Connector, coaxial power, 2.1 mm × 5.5 mm			plugz2go @ eBay
J3	Connector, coaxial power, 3.5 mm × 1.3 mm			plugz2go @ eBay
J8, J9	Header, 2x5 pin, 2.54 mm (0.1 in.)		various	eBay, etc.
JP1-JP12	Header, 1x3 pin, 2.54 mm (0.1 in.)		various	eBay, etc.
P8, P9	Connector, ribbon cable, 10 pin, 2.56 mm (0.1 in.)	FC-10P	various	eBay, etc.
R1, R2, R3, R4, R5	Resistor, 470 Ω, ¼ W, 5%		various	eBay, etc.
S1, S2, S3, S4	Switch, momentary contact		various	eBay, etc.
VR1, VR2, VR3, VR4	Diode, Zener, 1N4735A, 6.2 V, 1 W	1N4735A		Jameco
U1	DS75492, Hex driver	DS75492		Jameco
	Prototyping board, 5 cm × 7 cm			eBay, etc.
	Aluminum enclosure	EAS-100	LMB	LMB/Heeger

TABLE 12-1 Sequencer Parts List

Chapter 12: A Flexible Sequencer

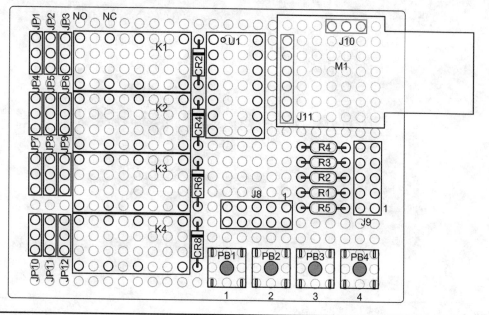

FIGURE 12-4   Sequencer parts placement.

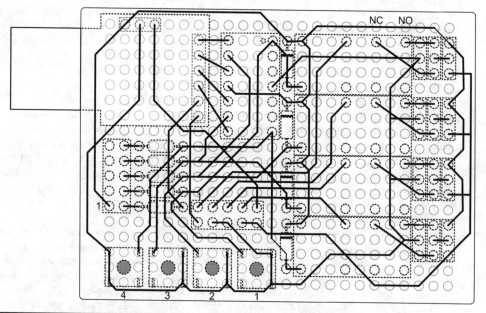

FIGURE 12-5   Sequencer wiring diagram.

FIGURE 12-6   Assembled sequencer board.

The assembled sequencer board is shown in Figure 12-6. You can clearly see the configuration jumpers on the left side of the shield, the Digispark at the upper right, the pushbuttons near the bottom-right for manual operation, and the two 10-pin headers connecting to the front and rear panels. It's a good idea to mark pin 1 of each header using a marker. We have permanently soldered the Digispark to the circuit board using header pins. If you are not comfortable mounting the Digispark in this manner, you can add header sockets to the Digispark to make it removable. (Don't forget, you could also program an ATtiny85 to do the job.)

One physical difference between the relay shield and this version of the sequencer is that the LEDs are mounted remotely on the front panel in order to be visible during operation. The LEDs are mounted on a smaller prototype board and connected to the main board with a ribbon cable. Figure 12-7 shows the front panel construction and how the prototype board is mounted. Mounting the LEDs is fairly simple. The first step is to drill the holes in the front panel for the LEDS and mounting screws. Next, place the LEDs in the appropriate locations on the prototyping board but do not solder them in place. The board is then mounted to the front panel and the LEDs slipped into their respective holes on the front panel. The LEDs are then soldered into place. (You could also mount the LEDs in rubber grommets on the front panel, but that is less robust than the approach used here.)

The rear panel is drilled for five RCA jacks and two coaxial power connectors. We used two different sized coaxial power connectors so they would not be mixed up in use. The rear panel is connected to the sequencer board using another piece of ribbon cable. We chose to use connectors on the ribbon cable for ease of disassembly; however, this is not required. The ribbon cables can be soldered in place if so desired. Figure 12-8 shows the rear panel of the sequencer. (You can often find old floppy drive ribbon cables for next to nothing at flea markets, hamfests, and even used items stores, like Goodwill or church-related shops.)

The sequencer is shown assembled in Figure 12-9. You can clearly see the front and rear panels and the main sequencer board and how they are interconnected. A labeler was used for the finishing touches on the assembly.

Chapter 12: A Flexible Sequencer 233

FIGURE 12-7 Front panel construction showing LEDs.

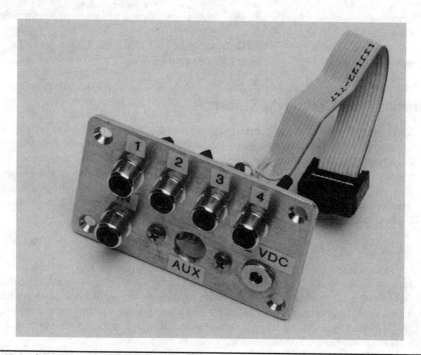

FIGURE 12-8 Sequencer rear panel construction.

FIGURE 12-9 Assembled sequencer.

Remember, before you ever apply power to the sequencer, you should check your wiring against the schematic and wiring diagrams. Use a continuity tester or multimeter for tracing the wiring. If you don't have access to these, use a yellow highlighter to trace a copy of the schematic. Once you are satisfied that the wiring is correct, you can proceed with applying power and testing the sequencer.

To operate the sequencer, 12 VDC is applied to the 12 V coaxial input pin and an input that is normally open to ground is connected to the Input RCA jack.

Programming and Testing the Sequencer

In this section, we load the sequencer program into the Digispark and begin testing the completed sequencer. You need to review the programming procedure described in Chapter 7 for the Digispark keyer. Some of the important points to remember are that you must use IDE 1.04 with the Digispark extensions installed. The new Digispark Pro, which offers more memory and I/O pins, does work with later versions of the Arduino IDE. You must use the Arduino IDE that has been modified by Digistump for use with the Digispark. The sequencer program source code for the Digispark is shown in Listing 12-1.

Initial Testing of the Sequencer

If you have assembled the sequencer into an enclosure similar to what we have used, you need to provide access to the USB connector on the Digispark. One other recommendation that comes from the Digispark Wiki site: use a USB hub to apply power to the Digispark via USB rather than a direct connection to the USB. In the event that there is a short circuit in your wiring, you run the risk of damaging the USB port on your computer. Using a USB hub may blow the hub away, but at least the computer's USB port is saved.

Now apply 12 VDC to the 12 V connector on the rear panel. If the Digispark you are using has never been programmed, none of the LEDs should be lit and none of the relays should have clicked. At this point, you should be able to check the relays and LEDs by pressing each pushbutton on the shield. If it all checks out, you can proceed with loading the sequencer program.

Loading the Sequencer Program and Testing

First, make sure that you have the 12 VDC power disconnected from your sequencer. Launch the Digispark IDE 1.04 with the Digispark extensions and load the sequencer sketch from Listing 12-1. Remember that you need to compile and upload with the Digispark *disconnected* from the USB port. The IDE prompts you when to connect the USB cable to the Digispark. Once the program is loaded, the IDE informs you that the load was successful.

To test the sequencer, apply a ground to the input pin (J1 in Figure 12-2). The four relays and corresponding LEDs will fire up in sequence. Releasing the switch causes the relays and LEDs to go off in the reverse sequence.

```
/******
Flexible 4-Port Sequencer   Version 1.0

Designed by Dennis Kidder W6DQ 5 Jan 2014

This sketch when used with the purpose-built hardware described in the
book, "Arduino Projects for Ham Radio," will provide a completely
configurable sequencer.  Two sets of two arrays configure the ON sequence
the OFF sequence and the delays between each step.  This allows the ON
and OFF sequence order to differ along with the ON and OFF times for each
step through sequence.
******/

/******
The sequencer uses the following Digispark digital IO pins:

        P0 - Relay K1
        P1 - Relay K2
        P2 - Relay K3
        P4 - Relay K4

        P3 - PTT Input Indicator LED
        P5 - PTT Input

The following four lines of code allow the user to configure the ON and OFF
sequence and delay times for each step.  The default configuration turns
the relays on in the sequence "1,2,3,4" (represented logically by digital
IO pins 0, 1, 2, and 4) and to turn them off in the reverse order "4,3,2,1."

The delay between steps is expressed in milliseconds and the default is 50.
The order of the values in the array corresponds to the order in which the
relays are turned ON or OFF.  Note that the first entry (default 40) takes
into consideration that the debounce delay of 10 mS must be included in the
total delay time, e.g. 50 mS (total) minus 10 mS (debounce delay) equals
40 milliseconds.  The remaining time delays remain unmodified.
******/
```

LISTING 12-1   Sequencer program for Digispark.

```
#define DEBOUNCEDELAY  10

int relayPin[] = {0,1,2,4};              // order to sequence relays on
int onDelay[] = {40,50,50,50};           // ON times in sequence

int relayPinOff[] = {4,2,1,0};           // order to sequence relays off
int offDelay[] = {40,50,50,50};          // OFF times in sequence

int inputInd = 3;                        // pin 3 drives the input indicator LED
int pttInput = 5;                        // pin 5 is the PTT input

int arrayElements = (sizeof(relayPin) / sizeof(relayPin[0]));  //  elements

void setup()
{

/******
The following for loop is used to setup each of the pins in the relayPin[]
array as an output.
******/

  for(int index = 0; index < arrayElements; index++)
  {
    pinMode(relayPin[index], OUTPUT);   //Set the relayPins as outputs
  }

  pinMode(inputInd, OUTPUT);             // PTT input indicator
  pinMode(pttInput, INPUT);              // PTT input

  digitalWrite(pttInput, HIGH);          // enable the internal pullup resistor

}

void loop()
{

/******
The following sequence calls the debounce() function to determine if there
Is a TRUE input on pttInput.

If pttInput is TRUE, the Input LED (inputInd) is lit, and the for loop uses
"index" to step through the onDelay[] and relayPin[] arrays, applying the
delay to the associated output pin.

When pttInput goes FALSE the Input LED is extinguished and the next for loop
Uses index to step through the offDelay[] and relayPin[] arrays, applying the
off delay for each relay in the order specified in the relayPin[] array.

******/
```

LISTING 12-1 Sequencer program for Digispark. (*continued*)

```
    if (debounce(pttInput) == true) {
      digitalWrite(inputInd, HIGH);
      for(int index = 0; index < arrayElements; index++)
      {
        delay(onDelay[index]);
        digitalWrite(relayPin[index], HIGH);
      }
    } else {
      digitalWrite(inputInd, LOW);
      for (int index = 0; index < arrayElements; index++)
      {
        delay(offDelay[index]);
        digitalWrite(relayPinOff[index], LOW);
      }
    }
}

/*****
  This function is a pretty standard debounce function so we can determine
  that the switch has really been pressed.

  Parameter list:
    int pin             the pin that registers the switch state

  Return value:
    boolean             true if the switch was pressed, false otherwise

*****/
boolean debounce(int pin)
{
  boolean currentState;
  boolean previousState;
  int i;

  previousState = digitalRead(pin);
  for (i = 0; i < DEBOUNCEDELAY; i++) {
    delay(1);      // small delay
    currentState = digitalRead(pin);   // Read it now
    if (currentState != previousState)
    {
      i = 0;
      previousState = currentState;
    }
  }
  if (currentState == LOW)
    return true;
  else
    return false;
}
```

LISTING 12-1 Sequencer program for Digispark. *(continued)*

Sequencer Code "Walk-Through"

The sequencer code from Listing 12-1 should be pretty straightforward to you by now. We define *DEBOUNCEDELAY* as 10 ms. We then set up four arrays: *relayPin[], onDelay[], relayPinOff[],* and *offDelay[]*, and populate them. As noted in the comments, the *relayPin[]* array sets the order in which the relays are turned on following the time delay set by *onDelay[]* for each relay. The first entry in *onDelay[]* corresponds to the first entry in *relayPin[]*. We also set up the pin numbers for the Input indicator LED and the PTT input pin. The following lines of code define the pin numbers for the PTT input (*pttInput*) and the PTT input indicator (*inputInd*):

```
int inputInd = 3;        // pin 3 drives the input indicator LED
int pttInput = 5;        // pin 5 is the PTT input
```

We have to be careful when defining the pins as inputs and outputs with the Digispark. There are multiple pins that serve multiple purposes. The USB interface is realized with P3 and P4. P3 causes problems for us under certain circumstances as it has a 1.5K pull-up resistor connected. Placing a load on P4 may cause the USB to not function. P1 has an on-board LED. If we use this as an output, we have to consider the extra load from the on-board LED. Fortunately, there is a circuit trace that can be cut to disable the LED, which we have done for this application. Cutting the trace to disable the LED is described in the Digispark Wiki:

http://digistump.com/wiki/digispark

In addition, the power indicating LED on the Digispark also has a trace that may be cut in order to reduce current consumption. P5 brings its own quirks to the table in that it is designed as a 3.3 V level output. While we can use P5 as an output, it seemed more reasonable to us to use it as the PTT input.

Next we determine the number of members in the arrays. When the compiler creates the array in memory, it allocates memory based on the number of elements and their data type. In this case, the members are all integers. This line of code:

```
int arrayElements = (sizeof(relayPin) / sizeof(relayPin[0]));
```

sets *arrayMembers* to the number of elements calculated to be in the *relayPin* array. *sizeof()* is actually a C operator used to determine the number of bytes occupied by an array or a variable data type. Although using *sizeof()* looks like a function call, it isn't. By dividing *sizeof(relayPin)* by *sizeof(relayPin[0])*, we know how many elements are in the array. All four arrays contain integers and by necessity, the same number of members and as a result, all four arrays are the same size. Using this approach means that we can change the array sizes or data types if need be and all of the loops that need to use the array size are automatically adjusted for us via the variable *arrayElements*.

In the *void setup()* section, we use the *relayPin[]* array to initialize each of the relay pins as an output. The *for* loop marches through each position in the array according to the value of *index*. Take a look at the *for* loop:

```
for(int index = 0; index < arrayElements; index++)
```

The value of index is incremented from "0" (zero) and used as a pointer to the array element, and then sets each element of the *relayPin* array as an *OUTPUT*. When the value of *index* increments and is no longer less than the number of elements in the array, the loop terminates.

We initialize the *inputInd* and the *pttInput* pins next. We use the last step in the sequence, a *digitalWrite()* to the input pin which may seem counter-intuitive but is most useful. By setting the input pin HIGH with the *digitalWrite()*, we enable an internal pullup resistor in the Digispark on the input pin, eliminating the need for an additional external pullup resistor.

The *void loop()* section of the program applies the associated on and off time delay to each relay in the sequence specified by the *relayPin[]* and *relayPinOff[]* arrays. We use the same debounce function that was used in the Real Time Clock/Timer sketch from Chapter 4. While the debounced *pttInput* pin is TRUE, we execute the *for* loop as we did above, incrementing through each element of the *onDelay* and *relayPin* arrays and following the specified delay, setting each pin HIGH, turning on the relay.

When *pttInput* goes FALSE, a second *for* loop increments through each element of the *offDelay[]* and *relayPinOff[]* arrays, and following the specified delay, sets each pin LOW, turning off each relay.

Modifying the Sequence Order and Delay Time

The sequencer program is designed to allow the user to easily modify the order in which the relays turn on and off, as well as the time delay between each step. There are only four lines of code that you might need to modify:

```
int relayPin[] = {0,1,2,4};           // order to sequence relays on
int onDelay[] = {40,50,50,50};        // ON times in sequence

int relayPinOff[] = {4,2,1,0};        // order to sequence relays off
int offDelay[] = {40,50,50,50};       // OFF times in sequence
```

The first line sets the order in which the relays are energized. It is highly likely that you would not need to modify this line, but you can if need be. (Remember that the initializer list uses logical pins, not the relay number.) The second line sets the delay in milliseconds between each relay being energized. Of course, the first entry is the time delay from when the input is detected. We have set this up so that each delay is individually configurable and corresponds to the relay order from the previous line.

So why is the first delay 40 ms instead of 50? The first entry should be reduced by 10 ms as dictated by the debounce delay. The debounce function introduces the time delay defined by *DEBOUNCEDELAY*, in this case, 10 ms. The debounce delay must elapse before the first entry of the on or off time delay arrays are executed. In other words, if the first entry is set to 0 (zero), the first relay is energized 10 ms following the detection of an input because the debounce delay does not provide a return until 10 ms have elapsed from a change of state of the input. As a result, since we want the default delay to be 50 ms, we must subtract the 10 ms from the first entry, hence a value of 40 ms.

The third and fourth lines are used to configure the order and time delays for de-energizing the relays. The default configuration de-energizes the relays in the opposite order to which they were energized. This can be changed to meet any special requirements you may encounter in your application. As we have shown in the default configuration, the first entry includes the 10 ms debounce delay.

Configuring the Jumpers for Different Situations

The sequencer is designed to fit into a lot of different configurations. In a typical ham station, we might encounter 28 V antenna relays as well as 12 V or even 115 VAC relays. We might

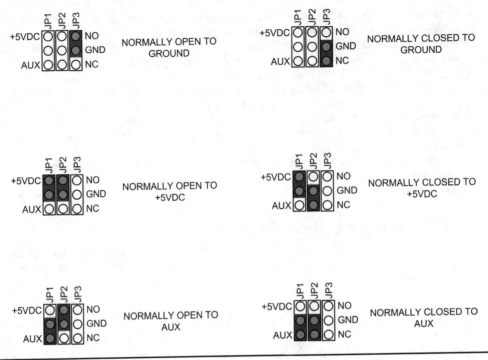

FIGURE 12-10 Jumper configurations.

have components that require a ground to be actuated or require switched 12 VDC input. The jumpers allow us to configure the output of the sequencer to accommodate many different requirements.

The RCA output jacks are connected to the center contact of each relay and each relay has a normally open (NO) and normally closed (NC) position. There are three possible selections for the NO or NC position: 1) Ground, 2) +12 VDC, or 3) AUX. AUX selects an arbitrary voltage applied to the AUX input on the rear panel. This could be 28 VDC (or AC). We would not recommend switching a 115 VAC circuit (even though the relays are rated for 1 A at 115 VAC), but would recommend using an external relay to switch the higher voltage. Figure 12-10 shows the six possible jumper positions for each output. Note that only the pins for relay K1 are shown. The other three relay's jumper pins are identical.

Modifying the Relay Shield from Chapter 11

To use the relay shield from Chapter 11, we make several simple additions. The additions consist of adding an input connection to the Arduino with which to trigger the sequencer. We have added another set of screw terminals for input interface, mounting a two-position screw terminal in the 40-pin header area of the shield. (The new terminal is near the right edge of the board shown in Figure 12-11.) One pin of the screw terminal is wired to ground and the other pin is wired to Digital IO pin 10. The new input detects a normally open circuit that would close to ground, such

Chapter 12: A Flexible Sequencer   241

FIGURE 12-11   New screw terminal added to the relay shield (far right).

as a push-to-talk (PTT) switch on a microphone. You can also add the LED used to indicate that an input is present. You can use Digital IO pin 13 to drive this LED. Be sure to add a 1K resistor in series with the LED. You may also wish to mount the LEDs remotely, rather than on the relay shield as it is shown in Chapter 11.

Figure 12-11 shows the location of the new screw terminals (compare to Figure 11-3). As this version is built as a shield, it is operated with an Arduino. The modified relay shield is shown being used with an Arduino Duemilanove, but the shield can be used with virtually any Arduino or compatible. **CAUTION**: When using the Mega shield with most Arduinos, be sure to cover the USB connector with insulating material, preferably two layers of black electrical tape. The Mega shield (and several others we might add) allows you to mount components in the area directly above the USB connector leading to the possibility of a short circuit if a component lead or wiring comes in contact with the USB connector.

Alternate Listing for the Relay Shield Sequencer

We have provided an alternate source code version of the sequencer program sketch for use with the modified relay shield and an Arduino. The alternate version is shown in Listing 12-2. The modifications are limited to changing pin assignments from what were used the Digispark version to those that are used with the Chapter 11 relay shield and an Arduino. Follow the same testing procedure for the modified relay shield that was used for testing the Digispark sequencer. Once you are satisfied that the shield is wired correctly you can proceed with loading the program in Listing 12-2 and continue with testing.

```
/******
Flexible 4-Port Sequencer  Version 1.0

Designed by Dennis Kidder W6DQ 5 Jan 2014

This sketch when used with the purpose-built hardware described in the
book, "Arduino Projects for Ham Radio," will provide a completely
configurable sequencer.  Two sets of two arrays configure the ON sequence
the OFF sequence and the delays between each step.  This allows the ON
and OFF sequence order to different along with the ON and OFF times for each
step through sequence.
******/

/******
The following four lines of code allow the user to configure the ON and OFF
sequence and delay times for each step.  The default configuration turns
the relays on in the sequence "1,2,3,4" (represented logically by digital
IO pins 0, 1, 2, and 4) and to turn them off in the reverse order "4,3,2,1."

The delay between steps is expressed in milliseconds and the default is 50.
The order of the values in the array corresponds to the order in which the
relays are turned ON or OFF.
******/

#define DEBOUNCEDELAY  10

int relayPin[] = {6,7,8,9};              // order to sequence relays on
int onDelay[] = {50,50,50,50};           // ON times in sequence

int relayPinOff[] = {9,8,7,6};           // order to sequence relays off
int offDelay[] = {50,50,50,50};          // OFF times in sequence

int inputInd = 13;                       // pin 13 drives the input indicator LED
int pttInput = 10;                       // pin 5 is the PTT input

int arrayElements = (sizeof(relayPin) / sizeof(relayPin[0])); //  elements

void setup()
{
  for(int index = 0; index < arrayElements; index++)
  {
    pinMode(relayPin[index], OUTPUT);
    pinMode(inputInd, OUTPUT);
    pinMode(pttInput, INPUT);

    digitalWrite(pttInput, HIGH);
  }
}

void loop()
{
```

LISTING 12-2 Sequencer program for Arduino.

```
    if (debounce(pttInput) == true) {
      digitalWrite(inputInd, HIGH);
      for(int index = 0; index < arrayElements; index++)
      {
        delay(onDelay[index]);
        digitalWrite(relayPin[index], HIGH);
      }
    }
    else {
      digitalWrite(inputInd, LOW);
      for (int index = 0; index < arrayElements; index++)
      {
        delay(offDelay[index]);
        digitalWrite(relayPinOff[index], LOW);
      }
    }
}

/*****
  This function is a pretty standard debounce function so we can determine
  that the switch has really been pressed.

  Parameter list:
    int pin              the pin that registers the switch state

  Return value:
    boolean              true if the switch was pressed, false otherwise
*****/
boolean debounce(int pin)
{
  boolean currentState;
  boolean previousState;
  int i;

  previousState = digitalRead(pin);
  for (i = 0; i < DEBOUNCEDELAY; i++) {
    delay(1);      // small delay
    currentState = digitalRead(pin);   // Read it now
    if (currentState != previousState)
    {
      i = 0;
      previousState = currentState;
    }
  }
  if (currentState == LOW)
    return true;
  else
    return false;
}
```

LISTING 12-2   Sequencer program for Arduino. *(continued)*

Conclusion

In this chapter we have constructed a device that can now be integrated into the user's station. It is a matter of determining the switching requirements of the attached devices and configuring the sequencer's jumper pins to match those requirements. One thing that has not been covered in this chapter is dealing with RF interference. It is always possible that stray RF signals can enter the sequencer's enclosure and cause indeterminate results. The aluminum enclosure we used does a pretty good job of shielding the contents, but in the event that RF does get in, then you might try adding a small capacitor across each input and output, and power connectors (.01 μF at 50 V should suffice). Other measures include reviewing the grounding on the station, using well-shielded RCA patch cables, and possibly adding clamp-on ferrite EMI suppressors to the cables. The combination of clamp-on ferrite and capacitor on each cable and jack should take care of any issues you may encounter.

Last but not least, this design is made to be expandable. If you should need more than four outputs, it is quite easy to add additional relays, up to six, without having to add another hex driver. Of course, if you need more relays, you cannot use the Digispark[1] version because of the limit number of I/O line. In such cases, you will have to use the Arduino shield version. Adding relays means adding elements to the arrays, making sure they are added in equal numbers to all of the affect arrays.

We are certain that you will discover new applications for the sequencer in your radio shack. It's up to you to explore the possibilities!

---

[1]Digistump has just come out with Digispark Pro, which ups the I/O pin count to 14 and increases the program space to 14 kb. There are some other enhancements, too, yet the size is still about the same.

CHAPTER **13**

Rotator Controller

One of Dennis's many interests in ham radio is operating "weak signal" VHF and UHF (V/UHF). And, as many V/UHF operators do, he does get involved in contesting. It is fun and when conditions are good, it can be downright exciting! Working a station across the country on 6 meters is an amazing thing when the conditions are good. HF is fun, but VHF, UHF, and microwave can be, for lack of a word, amazing! Working a station 600 miles away on 10 GHz SSB can change your life!

But, this book isn't about weak signal operating. Still, the project discussed in this chapter can enhance the ease of operating a station with any directional antenna on a rotator, and especially those stations that have multiple directional antennas on multiple masts or towers.

FIGURE 13-1   The Arduino Rotator Controller shown with a Yaesu G-800DXA Rotator.

Dennis's V/UHF station is set up to operate on many bands and the antennas are located on three different towers. If you do any V/UHF contesting, you know that many of the stations you work during the contest are set up the same way, on multiple bands. Each contact you make on a different band with the same station counts as points, so you always attempt to work that station on each band while you have them, and then QSY to the next band you both have.

When you have three towers with a plethora of antennas, it can be a handful to move all of those antennas in the direction of the station you want to work. Tuning, talking, logging, switching bands, and moving multiple antennas all at once can be a handful all right. Heck, it can be a handful with just one rotator to get the antenna around in the right direction.

This project automates at least part of that process. It provides your rotator with preset antenna headings that can be used at the touch of a button. It allows you to set an arbitrary heading, press a button, and the rotator turns to that heading for you.

If this sounds good to you, read on!

The Arduino Antenna Rotator Controller

The controller is designed so that it may be used with most of the popular antenna rotators. There are some rotators, of course, that won't work with this project, but they are few. Most rotators use a dual-winding motor to turn the antenna mast (one winding for each direction) and a potentiometer to sense the direction the antenna is pointed. Some rotators also include a solenoid-actuated brake to prevent the system from "windmilling" in a breeze. Still others use a single motor and reverse the polarity to change direction. Some rotators require a relay to handle the current of the motor switches. Another allows for driving the control head with open collector transistors, while even providing a 0–5 VDC voltage for azimuth, as well as an interface for controlling the speed of the motor.

Our controller doesn't replace your existing control box. Instead, the controller automates the control of the rotator, supplementing the existing control box. Ideally, we would like the Arduino Rotator Controller to be totally external to the rotator control box. In the case of the Yaesu rotators, this is possible as there are external control interfaces provided by the manufacturer. However, some rotators require that connections be made internally to the rotator control box.

But, we give fair warning at this point. ***Antenna rotator control boxes contain lethal voltages. Extreme care is required when implementing this project.*** Your control box has 115 VAC line voltage, as well as lower, high-current voltages present. Never open the control box unless you know exactly what you are doing and what you will encounter. We repeat, 115 VAC can be lethal. Never work on the box when it is plugged into the wall socket. Some connections to the Arduino Rotator Controller are made on the rear apron connector of the control box. If you understand and follow the instructions we provide, you will be successful.

In general, the type of rotator control box that this works with is any type that has two lever switches, one for clockwise and one for counterclockwise rotation; a lever switch that operates a brake (most rotators these days have a brake but older ones may not); and a panel meter that indicates beam heading. This encompasses a wide variety of rotators and controllers.

Supported Rotators

This project supports a number of different rotators from different manufacturers. Table 13-1 gives a list of the rotators the project supports. In general, most rotators provide the interfaces that are easily controlled or measured externally. Most rotators use a potentiometer as a voltage divider to determine the heading of the antenna. It is a simple matter to read the voltage across the potentiometer and then calculate the heading. Many rotators use dual-winding motors, as mentioned earlier, but a few, like those made by Yaesu, use a single motor and swap the supply

Mfg	Model	External Control	Heading	Connection
CDE, HyGain, Telex, MFJ	TR-44	Relay, see text	0–25 VDC	Screw terminals
	Ham-M			Screw terminals
	Ham-2	Relay	0–13 VDC	Screw terminals
	Ham-IV			Screw terminals
	T2X			Screw terminals
Yaesu	G-800DXA	Low level	0–4.5 VDC	Mini-DIN 6
	G-1000DXA			Mini-DIN 6
	G-2800DXA			Mini-DIN 6
	G-800SDX			Internal header
	G-1000SDX			Internal header

TABLE 13-1 Supported Rotators

voltage polarity to change direction. Fortunately, most of the Yaesu rotators include a connector for external control that is very easy to interface. It provides control of direction and speed as well as providing a voltage proportional to the heading.

The Arduino Rotator Controller uses two shields from previous chapters. We use the relay shield (with the addition of a header connector), and the panel meter shield. We added a new board that is the control panel for the controller. The control panel is not built on a shield but is assembled on the same style prototyping board we used for the Sequencer in Chapter 12 and is connected to the Arduino shield "stack" through a ribbon cable. The control panel contains a digital shaft encoder to set headings, switches to control functions, and an LCD for display of headings and functions.

Relay Shield

We use the Relay Shield from Chapter 11 with some consideration of parts to be used. The original relay shield was generic and used 5 V relays capable of switching up to an Amp or two. In order to switch the motor windings and brake of the rotator, the relay needs to be able to switch higher current. There is no reason why the relay shield as built in Chapter 11 couldn't be used with external relays to switch the higher current as described in that chapter. However, we decided to use a SPDT 12 VDC relay with contacts rated at least 10 A to simplify the wiring. The relay shield from Chapter 11 can be used to control the Yaesu rotators that include external control. The control voltage and current are quite low and are within the limits of the relays used in Chapter 11.

For rotators not listed in Table 13-1, don't despair. With schematics and a little bit of sleuthing, just about any rotator is controllable with the design presented in this chapter. The voltage range for the heading may vary in a different range. In that case, it is a simple matter of changing the scaling resistors as described in Chapter 5 when using the panel meter as a voltmeter.

Unfortunately, such design decisions can have undesirable consequences. In this case, while searching for a high current 5 VDC relay of a size where four would actually fit on a Mega prototyping shield, we could not find any that met our requirements (at a reasonable price or where we didn't have to purchase 5000 of them). Fortunately, we were able to find many 12 VDC relays. So, we settled on a Potter and Brumfield RT314012F because: 1), the contacts are rated at 16 A, 2), they are fairly inexpensive (under $3 each), and 3), they are readily available (Mouser Electronics carries them in stock at www.mouser.com). Four of these relays along with the driver circuit, four LEDs with current limiting resistors, and four screw terminals all fit nicely on a Mega

High Current Relay Shield				
Ref	Description	Part No	Mfg	Source
DS1-4	LED			
J1	10-pin header, 0.1-in. centers (2x5)			
JP1-4	2-pin header, 0.1-in. centers (2x4)			
K1-4	Relay, 12 VDC, DPDT, 16A/240 VAC, 12 VDC coil	RT314012F	Potter and Brumfield	Mouser
R1-4	470 Ω, ¼ W, 5%			
TS1-4	Screw Terminal, 3 position, 0.2 in. centers			
U1	ULN2068B, Quad Darlington switch		STI	Jameco
	Prototyping shield, Arduino Mega			

TABLE 13-2 Parts List for the High Current Relay Shield

prototyping shield, albeit tightly. There was even room left over to add the four jumpers to disable the relays as we included on the original relay shield in Chapter 11. Table 13-2 lists the parts needed for the high current relay shield. You may ask why there are four relays on the shield when only three are needed to control a rotator. The intent was for the new shield to duplicate the function of the relay shield in Chapter 11, which uses four relays. The fourth relay can be left off if you so desire, or it can be used to control another function. For instance, in the case of the early CDE rotators like the TR-44 and Ham-M rotators, power was only applied when the lever switch was moved by the operator. This means that the meter is also inoperative. We show a switch being added to turn on the power when the rotator and controller are in use. You could easily use the fourth relay to turn the rotator control head on instead of the switch, activating the relay when the Arduino is powered up.

Referring to the relay shield schematic shown in Figure 13-2, you should note that the 12 VDC supply for the relays is taken from the V_{IN} pin on the shield. This means that you need to provide 12 VDC to the Arduino power plug. The Arduino provides the supply voltage (minus one diode drop; about 0.6 V) as well as 5 VDC to the shield. Note also that only one of four relay driver circuits is shown in the schematic.

One additional change from the original relay shield must be made. The original shield uses a 75492 Hex driver. As time has passed from that original build, we ran into a problem obtaining these parts. As a result, we found a newer, better, part to use. These devices should be available for some time and are fairly inexpensive (about $2 online). We are now using a quad Darlington switch from STI designated ULN2068B. The ULN2068B can handle higher current, uses a Darlington output stage with an additional input driver (for operation with 5 V TTL logic levels), and even includes built-in suppressor diodes on the output for use with inductive loads such as a relay. This has reduced our part count on the shield by four diodes. The ULN2068B is an open collector output device, just like the 75492, so substitution is easy, except that the pinouts are quite different and it is in a 16-pin DIP package. Figure 13-3 shows the parts layout for the high current relay shield and Figure 13-4 shows the wiring. The completed shield is shown in Figure 13-5.

Panel Meter Shield

There is one consideration for the panel meter shield. Remember that the meter shield measures 0–1 mA DC. The potentiometers used in the rotators don't necessarily generate a 0–1 mA signal. Rather, they are typically a higher voltage that is scaled to be read by a 0–1 mA meter on the

Chapter 13: Rotator Controller   249

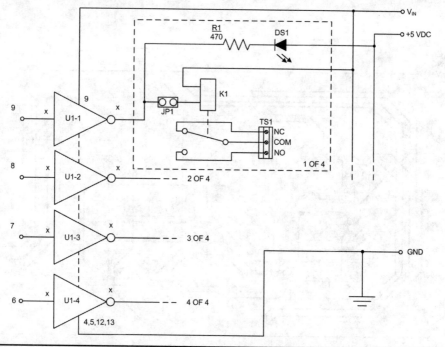

FIGURE 13-2   Updated relay shield with higher current capability.

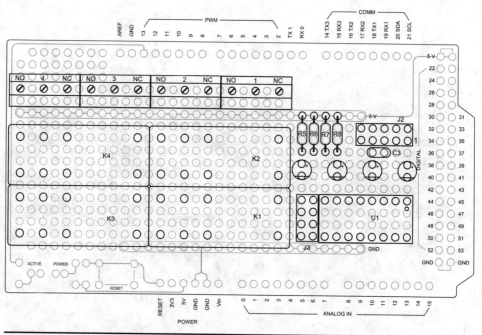

FIGURE 13-3   High current relay shield parts layout.

250 Arduino Projects for Amateur Radio

1 - SDA
2 - GND
3 - SCL
4 - GND
5 - GND
6 - NC
7 - V+
8 - V+
9 - CCW
10- CW

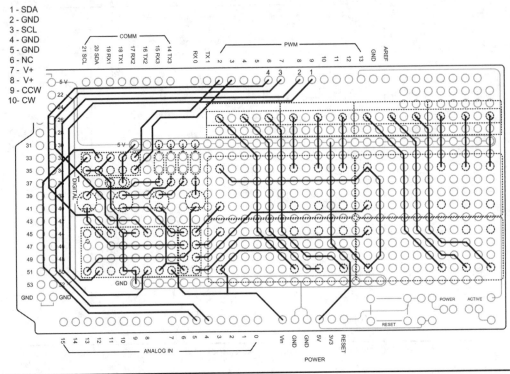

Figure 13-4 High current relay shield wiring.

Figure 13-5 Completed high current relay shield.

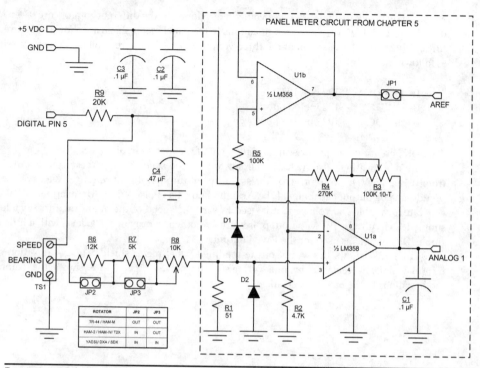

FIGURE 13-6 Schematic diagram of modifications to panel meter shield.

control box. Chapter 5 included a section on how to add scaling circuits to our panel meter and we make use of that information here. For the rotators we tested, it is a simple matter to measure the voltage from the heading potentiometer at the rotator control box and use that measurement to determine the scaling circuit values needed. Figure 13-6 shows the additional circuitry added to the Chapter 5 panel meter. The additional parts required are listed in Table 13-3. Note that we decided to use an LM358 Dual opamp rather than the LM324 used in Chapter 5 for the panel meter. Either is acceptable for this application. The design uses two opamps so the dual packge, LM358, is a good choice.

Chapter 5, Panel Meter, Added Components				
Ref	Description	Part No	Mfg	Source
C2, C3	Capacitor, 0.1 µF, 50 V Monolithic			Jameco, Radio Shack, others
C4	Capacitor, 0.47 µF, 50 V Monolithic			
JP1, 2, 3	2-pin header, 0.1-in. spacing			
R6	12 kΩ, ¼ W, 5%			
R7	5 kΩ, ¼ W, 5%			Jameco, Radio Shack, others
R8	10 kΩ, ¼ W, 5%			
R9	20 kΩ, ¼ W, 5%			
TS1	Screw Terminal, 3 position, 0.2-in. centers			

TABLE 13-3 Parts List for the Modified Panel Meter Shield

Referring back to Table 13-1, the voltages developed in the different rotators are 0–4.5 VDC, 0–13 VDC, and 0–25 VDC. The modifications to the panel meter shield are shown in Figure 13-6. JP1 and JP2 allow the selection of the three voltage ranges. The three voltage ranges are selected as follows:

0–4.5 VDC	JP2 out, JP3 in
0–13 VDC	JP2 in, JP3 out
0–25 VDC	JP2 out, JP3 out

The 10K potentiometer provides fine adjustment of the meter circuit.

The modifications also include a circuit to vary the speed on the Yaesu rotators. An input from 0.5 to 4.5 VDC to the rotator varies the speed through its entire range. We have used the pseudo-analog output capability of the Arduino known as Pulse Width Modulation, or PWM, to generate a variable DC voltage that is used to vary the speed of the Yaesu rotators. By adding a simple RC circuit, the PWM output pulses are smoothed to provide a DC level with a little bit of ripple. Digital pin 5 is used for the PWM output.

Figure 13-7 shows the wiring of the modified panel meter shield. The two-position screw terminal block is replaced by one with three terminals. Figure 13-8 shows the completed modifications to the panel meter shield.

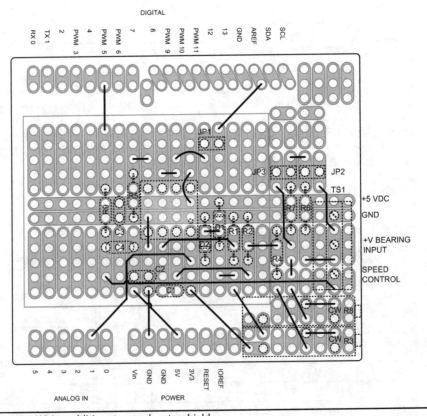

FIGURE 13-7 Wiring additions to panel meter shield.

FIGURE 13-8 Modifications completed to the panel meter shield.

The Control Panel

Some of our projects require more digital IO pins than some Arduinos can provide. For example, when you add the panel meter shields along with the relay shields and expect to have some controls to provide additional inputs, you find that there may not be enough pins! We want to use the relay shield to control an antenna rotator. This requires the use of the panel meter to sense the rotator's actual position, an LCD to display information, a shaft encoder to set headings and several switches that are used for other functions, such as storing and recalling a preset heading. Because we don't have enough IO pins to do all of this, we use a device called a "port expander." A port expander device uses the Arduino's built-in I2C bus to provide additional digital IO ports. We used the I2C bus previously in Chapter 4 to "talk" to the real time clock breakout board.

Recall that the I2C interface requires only two digital IO pins; one for clock and one for data. Using a port expander and adding power and ground, we can provide up to 16 external digital IO ports with only four wires! We constructed an external "control panel," providing an LCD display, five switches, and a shaft encoder (also with another switch), which is connected to the Arduino through the I2C interface. Note that the shaft encoder itself does not connect through the I2C bus but the shaft encoder switch does. We use hardware interrupts with the shaft encoder so it connects to the shield on two separate wires and connects to Arduino digital interrupt pins 2 and 3.

As mentioned earlier, the control panel board is built on a prototyping board rather than a shield. It is designed to be interconnected to the Arduino shield "stack" via a 10-pin ribbon cable so that the board may be mounted remotely. The control panel board uses a Microchip MCP23017 16-port expander chip with an I2C interface. The MCP23017 provides all the digital IO ports needed to support an LCD along with six momentary contact switches (five switches and the switch on the shaft encoder). The six momentary contact switches are used for controlling multiple functions. The schematic for the board is shown in Figure 13-9 and the parts required are listed in Table 13-4.

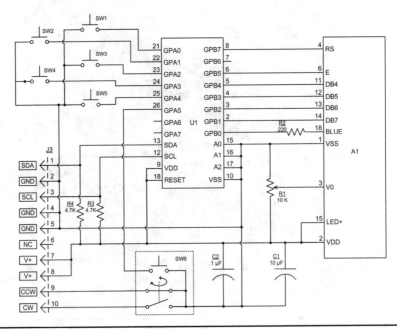

FIGURE 13-9 Schematic of the control panel board.

Front Panel Board

Ref	Description	Part No	Mfg	Source
A1	LCD, 2x16, HD44780		various	eBay, Maker Shed, Adafruit, SparkFun, Arduino Mall, etc.
C1	10 µF, Electrolytic, 15 VDC			eBay, Jameco, etc.
C2	0.1 µF, 50 VDC monolithic cap			eBay, Jameco, etc.
J3	10-pin header, 0.1-in. centers (2x5)			eBay, Jameco, etc.
R1	Pot, 10 kΩ, PC mount			eBay, Jameco, etc.
R2	220 Ω, ¼ W, 5%			eBay, Jameco, etc.
R3, 4	4, 7 kΩ, ¼ W, 5%			eBay, Jameco, etc.
SW1-5	Switch, momentary contact PC mount			eBay
SW6	Shaft encoder, with momentary switch		various	eBay, Suntek Store
U1	Port Expander, 16 port, I2C IF	MCP23017	MCP	eBay
	Socket, 28-pin DIP			eBay, Jameco
Misc parts				
	Protoyping PC board, 2 cm × 8 cm			eBay
P1, P3	10-pin plug, ribbon cable (2x5) qty: 2	FC-10P		eBay
Alternate		09185107803	Harting	Mouser, Allied
	Cable, ribbon, 6 conductor 0.05-in. pitch, 12-in. length		3M, others	eBay, old computer cables

TABLE 13-4 Parts List for the Control Panel Board

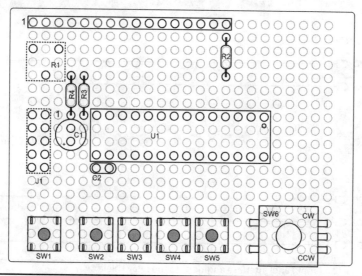

FIGURE 13-10 Control panel parts layout.

The control panel parts layout is shown in Figure 13-10. Note that there are components mounted on both sides of the board. The side shown in Figure 13-10 is the side facing the user and we refer to that as the front of the board. Components on the back of the board are shown with dashed lines.

The wiring diagram, shown in Figure 13-11, shows the back side of the board and the two components mounted on this side: the pot, R1, and the header plug, J1.

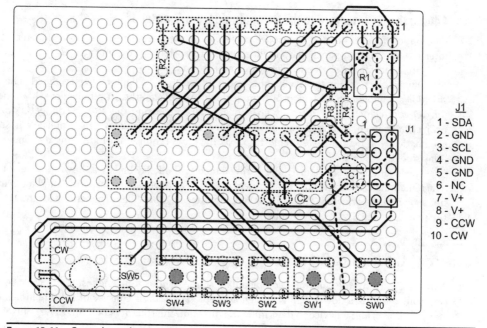

J1
1 - SDA
2 - GND
3 - SCL
4 - GND
5 - GND
6 - NC
7 - V+
8 - V+
9 - CCW
10 - CW

FIGURE 13-11 Control panel wiring diagram.

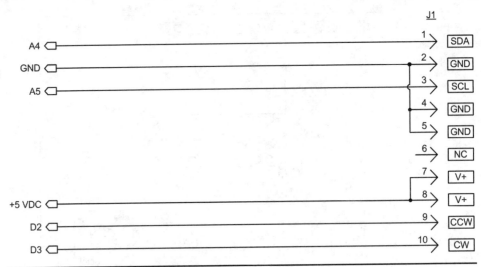

FIGURE 13-12 Front Panel interface additions to the relay shield.

Chapter 11, Relay Shield, Added Components				
Ref	Description	Part No	Mfg	Source
	Relay shield from Chapter 11			
J1	10-pin header, 0.1-in. centers (2x5)			

TABLE 13-5 Parts List for the Modifications to the Chapter 11 Relay Shield

Adding the I2C Interface to the Relay Shield from Chapter 11

One option for the Rotator Controller project is to use the relay shield from Chapter 11. An interface for the front panel board is added to the relay shield, providing connections for the I2C bus, interrupts for the shaft encoder, power, and ground.

Our I2C bus connector is a 10-pin header added to the Chapter 11 relay shield. The relay shield is assembled on a Mega prototyping shield. We placed the 10-pin header where the 40-pin IO connector would normally go. Since we are not using the 40-pin connector for this project, it seemed an ideal location. The added wiring for the front panel interface is shown schematically in Figure 13-12. Table 13-5 lists the parts required for modifying the Chapter 11 relay shield.

The location of connector J1 and wiring added to the Relay Shield is shown in Figure 13-13. The remaining circuitry for the Relay Shield is exactly as depicted in Chapter 11, and for clarity is not shown in the drawing. The Control Panel interface, J1, provides connections for the I2C bus, the shaft encoder, 5 VDC, and ground.

Connecting the Rotator Controller

Now that you have the controller completed, it is time to hook it up and test it out. Most rotators and control heads follow a similar design concept. The rotator consists of a bidirectional drive motor, a brake, and a device to determine the direction the rotator is pointing. The control head

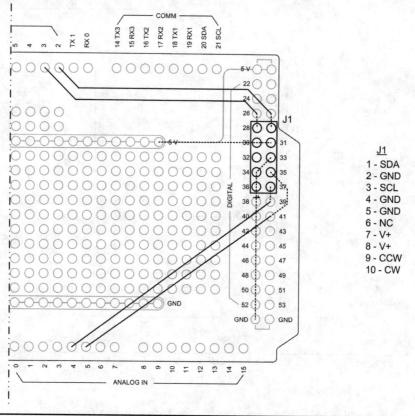

FIGURE 13-13 Adding the control panel interface to the Chapter 12 relay shield.

contains a power source, a direction indicator, and switches to control braking and direction of rotation. You need to have the schematic diagrams of your rotator and control head to connect the Rotator Controller. The following sections describe how to connect to the rotators we have listed. The methods described are adaptable to rotators that are not listed. Refer to the schematics of your rotator if it is not one of the models we list. One of the schematics presented here is likely similar to yours.

Early Cornell-Dublier Electronics (CDE) Models

Early CDE models include the TR-44 and Ham-M rotators. The TR-44 and Ham-M present a unique problem. As shown in Figure 13-14, the only time voltage is applied to the heading potentiometer is when the control levers are used. This led to a design change in later control heads where there is a power switch allowing the voltage to always be applied to the potentiometer. If you are using a TR-44 or Ham-M rotator and would like to use this design for control, the rotator's control box must be modified to always apply the voltage to the pot when it is in use, just as in the later CDE, HyGain, Telex, and MFJ models. Making these modifications requires opening the control box, adding some wiring, and in the case of the TR-44, an extra switch. (The fourth relay can be used here as well but will require an addition to the software to energize it.)

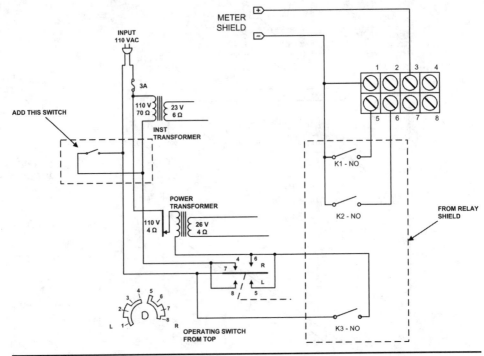

FIGURE 13-14 Interconnections for early CDE rotators: TR-44 and Ham-M.

Controlling the CDE rotators requires the higher current relays used on the Relay Shield described in this chapter or when using the Relay Shield from Chapter 11, external higher-current relays must be used.

Later Models from HyGain, Telex, and MFJ

These include the Ham 2, Ham IV, and Tailtwister T2X rotators. As shown in Figure 13-15, one set of connections is made inside the control box, placing relay K3 in parallel with the brake switch. The brake switch is the center switch of the three lever switches and is identified as S3 on their rotator schematic. The direction control relays are connected between the ground screw terminal (1) and the direction control screw terminals for clockwise/right (5) and counterclockwise/left (6). The voltmeter input (for heading) is connected between the ground terminal (1) and terminal 3.

Controlling this set of rotators also requires the higher current relays used on the relay shield described in this chapter, or when using the relay shield from Chapter 11, external higher-current relays must be used.

Recent models of the HyGain control heads use an 8-pin Cinch-Jones connector, rather than a barrier terminal strip, to connect to the rotator. In this case, we used a chassis-mounted connector on the rear apron to make the internal connections to the control head. The connector used is a Molex 03-09-1061 receptacle, rated at 11 A per pin, more than enough to operate our rotator. If you do use a different connector, make sure that it has ample current rating for your application. Figure 13-16 shows our installation on a HyGain TailTwister rotator control head.

Chapter 13: Rotator Controller 259

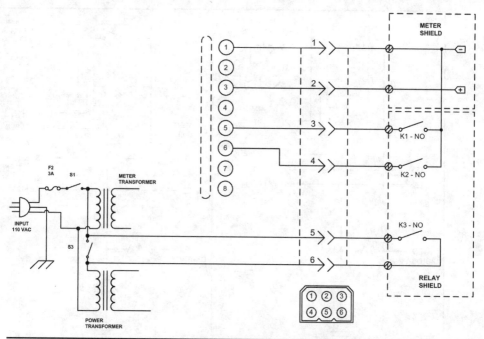

FIGURE 13-15   Interconnections for later rotators: Ham-2, Ham-IV, and T2X.

FIGURE 13-16   Installing the rotator controller interface on a late model HyGain TailTwister.

Yaesu Models G-800SDX/DXA, G-1000SDX/DXA, and G-2800DXA

These Yaesu models include a rear-panel connection for external control. The models with the "DXA" suffix use a 6-pin Mini-DIN connector for connections. The connections are described in Figure 13-17A. External connections to the rotators with the SDX suffix can be accessed through

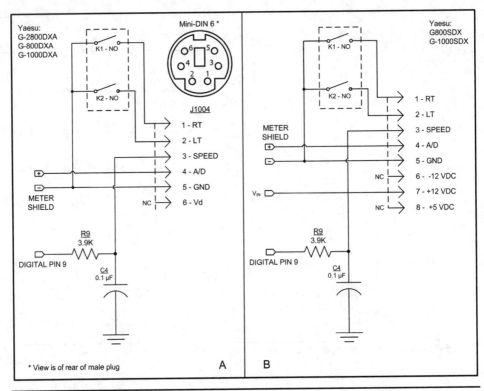

FIGURE 13-17   Interconnections for Yaesu DXA and SDX rotators.

a grommeted hole on the rear panel. There is an 8-pin header on the circuit board with connections as described in Figure 13-17B.

In all of these examples of Yaesu rotators, the relay shield from Chapter 11 is sufficient for control without changing the relays.

Software

Actually, there are two pieces of software for this project. The first piece of software is the Arduino sketch used to control the hardware and the rotator. The second piece of software is a Windows program that allows you to enter in your QTH, and, once entered, it determines your longitude and latitude. From there, you can enter any of 330 amateur radio call sign prefixes from around the world and the program gives you a beam heading for that prefix. You can also print the list out so you can use the heading specific to a given location without the use of a computer.

Arduino Beam Heading Software

Listing 13-1 presents the code for the software that controls the hardware. There are a large number of *#define* preprocessor directives, many of which are shared with the DDS VFO discussed in Chapter 16. The reason is because the VFO and the rotator control use virtually the same control shield for the LCD display, encoder, and switches. We concentrate on the directives that are germane to this chapter.

Chapter 13: Rotator Controller

The first important *#define* is for INITIALIZEEEPROM, which appears around line 16 in Listing 13-1. The first time you run the software, you should compile this code with the comment characters (//) removed so the directive is compiled into the program. The reason for uncommenting the line is so the following two statement blocks are compiled into the program.

```
#ifdef INITIALIZEEEPROM
int headings[]      = {      90,    120,    150,    180,    125,    75};
#else
int headings[6];
#endif
// Later in setup() you'll find:
#ifdef INITIALIZEEEPROM
  for (int i = 0; i < 6; i++)
    writeEEPROMRecord(headings[i], i);
#endif
```

By uncommenting the *#define* preprocessor line, you initialize several EEPROM values that you can use to test the software. Once you have run the program one time, comment out the *#define* INITIALIZEEEPROM line so that it is no longer compiled into the program. When you run the program a second time, you can change our test values to whatever you wish them to be. We explain the process a little later in this chapter.

Also found in the *setup()* function is the statement:

```
typeRotator = TEST;             // *** User sets appropriate type here ***
```

Again, the *TEST* symbolic constant can be used to test the code in general terms. You should, however, replace the *TEST* symbolic constant with the name of your rotator (e.g., *CDE*, *HYGAIN*) from the list of rotators described in the list of symbolic constants labeled Rotator Type. The rest of the *setup()* function code initializes a number of working variables for the program and should look pretty familiar to you by now.

```
/*
Jack Purdum and Dennis Kidder:
   Rev 1.4:   May 3, 2014
*/

#include <pins.h>
#include <avr/pgmspace.h>
#include <EEPROM.h>
#include <LiquidCrystal.h>
#include <rotary.h>
#include <Wire.h>
#include <Adafruit_MCP23017.h>
#include <DDSLCDShield.h>
#include "W6DQ_DDS_VFO.h"

//#define INITIALIZEEEPROM 1       // Uncomment 1st you run the software
```

LISTING **13-1**   Rotator control software.

```
//#define DEBUG 1

#define HowBigIsArray(x)        (sizeof(x) / sizeof(x[0]))

#define ROTATORSENSORPIN    A1
#define PWMPIN              5
#define BRAKERELAYPIN       7
#define CCWRELAYPIN         8
#define CWRELAYPIN          9

#define READDELAY           500     // Enough time tfor beam heading reading
#define ABORTDELAY          2000    // Delay after abort
#define DELTAMOVE           10      // A minimum movement in rotating the beam
#define BRAKEDELAY          100     // Wait for brake release
#define DUTYCYCLE           240
#define YAESUCROSSOVER      816     // analog value when past 360 degrees
#define YAESUNORTH          408     // linear scale for North

//===================== Rotator Type =========================================
#define TEST                0
#define CDE                 1
#define HYGAIN              2
#define TELEX               3
#define MFJ                 4
#define YAESU               5

//===================== Switch configuration =================================
#define STARTSWITCH         1       // SW1 on board
#define MEMORYSWITCH1       2       // SW2
#define MEMORYSWITCH2       4       // SW3
#define MEMORYSWITCH3       8       // SW4
#define MEMORYSWITCH4       16      // SW5
#define MEMORYSTORE         32      // SW6, encoder switch

//===================== LCD configuration ====================================
#define LCDCOLS             16
#define LCDROWS             2
#define CURRENTHEADINGOFFSET 8      // Place current heading in column 8, row 0
#define DEGREESOFFSET       12      // Place degrees in column 12, row 1
#define GOTOOFFSET          8       // Place the ending heading

//===================== Encoder configuration ================================
#define ENCODERASLEEP       0       // The EEPROM encoder is just tuning
#define ENCODERREVIEW       1       // Scroll thru user-stored EEPROM freq
#define ENCODERSTORE        2       // Store a new frequency in user EEPROM
#define ENCODERCHANGED      3

//===================== Beam configuration ===================================
#define BRAKERELEASEDELAY   5000    // Five second delay after brake release
```

LISTING 13-1   Rotator control software. *(continued)*

```c
#define BEAMSTABLE           0
#define BEAMINTRANSIT        1
#define BEAMABORT            2

#define MOVEINITIALIZATION   0    // Prepare to move beam
#define MOVEINMOTION         2    // The beam is in motion
#define MOVEATREST           5    // Movement has stopped, beam is locked

#define MOVEBEAMUP           1
#define MOVEBEAMNOWHERE      0
#define MOVEBEAMDOWN        -1

//==================== Heading configuration ==============================
#define CURRENTHEADING       0    // Holds the current heading being used
#define MEMORYHEADING1       1    // Each tied to its switch
#define MEMORYHEADING2       2
#define MEMORYHEADING3       3
#define MEMORYHEADING4       4
#define MANUALMOVE           5

#define ROTATIONIDLE         6
#define ROTATIONCW           7
#define ROTATIONCCW          8
#define ABORTROTATOR         9

char statusField[][10] = {"CUR ", "MEM1", "MEM2", "MEM3", "MEM4", "MAN ",
                          "IDLE", "CW  ", "CCW ", "ABRT"};

#ifdef INITIALIZEEEPROM
int headings[]      = {90, 130, 200, 280, 325};   // Trial headings, run once
#else
int headings[5];                                   // Used rest of the time
#endif

int typeRotator;
int statusFieldIndex;
int currentGoto;
int beamStatus;           // What's the beam doing?
int beamDirection;
int beamMoveStatus;       // Where we are in move sequence
int linearGoto;
int linearCurrent;
int memorySelected;
int initialHeading;

int mapCurrent, mapGoto;     // Used to find rotations in CalculateDirection()
int convertCurrent, convertGoto;
int lastDisplay;
```

LISTING 13-1   Rotator control software. (*continued*)

```
volatile int encoderStatus;
volatile int currentHeading;
volatile int oldHeading;


Rotary myRotary = Rotary(2,3);            // Rotary encoder pins.  Must be
                                          // interrupt pins.
DDSLCDShield lcd = DDSLCDShield();

//==================== setup() function ===================================
void setup() {
  //analogReference(EXTERNAL);

   //These two lines are run once to establish some test records in EEPROM.
#ifdef INITIALIZEEEPROM
  for (int i = 0; i < HowBigIsArray(headings); i++)
    writeEEPROMRecord(headings[i], i);
#endif

#ifdef DEBUG
Serial.begin(115200);         // Turn on serial link
#endif

  pinMode(ROTATORSENSORPIN, INPUT);
  typeRotator = YAESU;            // *** User sets appropriate type here ***
  //typeRotator = HYGAIN;         // *** User sets appropriate type here ***

  ReadBeamHeading();              // Get the initial beam heading
  MapLinearCurrentToCurrentHeading();   // Fills in linearCurrent

  currentGoto = currentHeading = initialHeading;

  ReadEEPROMArray();              // Read the headings[] array from EEPROM

  lcd.begin(LCDCOLS, LCDROWS);    // Display LCD
  InitializeDisplay();

  beamStatus = BEAMSTABLE;
  beamDirection = MOVEBEAMNOWHERE;
  beamMoveStatus = MOVEATREST;
  encoderStatus = ENCODERASLEEP;

  memorySelected = MEMORYHEADING1;
  statusFieldIndex = ROTATIONIDLE;
  DisplayCurrentHeading();                // Show all three fields
  DisplayDegrees(ROTATIONIDLE);
  DisplayGoto();


  cli();                                  // Disable interrupts
  PCICR |= (1 << PCIE2);                  // ISR for encoder
```

LISTING 13-1   Rotator control software. (*continued*)

```
    PCMSK2 |= (1 << PCINT18) | (1 << PCINT19);
    sei();                                    // Enable interrupts
}


//===================== loop() function ====================================
void loop() {
  uint8_t buttons = lcd.readButtons();       // See if they pressed a button
  delay(BOUNCEDELAY);

  ReadBeamHeading();
  if (lastDisplay != currentHeading) {       // Done to reduce flicker
    DisplayCurrentHeading();                 // Show we're making progress...
    lastDisplay = currentHeading;
  }

  if (buttons > 0 && beamMoveStatus == MOVEINMOTION) {  // Abort beam move
    AbortBeamMove();       // Beam is in transit, but they pressed a button.
    buttons = 0;
  }
                           // The beam is being moved...
  if (beamStatus == BEAMINTRANSIT && beamMoveStatus == MOVEINMOTION) {
    ReadBeamHeading();                  // Sets linearCurrent by reading beam sensor
    currentHeading += beamDirection;
    MoveBeam();
    if (currentHeading > 359)           // Handle wrap-around
      currentHeading = 0;
    if (currentHeading < 0)
      currentHeading = 359;
    if (currentHeading == currentGoto) {  // Are we there??
      StopRotator();                      // Stop rotator sequence
      beamStatus = BEAMSTABLE;            // We're done
      beamMoveStatus = MOVEATREST;
      encoderStatus = ENCODERASLEEP;      // Encoder not being moved
      DisplayDegrees(ROTATIONIDLE);       // Rotator idle
    }
  }
                   // Did they pressed a button, but not in ABORT state?
  if (buttons && beamStatus != BEAMABORT) {
    ButtonPush(buttons);                   // Yep...
  }

  if (encoderStatus == ENCODERCHANGED) {
    statusFieldIndex = MANUALMOVE;
    DisplayDegrees(statusFieldIndex);      // Setting the heading manually
    DisplayCurrentHeading();               // Show we're making progress...
    DisplayGoto();
    encoderStatus = ENCODERASLEEP;
  }
```

LISTING 13-1   Rotator control software. (*continued*)

```
      // Figure out display fields...
  if (oldHeading != currentHeading && encoderStatus != ENCODERASLEEP) {
    oldHeading = currentHeading;
    DisplayCurrentHeading();
  }
  if (oldHeading != currentGoto && beamStatus != BEAMSTABLE) {
    oldHeading = currentGoto;
    DisplayGoto();
  }

}

//=================== Internal functions ================================

/*****
  This function takes the value of currentGoto and maps it to linearGoto

  Argument list:
    void

  Return value:
    void
*****/
void SetLinearGoto()
{
  linearCurrent = analogRead(ROTATORSENSORPIN);  // Where's beam at start?
  MapLinearCurrentToCurrentHeading();
        // Map linear to current...
  currentHeading = initialHeading;
        // ...and copy
  if (typeRotator != YAESU) {
    if (currentGoto >= 180)           // Degrees to rotator coordinates
      linearGoto = map(currentGoto, 180, 360, 0, 512);
    else
      linearGoto = map(currentGoto, 0, 179, 512, 1024);
  } else {
    if (currentHeading == currentGoto)
      return;
                                      // Beam is in overlap, but Goto in range
    if (linearCurrent >= YAESUCROSSOVER && currentGoto <= 270) {
      linearGoto = map(currentGoto, 180, 270, YAESUCROSSOVER, 1023);
    } else {
      if (linearCurrent >= YAESUCROSSOVER && currentGoto > 270) {
        linearGoto = map(currentGoto, 0, 360, 0, YAESUNORTH);
      } else                          // Beam between 180-360, Goto > 180
      if (linearCurrent <= YAESUNORTH && currentGoto >= 180) {
        linearGoto = map(currentGoto, 180, 360, 0, YAESUNORTH);
                                      // Beam < 360, Goto > 0 && < 180
      } else if (linearCurrent <= YAESUNORTH && currentGoto <= 180) {
        linearGoto = map(currentGoto, 0, 180, YAESUNORTH, YAESUCROSSOVER);
```

LISTING 13-1   Rotator control software. (*continued*)

```
                                            // Beam > 0, Goto > 0 && < 180
        } else if (linearCurrent > YAESUNORTH && currentGoto <= 180) {
          linearGoto = map(currentGoto, 0, 180, YAESUNORTH, YAESUCROSSOVER);
        } else if (linearCurrent >= YAESUNORTH && currentGoto >= 180) {
          linearGoto = map(currentGoto, 180, 450, YAESUCROSSOVER, 1023);
        }
      }
    }
  }
  DisplayGoto();
}
/*****
  This function reads the current heading for the beam and assigns that
  reading into linearCurrent

  Argument list:
    void

  Return value:
    void
*****/
void MapLinearCurrentToCurrentHeading()
{
  if (typeRotator == YAESU) {
    initialHeading = map(linearCurrent, 0, 1023, -180, 270);
    if ( initialHeading < 360 ){
      initialHeading = map(initialHeading, -180, -1, 180, 359);
    }
    if (initialHeading > 359) {
      initialHeading = initialHeading - 360;
    }
  } else {
    initialHeading = map(linearCurrent, 0, 1023, -180, 180);
    if (initialHeading <= 0 ){
      initialHeading = map(initialHeading, -180, 0, 180, 360);
    }
  }
}
/*****
  This function reads the current heading for the beam and assigns that
  reading into linearCurrent

  Argument list:
    void

  Return value:
    void
*****/
void ReadBeamHeading()
{
  linearCurrent = analogRead(ROTATORSENSORPIN);      // Where beam is pointing
```

LISTING 13-1 Rotator control software. (*continued*)

```
    MapLinearCurrentToCurrentHeading();        // Map for display update
    currentHeading = initialHeading;
}
/*****
    This function is used to map the rotator sensor readings to the current
    beam heading to the new beam heading.

    Argument list:
      void

    Return value:
      void
*****/
void CalculateDirection()
{
  if (linearCurrent > linearGoto)
    beamDirection = MOVEBEAMDOWN;
  else
    beamDirection = MOVEBEAMUP;

  beamStatus = BEAMINTRANSIT;
  beamMoveStatus = MOVEINITIALIZATION;

}


/*****
    This function is used to prepare the rotator for moving the beam.

    Argument list:
      void

    Return value:
      void
*****/
void InitializeMoveSequence()
{
  if (typeRotator != YAESU) {              // Release brake for all but Yaesu
    digitalWrite(BRAKERELAYPIN, HIGH);     // Release brake
    delay(BRAKEDELAY);
  }
  if (beamDirection < 0) {                 // Move CCW by energizing relay
    digitalWrite(CCWRELAYPIN, HIGH);       // Start moving CCW
    digitalWrite(CWRELAYPIN, LOW);         // Just playing it safe
  } else {
    digitalWrite(CWRELAYPIN, HIGH);        // Start moving CW
    digitalWrite(CCWRELAYPIN, LOW);        // Still playing it safe
  }
  beamMoveStatus = MOVEINMOTION;
}
```

LISTING 13-1 Rotator control software. *(continued)*

```c
/*****
    This function is used with the Yaesu rotator, since it wants to slow as it
    approaches the correct heading.

    Argument list:
        void

    Return value:
        void
*****/
void MoveBeam()
{
  int headingDifference;
  if (typeRotator == YAESU) {       // How much left to turn?
      headingDifference = abs(currentHeading - currentGoto);

      switch (headingDifference) {
        case 10:                                        // Coast to a stop...
        case 9:
          analogWrite(PWMPIN, 240);
          break;
        case 8:
        case 7:
          analogWrite(PWMPIN, 184);
          break;
        case 6:
        case 5:
          analogWrite(PWMPIN, 128);
          break;
        case 4:
        case 3:
          analogWrite(PWMPIN, 72);
          break;
        case 2:
        case 1:
        case 0:                                         // Turn off relays
          if (beamDirection < 0) {                      // Move CCW
            digitalWrite(CCWRELAYPIN, LOW);
          } else {
            digitalWrite(CWRELAYPIN, LOW);    //Move CW
          }
          currentHeading = currentGoto;
          break;
        default:              // All other values, move fast
          analogWrite(PWMPIN, DUTYCYCLE);
          break;
      }
  }
  beamMoveStatus = MOVEINMOTION;
}
```

LISTING 13-1   Rotator control software. (*continued*)

```
/*****
  This function is used to stop the rotator according to the needs of the
  specific rotator being used.

  Argument list:
    void

  Return value:
    void
*****/
void StopRotator()
{
  digitalWrite(CWRELAYPIN, LOW);              // Won't hurt to turn them both off
  digitalWrite(CCWRELAYPIN, LOW);
  if (typeRotator != YAESU) {
    delay(BRAKEDELAY);                        // Wait for it to settle down...
    digitalWrite(BRAKERELAYPIN, HIGH);        // Lock the beam in place
  }
  oldHeading = currentHeading = currentGoto;  // All the same now
  beamStatus = BEAMSTABLE;
  beamMoveStatus = MOVEATREST;
  encoderStatus = ENCODERASLEEP;
  DisplayCurrentHeading();                    // We're there
  statusFieldIndex = ROTATIONIDLE;            // Setting the heading manually
  DisplayDegrees(ROTATIONIDLE);               // Rotator idle

}
/*****
  This function is called if the beam is being moved but the user presses any
  other switch. This is interpreted as an Abort of the beam movement.

  Argument list:
    void

  Return value:
    void
*****/
void Abort()
{
  DisplayDegrees(ABORTROTATOR);
  oldHeading = currentGoto = currentHeading;
  if (beamStatus == BEAMINTRANSIT) {
    StopRotator();
  }
  beamStatus = BEAMABORT;
  beamMoveStatus == MOVEATREST;
  DisplayCurrentHeading();
  DisplayGoto();
}
```

LISTING 13-1 Rotator control software. (*continued*)

```c
/*****
  This function updates the affected fields when a memory stored heading is
  invoked

  Argument list:
    void

  Return value:
    void
*****/
void UpdateMemoryHeadingDisplays()
{
  DisplayDegrees(statusFieldIndex);
  currentGoto = headings[statusFieldIndex];
  DisplayGoto();
  DisplayCurrentHeading();
}
/*****
  This function is used to display the current heading

  Argument list:
    void

  Return value:
    void
*****/
void DisplayCurrentHeading()
{
  if (currentHeading < 0)
    currentHeading = 360 + currentHeading;
  lcd.setCursor(CURRENTHEADINGOFFSET, 0);
  lcd.print("   ");
  if (currentHeading > 99)                              // Right-justify in field
    lcd.setCursor(CURRENTHEADINGOFFSET, 0);
  else if (currentHeading > 9)
     lcd.setCursor(CURRENTHEADINGOFFSET + 1, 0);
    else
     lcd.setCursor(CURRENTHEADINGOFFSET + 2, 0);
  lcd.print(currentHeading);
}

/*****
  This function is used to display the degrees

  Argument list:
    int which     the index into the statusField array

  Return value:
    void
*****/
```

LISTING 13-1 Rotator control software. (*continued*)

```
void DisplayDegrees(int which)
{
  lcd.setCursor(DEGREESOFFSET, 1);
  lcd.print("   ");
  lcd.setCursor(DEGREESOFFSET, 1);
  lcd.print(statusField[which]);
}

/*****
  This function is used to display the Go To heading

  Argument list:
    void

  Return value:
    void
*****/
void DisplayGoto()
{
  if (currentGoto < 0)
    currentGoto = 360 + currentGoto;

  lcd.setCursor(GOTOOFFSET, 1);
  lcd.print("   ");
  if (currentGoto > 99)                          // Right-justify in field
    lcd.setCursor(GOTOOFFSET, 1);
  else if (currentGoto > 9)
      lcd.setCursor(GOTOOFFSET + 1, 1);
    else
      lcd.setCursor(GOTOOFFSET + 2, 1);
  lcd.print(currentGoto);
}


/*****
  This function is used to set the LCD display fields when the system is
  first powered up.

  Argument list:
    void

  Return value:
    void
*****/
void InitializeDisplay()
{
  lcd.setCursor(0,0);
  lcd.print("CURRENT:     DEG");
  lcd.setCursor(0, 1);
  lcd.print("   GO TO:");
}
```

LISTING 13-1   Rotator control software. *(continued)*

```
/*****
  This function returns the stored heading.

  Argument list:
    int which        which of the 5 available headings to return

  Return value:
    int              the heading or -1 on error
*****/
int ReadHeading(int which)
{

  if (which >= HowBigIsArray(headings))
    return -1;
  return headings[which];
}

/*****
  This function writes the current contents of the headings[] array to EEPROM.

  Argument list:
    void

  Return value:
    void

*****/
void ReadEEPROMArray()
{
  int i;

  for (i = 0; i < HowBigIsArray(headings); i++) {
    headings[i] = readEEPROMRecord(i);
  }
}
/*****
  This method is used to read a record from EEPROM. Each record is 4 bytes
  (sizeof(unsigned long)) and is used to calculate where to read from EEPROM.

  Argument list:
    int record          the record to be read. While tuning, it is record 0

  Return value:
    unsigned long       the value of the record,

  CAUTION:  Record 0 is the current frequency while tuning, etc. Record 1 is
            the number of stored frequencies the user has set. Therefore, the
            stored frequencies list starts with record 23.
*****/
```

LISTING 13-1 Rotator control software. (*continued*)

```
unsigned int readEEPROMRecord(int record)
{
  int offset;
  union {
    byte array[2];
    int val;
  } myUnion;

  offset = record * sizeof(int);

  myUnion.array[0] = EEPROM.read(offset);
  myUnion.array[1] = EEPROM.read(offset + 1);

  return myUnion.val;
}


/*****
  This method is write the heading data to EEPROM

  Argument list:
    int heading        the heading to save
    int record         the record where to store the heading

  Return value:
    void
*****/
void writeEEPROMRecord(int heading, int record)
{
  int offset;
  union {
    byte array[2];
    int val;
  } myUnion;
                       // No negative headings or greater than 360 degrees
  if (heading < 0 || heading > 360) {
    Error();
    return;
  }
  myUnion.val = heading;
  offset = record * sizeof(int);

  EEPROM.write(offset, myUnion.array[0]);
  EEPROM.write(offset + 1, myUnion.array[1]);
}


/*****
  This function is a general purpose error indicator and can be used for any
  error condition, being it improper input or electronic malfunction. The
  word ERR is display in the DEG position on the LCD display.
```

LISTING 13-1   Rotator control software. *(continued)*

```
  Argument list:
    void

  Return value:
    void
*****/
void Error()
{
  lcd.setCursor(DEGREESOFFSET, 1);
  lcd.print("ERR");
  delay(2000);
}

/*****
  This function is called if the user has started a beam move but decides to
  cancel it.

  Argument list:
    void

  Return value:
    void
*****/
void AbortBeamMove()
{
    Abort();
    delay(ABORTDELAY);                     // Let them see ABRT message
    beamStatus = BEAMSTABLE;               // We're done
    beamMoveStatus = MOVEATREST;
    statusFieldIndex = ROTATIONIDLE;
    encoderStatus = ENCODERASLEEP;
    oldHeading = currentHeading = currentGoto;   // All the same now
    DisplayCurrentHeading();                     // We're there
    DisplayDegrees(ROTATIONIDLE);

}

/*****
  This is the Interrupt Service Routine that is executed anytime the rotary
  encoded is turned.

  Argument list:
    PCINT2_vect     the vector for the interrupt routine

  Return value:
    void
*****/
ISR(PCINT2_vect) {
  unsigned char result = myRotary.process();
  if (beamStatus == BEAMINTRANSIT)     // No change during transit; only abort
```

LISTING 13-1 Rotator control software. (*continued*)

```
      return;
    switch (result) {
      case 0:                              // Nothing done...
        return;

      case DIR_CCW:                        // Turning Clockwise, higher frequencies
        currentGoto++;
        if (currentGoto >= 360)
          currentGoto = 0;
        break;

      case DIR_CW:                         // Turning Counter-Clockwise, lower frequencies
        currentGoto--;
        if (currentGoto < 0)
          currentGoto = 359;
         break;
      default:
         // Should never be here
        break;
    }
    encoderStatus = ENCODERCHANGED;   // They are rotating the encoder shaft
}

/*****
  This method is used to process the various button presses by the user.

  Argument list:
    uint8_t button    Will be either BANDUPMODE or BANDDOWNMODE

  Return value:
    void
*****/
void ButtonPush(uint8_t button)
{
  switch (button) {
    case STARTSWITCH:                 // Start rotator sequence
      ReadBeamHeading();              // Sets linearCurrent
      SetLinearGoto();                // Sets linearGoto from degrees to linear
      if (currentHeading == currentGoto) {
        DisplayDegrees(ROTATIONIDLE);
        break;
      }
      CalculateDirection();           // Which way to turn beam
      InitializeMoveSequence();
      if (beamDirection < 0)
        DisplayDegrees(ROTATIONCCW);
      else
        DisplayDegrees(ROTATIONCW);
      break;

    case MEMORYSWITCH1:   // Store current SWITCH heading in Memory location 1
```

LISTING 13-1   Rotator control software. *(continued)*

```
          memorySelected = statusFieldIndex = MEMORYHEADING1;
          UpdateMemoryHeadingDisplays();
          break;

      case MEMORYSWITCH2:  // Store current heading in Memory location 2
          memorySelected = statusFieldIndex = MEMORYHEADING2;
          UpdateMemoryHeadingDisplays();
          break;

      case MEMORYSWITCH3:  // Store current heading in Memory location 3
          memorySelected = statusFieldIndex = MEMORYHEADING3;
          UpdateMemoryHeadingDisplays();
          break;

      case MEMORYSWITCH4:  // Store current heading in Memory location 4
          memorySelected = statusFieldIndex = MEMORYHEADING4;
          UpdateMemoryHeadingDisplays();
          break;

      case MEMORYSTORE:                          // Encoder switch
        if (memorySelected < 1 || memorySelected > 4) {
          Error();
          break;
        }
        headings[memorySelected] = currentGoto;            // Assign new heading
        writeEEPROMRecord(currentGoto, memorySelected);    // Write it to EEPROM
        beamStatus = BEAMSTABLE;                           // Beam is not moving
        break;

      default:      // ERROR
        Error();
        break;
    }
  }
}
```

LISTING 13-1   Rotator control software. (*continued*)

Moving the Beam

The *loop()* function constantly polls the controller shield hardware looking for a switch press or encoder rotation. The *readButtons()* method of the *lcd* object monitors the switches and, when one is sensed, *ButtonPush()* is called. There are five switches on the controller shield plus the encoder switch, as shown in Figure 13-10. SW1 is used to activate the rotator and move to a new heading. The next four switches (SW2-SW5) control the EEPROM memory addresses (MEM1-MEM2) where you store the headings that you wish to save for later recall. For example, if you look near the top of Listing 13-1, you'll find the statement:

```
int headings[]     = {    90,    120,    150,    180,    125};
```

which are the sample headings that get stored when you first compile the program. (That's what the symbolic constant *INITIALIZEEEPROM* was all about.) Figure 13-18 shows the three fields

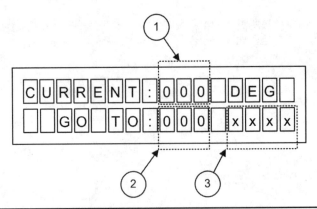

FIGURE 13-18   The three LCD display fields.

used on the LCD display. The second time you run the program, you will see the LCD display with the CURRENT and GO TO fields filled in with the number 90 and the DEG field displays CUR. The interpretation of the three fields is that the beam is currently on a heading of 90 degrees, the position to move the beam is to heading 90 degrees, and that is the current position. If you look at the *headings[]* array, you can see that its value is 90, which is the number you are seeing in Fields 1 and 2 in Figure 13-18. Therefore, *headings[0]* corresponds to the current heading.

If you press switch SW2, that switch is tied to MEM1, or memory heading 1 as stored in EEPROM. If you look at the test headings we wrote to *headings[]* the first time you ran the program, you can see that *headings[1]* has the value 120. The instant you press SW2, the GO TO field is filled in with 120 and the DEG field has MEM1 displayed immediately below it in Field 3. The CURRENT field still shows 90. The reason is because that's the current heading of the beam.

If you wish to move the beam from its CURRENT heading (90) and GO TO the heading stored in MEM1 (120), touch the Rotate button (SW1). As soon as the rotator brake is released and the beam starts to move, the CURRENT field starts to count up from 90 and stops counting when it reaches 120. In the sequence you just performed, you moved the beam from its current heading of 90 degrees to a new heading of 120 degrees, which you had previously stored in EEPROM at position MEM1.

When you performed the move from 90 to 120 degrees, the call to *ButtonPush()* used a *switch-case* statement block to: 1) set the proper parameters for the calls to *DisplayDegrees()*, 2) which updates the DEG Field, *DisplayGoto()*, 3) which updates the GO TO Field, and 4) call *DisplayCurrentHeading()*, which updates the CURRENT heading field. The call to *CalculateDirection()* determines the best way to reach the new heading and *MoveBeam()* controls how the beam is actually moved.

If you press switches SW3-SW5, you call up the headings stored in EEPROM at locations MEM2 (150), MEM3 (180), and MEM4 (125). The instant you touch one of the heading switches, its associated heading immediately appears in the GO TO field and the DEG field is updated to reflect the memory heading you selected (e.g., MEM3). The CURRENT field does not change, as it is always used to show the current heading of the beam. If you had touched SW4, CURRENT would still be at 120, GO TO immediately changes to 180, and DEG changes to MEM3. Touch the Rotate switch, SW1, and the beam starts to move to the new heading and the CURRENT field is updated accordingly.

Setting a New Heading

Obviously you need the ability to move the beam to more than four headings. Suppose you had previously moved the beam to MEM3 with a heading of 180 degrees. Now you want to move it to 160 degrees, which is not one of your stored headings. Simply rotate the encoder shaft counter-clockwise (CCW) and the GO TO field displays the numbers decreasing from 180 down to 160 as you rotate the encoder shaft in the CCW direction. Once you have set the GO TO field to 160, touch the Rotate switch (SW1) and the CURRENT field decreases from a heading of 180 to 160 as the beam turns to the new heading.

Storing a New Heading in EEPROM

Suppose you wish to replace one of the previously stored headings with a new one. To change any MEMn heading, press the switch associated with that memory heading. The code immediately updates the GO TO field with the heading that is currently stored at that address and the DEG field changes to display MEMn. Now rotate the encoder shaft to the new heading you wish to save as seen in the GO TO field. When the desired heading is set in the GO TO field, press the encoder shaft to engage the switch that is built into the encoder. The new heading is now stored in MEMn.

For example, suppose you wish to change MEM4 from its current value of 125 to its replacement value of 110. First, press the SW5 switch, which immediately updates the DEG field to MEM4 and the GO TO field to 125. Now rotate the encoder shaft in the CCW direction until you see 110 in the GO TO field. Now press the encoder shaft switch. The new heading is now stored in MEM4. You can verify the change by pressing some other MEMn switch, which changes the DEG and GO TO fields, and then press the MEM4 switch. You should see the new heading you just entered.

World Beam Headings

The program discussed in this section can be used to create a list of beam headings for 330 call districts around the world. There is good news and bad news associated with this program. First, the good news.

The program first asks you to enter the location of the antenna for which the headings are to be calculated. Once that location is entered, the program then can be used to determine the beam heading for any one of the 330 worldwide call districts as determined from the QTH location that was entered.

The bad news is actually twofold: 1) the program only runs on Windows, and 2) the program must be run on a system that is connected to the Internet. The reason for the first limitation is that the code was developed with Microsoft's Visual Studio using C#. The second limitation arises because the code uses the Google Maps Application Programming Interface (API) as a Web service to determine the longitude and latitude coordinates for the location you type into the program.

Finding the Coordinates for a QTH

Using the Google Maps API has a number of advantages, but the biggest in this instance is the ability to determine the exact coordinates of wherever you might choose to operate. For example, suppose you wanted to operate a Field Day station from the Wayne National Forest in southern Ohio. Even though you don't have an address *per se* for the forest, you can easily determine its coordinates using Google maps.

First, load Google maps into your browser using the https://maps.google.com URL. When the map program finishes loading, drag and center the map on the approximate location where you're going to set up your Field Day operation. For our example, when you see Wayne National Forest displayed on the map, click on your best approximation of where you will be operating.

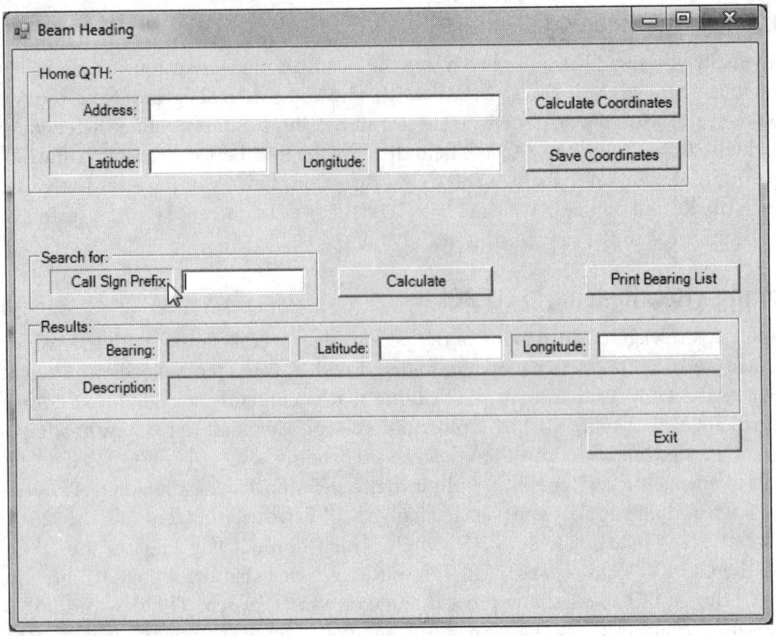

FIGURE 13-19 World beam headings initial screen.

Now right-click on that location and the address bar in the upper-left side of the display pops up with the address of the location you selected (e.g., 27515 Ohio 7, Marietta, OH). Immediately below that address are the longitude and latitude for the location you selected (39.592886, −82.202334). You can use either the address that is displayed or the exact coordinates as input into the Windows program.

Figure 13-19 shows how the World Beam Headings program appears when you first load and run it. You can enter the address or coordinates into the textboxes shown in the Home QTH group box. In Figure 13-20, we got lazy and just entered Wayne National Forest into the Address textbox and clicked the Calculate Coordinates button, and the coordinate textboxes were filled in from the coordinates supplied from the Google Maps API. If you click on the Save Coordinates button, the information you see displayed is saved to a disk file so that the next time you run the program, those coordinates are automatically filled in. This means that, if you reload the program a second time, you don't need to go through the steps to determine the coordinates used in the program.

Finding a Beam Heading

Suppose you hear the call T2B on the air and you want to set your beam to the heading for T2B. Since most people don't know where a T2 ham is operating from, you can type "T2" into the Call Sign Prefix and click the Calculate button. The display immediately changes to that shown in Figure 13-21. The program tells you that the beam heading from the Wayne National Forest to Funafuti, Tuvalu, is 268.601478829149. Because the Rotator control program doesn't have that much granularity for a beam heading, use the encoder to rotate to a value of 269 and press the Move switch. Your beam is now pointing toward Funafuti, which is the capital city of Tuvalu.

This example points out a number of important factors about the call search program. First, once you've set the Home QTH parameter, the beam heading look-up process is extremely fast. Simply enter the prefix and click Calculate. Second, in all cases, the heading is calculated for the capital city of the call area being searched. Our feeling is that there is no need to set the exact

Chapter 13: Rotator Controller 281

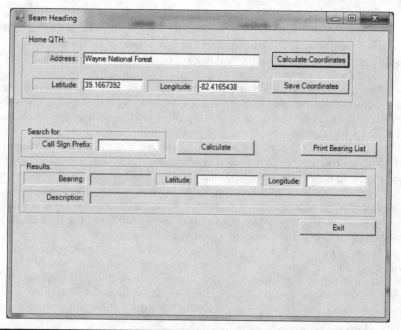

FIGURE 13-20   Display after entering Wayne National Forest and clicking Calculate Coordinates.

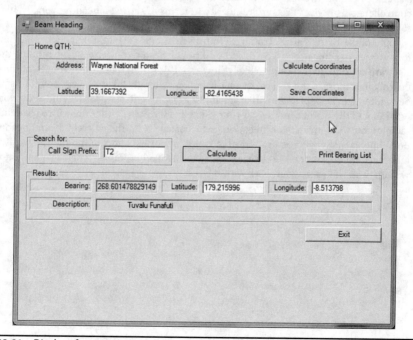

FIGURE 13-21   Display after entering a Call Sign Prefix and clicking Calculate.

coordinates for the QTH of the station you are calling. There are two exceptions to this rule. The first are call prefixes for Canada. Because of their close proximity to the United States, coordinates are based on the capital cities of the Canadian provinces. Likewise, any US call points to Washington, DC. If you have problems figuring out the approximate heading for a US call prefix, you need to spend a little more time with a map.

Tom Epperly (NS6T) has a nice web site (http://ns6t.net/azimuth/azimuth.html) that allows you to create a custom world map centered on your own QTH coordinates. Once you type in your longitude and latitude (which you obtained earlier from Google maps), the map is drawn with the compass points clearly marked on the edge of the map. You can print the map, which gives you a better idea of where the output of our World Beam Heading program is pointing your beam.

A side benefit of this example is that you now know approximately where Tuvalu is … somewhere in the Pacific. Why is this important? Well, first, is because most people don't have a clue where Tuvalu is. Second, Funafuti is a really awesome name for a city. Third, because of its location relative to the International Dateline, Tuvalu is the first country in the world to celebrate New Year's Day. You'd be surprised how often people ask: "I wonder what country is the first to celebrate New Year's Day?" Now you know, which may give you Phone-a-Friend status.

Finally, if you wish, you can click on the Print Bearing List, which prompts you to enter a file name to hold the output data. This action creates a text file that contains all 330 call prefixes, the country and capital city, and a beam heading relative to the QTH displayed on the screen at the time the list is printed. You can print this list out with any text reader and have a heading list that you can use when a computer isn't handy. The list is printed in ascending sorted order.

Conclusion

In this chapter you built a shield that can help you determine, set, and rotate an antenna beam to a specific heading. We feel that this beam control system is a little easier to use than some beam controllers plus it gives greater granularity in terms of setting a heading. Finally, it allows you to easily store up to four headings that you use often, perhaps to check into your favorite net.

Of course, there is always room for improvement. Some might find it useful to have more than four preset headings. You could add additional heading switches and modify the hardware/software accordingly. More than likely, however, this would require you to mount the switches "off board" since most of the real estate on the board is already taken. A very ambitious project would be to interface the controller with a small LCD graphics display. This would allow you to display a polar plot of the world centered on your QTH like the NS6T program, with a shaded "wedge" that radiates outward from your location in accordance to the current beam heading. Another possibility is that you have a rotator that is not on our list and you modify the hardware/software interface for that rotator. We're sure you can think of other improvements you might make. Again, we hope you will share your work with the rest of us.

CHAPTER 14

A Directional Watt and SWR Meter

When Dennis was back in engineering school, one of his passions was fast cars. Light weight, lots of horsepower; that was the ticket (...often literally!). But, a little too heavy on the throttle and all that power was wasted when the tires broke loose and would spin freely. So, what does this have to do with radio? Being a "radio guy," Dennis realized that he needed a better match between the car and the pavement; a lower SWR! One solution: wider tires!

Way back in 1840, a Russian, Moritz von Jacobi, determined that the maximum power transfer in a circuit occurred when the source resistance was equal to the load resistance. The maximum power transfer theorem, as we know it today, is also referred to as "Jacobi's Law." Jacobi was talking about DC circuits but this law applies to AC circuits just as well, although there is now a reactive component added to the equation.

In radio, we understand this to mean that there should be a good match between the source (a transmitter) and the load (an antenna); where the rubber meets the road, so to speak. When the transmitter and antenna are "matched" in impedance, we have the maximum power transfer between the transmitter and the antenna. If they are not matched, we radiate less power from the antenna, more power is wasted in the transmitter, and, like Dennis's spinning tires, the result just might be a lot of smoke!

This is where the SWR meter comes in. How do you know if you have a good match between the transmitter and antenna? One way is to use an SWR meter. SWR meters come in all shapes and sizes, and all ranges of sensitivity. For this project, we have built a meter that is scalable in size to handle different power requirements.

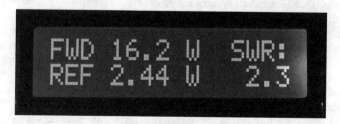

FIGURE 14-1   One of four display modes of the Directional Coupler/SWR Indicator.

283

SWR and How It Is Measured

Standing Wave Ratio, or SWR, is a measurement of the impedance match between a source and a load, in our case, a transmitter and an antenna *system*, and is expressed as a ratio, that is, "1.5:1 SWR" (read as "one point five to one"). We use the term *system* here because the load is comprised of much more than just the antenna. The load consists of the antenna, of course, but also includes the feedline and possibly an impedance matching network (i.e., an antenna tuner), connectors, switches, and the other components that comprise the actual load seen by the transmitter. To fully discuss the details of SWR is beyond the scope of this chapter. There are many good resources on the Internet and through various publications. A search on Wikipedia turns up an excellent discussion of SWR (http://en.wikipedia.org/wiki/Standing_wave_ratio). Simply stated, SWR is the ratio between the forward traveling wave on the feedline and the reflected traveling wave, the latter coming from any mismatch in the system.

Obtaining the Antenna System SWR

There are a number of ways to determine SWR either as the Voltage SWR or VSWR, or by measuring the forward and reflected *power* in the signal and deriving the SWR from their ratio. The measurement of RF power is a useful tool in troubleshooting and testing transmitters. Knowing both the SWR and the forward and reflected power is very useful as a measure of how well an antenna system is performing. There are many commercially manufactured SWR and Watt meters in the amateur market, but we chose to build one using an Arduino. There are a number of reasons we chose to build: 1) because we always like building things from scratch and maybe learning something in the process, 2) the design is tailorable to different applications, and 3) it's often less expensive to build your own equipment. For this project we targeted the QRP community by limiting the meter's maximum power handling to 20 W. However, we provide details on how to modify the design to increase the maximum power handling capability.

The SWR value is derived by measuring the forward and reflected power on the transmission line in our system. We use a very clever device called a "Stockton Bridge," named after David Stockton, G4NZQ, the chap who first came up with this configuration for a directional coupler. A typical directional coupler is depicted in Figure 14-2 as a 4-Port "Black Box." We refer to the depiction as a "Black Box" because describing the mechanism by which it works is beyond the scope of this chapter. However, we are able to accurately describe the coupler's behavior. If you are interested in learning more about directional couplers, the Web is a good source. For the sake of brevity, we refer to the Stockton Directional Coupler as merely the "coupler."

Looking at the "FORWARD" drawing in Figure 14-2, a small portion of the power entering Port A is redirected to Port D while the majority of the power is passed through to Port B. At the same time, Port C receives virtually none of the power entering Port A. The coupler is also symmetrical. If we turn the coupler around and use Port B as the input, as depicted in the "REFLECTED" drawing, most of the power entering Port B is passed through to Port A while a small portion of the power is redirected to Port C, and Port D receives none of the power entering Port B. The ratio of the input power either on Port A or B with respect to the "sampling" ports (Port D and C, respectively) is called the "coupling factor." The reduction in power between Ports A and B because of a small portion of the power being redirected is called "insertion loss." And while the actual level is very small, the amount of power redirected between Port A and C or Port B and D is known as "isolation." A fourth factor, "directivity," is the difference between the coupling factor and isolation of Port C or D. We refer to Ports C and D as the sampling ports.

It should be apparent now that the coupler we have described could be a useful device for measuring forward and reflected power. With an unknown power level applied to Port A from a

Chapter 14: A Directional Watt and SWR Meter 285

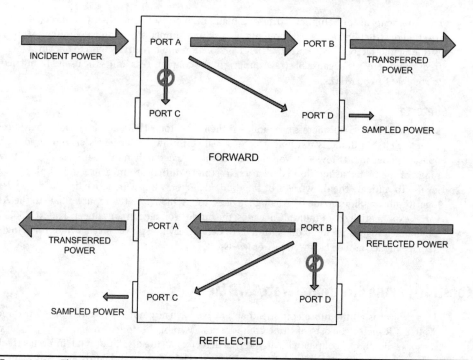

FIGURE 14-2   The Stockton Directional Coupler as a 4-Port Black Box.

transmitter, and knowing the coupling factor, we measure the power at Port D and calculate the input power at Port A. In a similar manner, we measure the power at Port C and determine the reflected power from the antenna system. From this we calculate the SWR using the following formula:

$$SWR = \frac{1+\sqrt{\frac{P_R}{P_F}}}{1-\sqrt{\frac{P_R}{P_F}}}$$

where P_F is the measured forward power and P_R is the measured reflected power.

The four factors: 1) isolation, 2) coupling factor, 3) insertion loss, and 4) directivity, all play a part in how well the coupler performs. We want to minimize insertion loss and maximize isolation while setting the coupling factor such that we have a useful output for measurement. Reducing the coupling factor means that more of the input power appears on the sampling port, which increases the insertion loss. Poor directivity means that a portion of the reflected power appears on the forward sampling port. Coupler design becomes a trade-off of the four factors.

We have designed our coupler with a nominal coupling factor of 30 dB, meaning that 1/1000 of the power on the input ports appears at the sampling ports. The sampled power level using 30 dB of coupling factor is still sufficient that we are able to measure the power level without difficulty. The insertion loss is approximately equal to the input power minus the amount appearing on the sampling port; in other words, it is negligible. We want the isolation to be high in order to maintain directivity. Lowering the directivity means that errors are introduced into the measurements at the sampling ports. We maintain high isolation (therefore directivity) through careful construction of the components we use and proper shielding to eliminate stray signals. While all four factors are frequency sensitive, isolation and directivity tend to become worse as the frequency increases because of stray coupling in the coupler. This limits the useful frequency range to roughly 1 to 60 MHz.

Detectors

Most inexpensive SWR meters use diodes as their detectors. There are drawbacks to using a diode detector. First, a diode is not going to be sensitive to low power levels encountered with QRP operation. Second, at low voltage values diodes are exceedingly nonlinear, meaning that over the range of the meter using diode detectors, it is more difficult to measure and display low power levels. To mitigate the drawbacks of the diode detector, we use an active device in our design: a logarithmic amplifier made by Analog Devices, the AD8307. As the input voltage to the AD8307 varies exponentially over multiple decades, the output voltage remains linear. The AD8307, having a dynamic range of up to 90 dB, allows us to detect very low power levels while still being able to manage higher, as well as lower, power levels.

Constructing the Directional Watt/SWR Meter

This project is a little more complex than any we have attempted up until now. Our directional Watt/SWR meter consists of three components assembled into two major parts: 1) the remote sensor (a directional coupler with an internal detector), 2) the display unit (an interface shield with two op-amp DC amplifiers), and 3) an LCD shield. The software and hardware are designed to use either the LCD shield from Chapter 3 or an alternative design shown in this chapter that includes an RGB backlit display. Using the RGB backlit display offers certain advantages that we detail later in the chapter. The components of the SWR/Wattmeter are shown in Figure 14-3 along with the

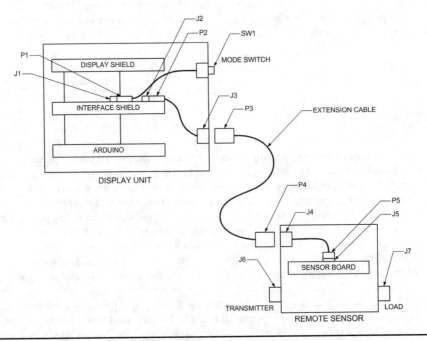

FIGURE 14-3 Components of the SWR/Wattmeter.

interconnecting cables, identifying the jack and plug designations. The remote sensor contains the directional coupler and the two log amplifiers. The display assembly contains the Arduino, an interface shield, and the LCD shield.

Construction of each of the components is described in the following sections. The overall parts list for the project is shown in Table 14-1.

Ref	Description	Part No	Mfg	Source
C1, 2, 4, 5, 6, 8, 13, 14, 15	Capacitor, .001 µF, 50 V Monolithic			Jameco, Radio Shack, others
C3, 7, 9, 10, 11, 12	Capacitor, 0.1 µF, 50 V Monolithic			
L1, L2	Fixed Inductor, 33 µH ¼ W			Mouser, eBay
R1, R2	50 Ω, 1 W 5% Carbon Comp (see text)	R-A100		Antique Electronics Supply
R3, R7	2.7K, ¼ W, 5%			Jameco, Radio Shack, others
R4, R8	82K, ¼ W, 5%			
R5, R9	50K multiturn potentiometer	3006P-204	Bourns, others	eBay, Mouser
R6, R10	33K, ¼ W, 5%			Jameco, Radio Shack
R11	100K, ¼ W, 5%			
T1, T2	Transformer, RF, 32t (see text)	FT-50-67	Amidon	Amidon, Alltronics
U1, U2	AD8307A Log Amplifier SOIC 8	AD8307A	Analog Devices	eBay
U3	LM324N, Quad Operational Amplifier	LM324N	various	Jameco, Mouser, eBay
J1	4-pin header, 0.1-in. centers, RT angle			eBay
J2	2-pin header, 0.1-in. centers, RT angle			
J3, J4	4-pin mic connector, male, chassis mount	61-624	Philmore	Intertex Electronics
J5	6-pin header, 0.1-in. centers (2x3)			
J6, J7	SO-239 UHF female bulkhead mount, "EZ Mount" or equivalent		NorCal	High Sierra Communications Products
JP1	2-pin header, 0.1-in. centers			eBay
P1	2-pin housing	22-01-2021	Molex	Mouser
	Crimp terminal for crimp housing (qty 2)	08-50-0114		
P2	4-pin housing	22-01-2041	Molex	Mouser
	Crimp terminal for crimp housing (qty 6)	08-50-0114		
P3, P4	4-pin mic connector, female plug	61-604	Philmore	Intertex Electronics
P5	6-pin plug, ribbon cable (2x3)	FC-6P		eBay
P5 (Alt)		09185067803	Harting	Mouser, Allied

TABLE 14-1 SWR/Directional Wattmeter Parts List

Ref	Description	Part No	Mfg	Source
	Cable, 4 conductor shielded, 22-24 AWG			Jameco
	Cable, ribbon, 6 conductor 0.05-in. pitch, 12-in. length		3M, others	eBay, old computer cables
	Coax, 12-in. length	RG-8X		
	Chassis box, 5¼ in. × 3 in. × 2 ⅛ in., unfinished Aluminum, "Tite-Fit"	No. 780	LMB	Mouser, Alltronics
	Chassis box, extruded Aluminum, 3.12 in. × 5.0 in. × 1.85 in.	EAS-200	LMB	Mouser, Alltronics
	Insulated standoffs ⅜ in. (2) (see text)	572-4863-01	Cambion	eBay
	Insulated spacers, ¼ in. (2)			Jameco, eBay, others
	Machine screw, 4-40 × ¼ in. (qty 12)			
	Hex nut, 4-40 (qty 10)			
	Solder lug, 4-40 (qty 12)			
	R3 Prototyping shield			
	Protoyping PC board, 2 cm × 8 cm			eBay
	SOIC 8 carrier (qty 2)			eBay
	Shielding material (Aluminum, PC board material, etc.)			eBay
	LCD Shield			Chapter 3
Alternative RGB backlit display:				
	LCD with RGB backlight			Adafruit, eBay, others
R12	10K Ω potentiometer			
R13, R15	220 Ω, ¼ W, 5% resistor			
R14	330 Ω, ¼ W, 5% resistor			
	Header pins (as required)			

TABLE 14-1 SWR/Directional Wattmeter Parts List. *(continued)*

Design and Construction of the Directional Coupler/Remote Sensor

Figure 14-4 shows the coupler/sensor design schematic. The directional coupler consists of two identical transformers, T1 and T2, with 50 Ω terminating resistors, R1 and R2. R1 and R2 reflect the characteristic impedance of the system transmission line. We chose 50 Ω as this is the most common type of coaxial cable used in amateur antenna systems. If your system uses a different impedance, the values of R1 and R2 should be adjusted accordingly.

U1 and U2 are AD8307 log amplifiers and they form the detector portion of our design. The AD8307 has a maximum input power of 15 dBm. Combining this with the coupling factor of 30 dB, the maximum permissible input power to the coupler is 15 dBm plus 30 dB or 45 dBm. This is roughly 20 W. If you wish to measure higher power levels, an attenuator must be used in

Chapter 14: A Directional Watt and SWR Meter

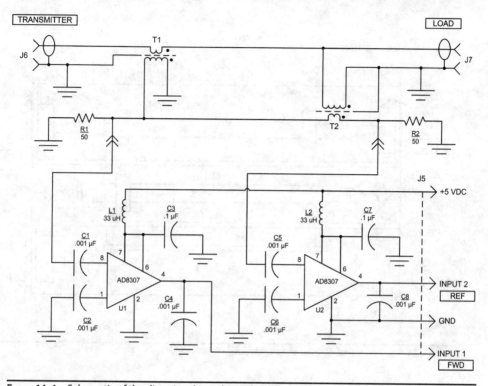

FIGURE 14-4 Schematic of the directional coupler and sensor.

front of pin 8 of the detector. Remember that the attenuator must reflect the characteristic impedance of the transmission line in the system. In the case of our system, the attenuator would have an input impedance of 50 Ω.

Our directional coupler/sensor assembly is housed in an LMB "Tite-Fit" chassis, Number 780, available from Alltronics in Santa Clara, CA (www.alltronics.com) or from Mouser Electronics in Mansfield, TX (www.mouser.com). The cost from either supplier is roughly the same, around $7.50 plus shipping and applicable taxes and shipping. The general layout of the coupler is shown in Figure 14-5.

Figure 14-6 shows how the holes for the coax connectors and interface cable connector were placed on the ends of the chassis. We used a ⅝-in. Greenlee chassis punch to make the holes for the connectors. An alternative to a chassis punch is a ⅝-in. twist drill or a stepped drill, but the chassis must be very securely held down to prevent the drill from "walking." A good technique for drilling holes in a chassis like this is to make up a paper template and tape it to the chassis. Punch the drill locations using a center punch and then remove the paper.

The coupler uses two lengths of RG-8X. Cut the coax into two identical pieces per the dimensions as shown in Figure 14-7. The shield braid is exposed at one end only. Take a 3-in. piece of hookup wire and make three wraps around the exposed braid on the coax and then solder the connection. Transformers T1 and T2 are each wound on an Amidon FT-50-67 Ferrite core. The ferrite cores

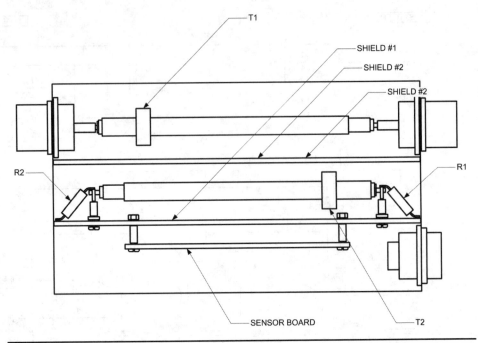

FIGURE 14-5 Directional coupler chassis layout.

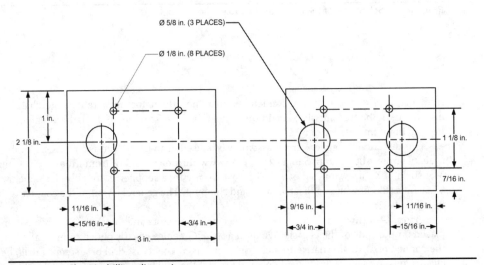

FIGURE 14-6 Chassis drilling dimensions.

are available from Alltronics. Alternatively, the ferrite cores are available directly from Amidon (www.amidon.com).

Cut a 3-ft length of AWG24 enamel coated wire. Tightly wind 32 turns on the ferrite cores and then twist the free ends of the enameled wire together. Each time you pass the wire through the core counts as one turn. Spread the windings out as evenly as possible. The completed

Chapter 14: A Directional Watt and SWR Meter 291

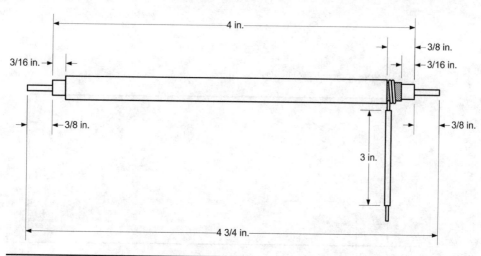

FIGURE 14-7   Preparing the coax for the coupler.

transformers should look like Figure 14-8. Slip one core over each the two coax pieces you have prepared. It takes some pressure to push the core over the coax as it is a tight fit. The finished transformer core and coax should look like Figure 14-8.

The coupler includes two shields made from copperclad circuit board that are cut to 5 in. by 1½ in. We used this material because it is convenient to work with, but you might use other materials such as aluminum. The shields are shown in Figure 14-5 separating the two pieces of coax and the sensor board. Make sure, if you use doubled-sided copperclad board, that the two sides are connected together. We wrapped small wires around the edges and soldered them in place. Not having both sides connected did cause some confusing results early in the build. Floating grounds are not a good thing around RF.

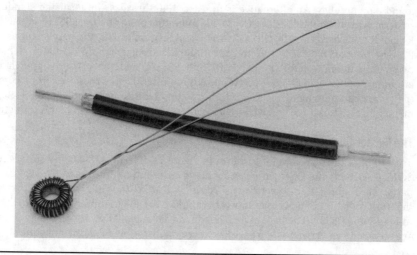

FIGURE 14-8   Ferrite cores after winding and the coaxial transformer.

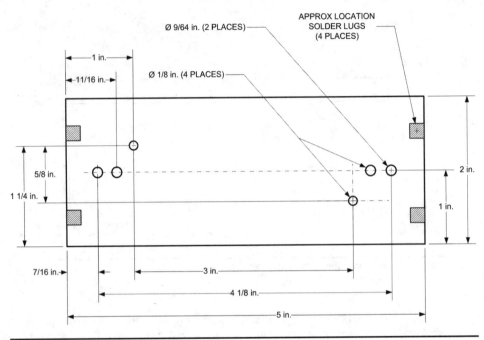

FIGURE 14-9 Drilling guide for the shields.

A drilling guide is shown in Figure 14-9. Drill holes in one shield to allow the wires from the transformer cores to pass through to the next "compartment." We used Teflon tubing to insulate the wires from the shield. Drill holes in the second shield to mount the sensor board (where the holes are placed depends on the choice of prototyping board used for the sensor board). Drill two mounting holes for the insulated standoffs used to hold the second coax assembly. We used a Cambion part for our coupler but these standoffs are difficult to find. We found ours on eBay. An alternative is to use a 4-40 thread, ⅜-in. insulated standoff with two solderlugs to attach the coax and to provide mounting points for the 50 Ω terminating resistors, R1 and R2. We used two 100 Ω ½ W carbon composition resistors in parallel for each of R1 and R2. The resistors we used came from Antique Electronics Supply located in Tempe, Arizona (www.tubesandmore.com). Resistors R1 and R2 are soldered to the copperclad board for grounding as shown in Figure 14-5. While it is not essential that these be carbon comp resistors, they *must* be noninductive. Carbon film resistors tend to be noninductive. Metal oxide and metal film resistors are trimmed with a laser-cut spiral making them highly inductive.

The Sensor Board

The layout of the sensor board is shown in Figure 14-10. The AD8307 is a surface mount 8-pin SOIC ("SOIC" is the acronym given to the "Small Outline Integrated Circuit" surface mount parts). The 8-pin SOIC is sometimes called an "SOIC-8" part. While the SOIC-8 part is very small, the truth is that it is much easier to work with than you might think. We used an SOIC-8 carrier to hold the AD8307. The bare carrier board and mounted AD8307 are shown in Figure 14-11. We obtained the carrier board from eBay by searching for "SOIC-8 carrier board."

Chapter 14: A Directional Watt and SWR Meter

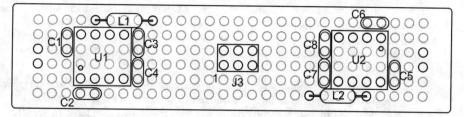

TOP VIEW

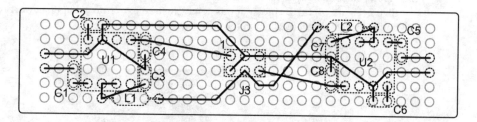

BOTTOM VIEW

FIGURE 14-10 Sensor board layout and wiring.

FIGURE 14-11 SOIC 8 carrier board and headers.

Soldering SOIC Components

Soldering the SOIC part to the carrier is easy if you follow the steps outlined below. Dennis has Dave Glawson, WA6CGR, to thank for teaching him this technique many, many years ago. Prior to Dave's instruction, Dennis shied away from projects using surface mount parts (SMT).

Today, he doesn't give it a second thought. Hopefully, this technique is as beneficial to you as it was to him. Once you lose the fear of SMT, a whole new world of project possibilities opens up! There are two cautions though: 1) make sure that you have pin 1 on the SOIC package aligned with pin 1 on the carrier and 2) do not use excessive heat.

1. Apply a small amount of solder to one of the corner pads on the carrier (Figure 14-12). If you have some liquid flux, feel free to use it as well, but it is not essential.
2. Hold the SOIC part on the carrier and heat the pin you soldered in Step 1. A pair of tweezers is great for putting the part in place. A toothpick is a useful "tool" to hold the part in place (Figure 14-13).
3. Apply solder liberally to the remaining pins starting opposite the pin you just soldered in Step 2 (Figure 14-14). Be messy at this point. No need to overdo it, but don't worry about solder bridges. We'll fix them in the next step.

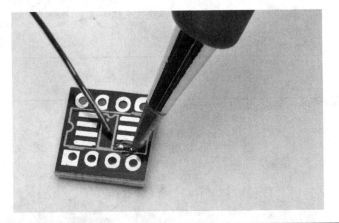

FIGURE 14-12 Apply solder to one corner pad.

FIGURE 14-13 Apply the part and heat the pad you soldered.

4. Remove the excess solder with solder wick (Figure 14-15).
5. Solder two 4-pin headers to the carrier (Figure 14-16). We use a solderless breadboard to hold the header pins while soldering.

The finished carrier should look like Figure 14-17.

FIGURE 14-14 Slather the pads with solder. You don't need to be neat!

FIGURE 14-15 Remove the excess solder with solder wick.

FIGURE 14-16 Soldering the 4-pin headers to the carrier.

FIGURE 14-17 Voila! You just soldered your first SMT part!

Final Assembly of the Coupler/Sensor

Now it's time to put the pieces of the directional coupler/sensor together. We mounted the copperclad shields to our chassis using solder lugs. The solder lugs were screwed to the chassis with 4-40 x ¼ in. machine screws and then soldered to the copperclad board. The wires from the transformers were passed through the holes in the shield separating the two pieces of coax and slipped through a short length of Teflon tubing. Be sure to observe the polarity of the windings; the wires must be attached exactly as shown in the schematic. Reversing a winding will actually do nothing more than reverse the Transmitter and Load ports, as well as the Forward and Reflected Power ports of the coupler. The sensor board should be mounted to its shield and wires routed through the shield to the standoff insulators holding the coax and terminating resistors using Teflon insulation over the wire to protect it from touching the shield. The ground wires from the sensor are soldered to the copperclad shield. We found that it was easier to mount the shields in the chassis before mounting the SO-239 connectors.

Make up a short jumper (about 4 in. in length) as shown in Figure 14-18 using six-conductor ribbon cable between the 4-pin mic connector and one of the 6-pin header plugs (FC-6P). The FC-6P connectors are easily found on eBay. They are also available as Harting part number 09185067803 from Mouser or Allied Electronics. We use a small vise to crimp the FC-type connectors onto the ribbon cable, but a pair of pliers also works. Be careful so as not to damage the connector. Regarding ribbon cable, scraps of ribbon cable are found in many places, most common being old computer disk drive cables. These tend to be 40 or 50 conductors, but it is easy to strip off as many conductors as you need using either a pair of flush cutters or an Exacto knife to separate the number of conductors you need from the rest. It is a simple matter of just "peeling" the six conductors away from the rest of the cable. Use an old pair of scissors to cut the cable to length.

Three 0.001 µF monolithic capacitors are used to bypass the power and sensor signal leads on the chassis connector. Solder one side of each capacitor to pins 1, 3, and 4 of the mic connector along

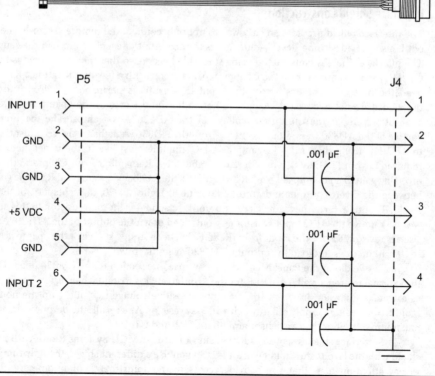

FIGURE 14-18 Sensor jumper cable.

FIGURE 14-19   The completed directional coupler/sensor assembly.

with the six leads of the ribbon cable. We used a leftover solder lug from one of the SO-239 coax connectors as a tie-point for the ground leads of the capacitors. The FC-6P connector and ribbon cable are small enough to fit through the ⅝-in. mounting hole for the 4-pin mic connector. Do not solder the capacitors to the solder lug until the cable has been passed through the chassis and the mic connector is tightened down. When you attach the FC-6P connector to the sensor board be sure to observe the location of pin 1. Pin 1 on an FC connector is marked with a little triangle on one side near the end. A close-up of the coupler/sensor assembly is shown in Figure 14-19.

Interface Shield Construction

The Interface Shield provides an easy way to not only connect the remote sensor to the Arduino, but it also provides some "level" adjustments to the output of each log amp on the sensor board. The interface shield is constructed using stackable headers so that the LCD shield is inserted on top. We used the circuit from the Chapter 5 panel meter with a few modifications. First, we don't need the protection diodes across the input. Since this is permanently wired into a circuit, the diodes are not necessary. Second, we have adjusted the gain of the opamp to better suit the application, amplifying the output voltage of the AD8307A log amps. The maximum output voltage of the AD8307A is about 2.6 VDC (from the AD8307A datasheet). We adjust the 2.6 VDC to 3.6 VDC using the LM 324 opamp as a DC amplifier, just like in the panel meter. The gain required is Vout (3.6) divided by Vin (2.6) or about 1.4. Remember from Chapter 5 that the non-inverting gain of an opamp is $1 + R_F/R_G$ where R_F is the feedback resistor and R_G is the gain resistor between the inverting input and ground. Looking at Figure 14-20, the schematic of the interface shield, R5 and R6 form the feedback resistor for U1c and R4 is the gain resistor. In this application, we also remove the 51 Ω input resistor as would load down the output of the AD8307A. We leave a series resistor, R3, for the input from the sensor. The input resistance of the opamp is quite high, so the circuit does not load the output of AD8307 significantly.

Just as we did in the Panel Meter design, we use one section of the LM324 opamp to provide the external reference voltage, AREF, for the Arduino analog to digital converter. U1d is configured as a voltage follower, driven to the maximum positive voltage by pulling up the non-inverting input to the positive supply rail through a 100 kΩ resistor. As a result, the output is the same as the maximum output of the two sensor amplifiers, U1b and U1c.

The interface shield was constructed using an Omega MCU Systems ProtoPro-B prototyping shield with the layout shown in Figure 14-21. If you use a different shield, the only thing you need to pay attention to is that you provide access to the adjustment potentiometers, R5 and R9. We show them mounted to the edge so that they are adjustable once the LCD shield is in place.

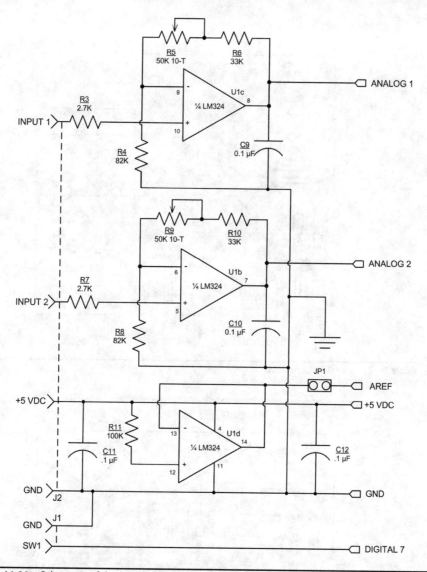

FIGURE 14-20 Schematic of the interface shield.

Figure 14-22 shows the wiring side of the interface shield. The completed interface shield is shown in Figure 14-23 mounted on an Arduino Dumilanove.

LCD Shield Options

While the LCD shield from Chapter 3 works just fine with our circuit, we decided to add a little feature by using an LCD with a multicolor backlight. We use the colors as a warning indicator as SWR is getting higher. White backlight means that everything is fine and that the SWR is within acceptable limits, usually below 2 to 1. As the SWR increases above 5:1 the display turns yellow. Above 10:1 it becomes red. The limits are all user-adjustable parameters set in software

300 Arduino Projects for Amateur Radio

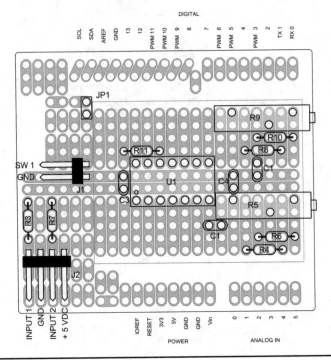

FIGURE 14-21 Interface shield layout.

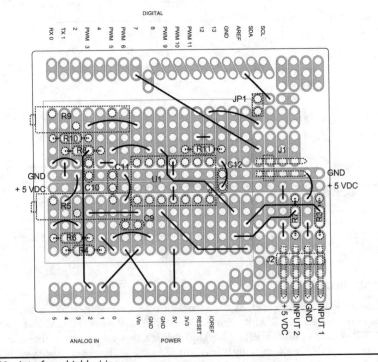

FIGURE 14-22 Interface shield wiring.

FIGURE 14-23   The completed interface shield mounted on the Arduino.

and are discussed in the software section of this chapter. The RGB LCD uses an 18-pin header. We added two additional resistors and along with the original 220 Ω resistor that was tied to ground, all three are now tied to digital output pins so that they may be turned on and off. Setting each of the three digital outputs (8, 9, and 10) LOW turns each of the three backlight LEDs on. Figure 14-24 shows the schematic of the new RGB LCD shield and Figure 14-25 shows the wiring side of the shield.

An easy way to make a quick check that your LCD shield is working correctly is to load the "Hello world" sketch from Chapter 3 and verify that the display is updating. However, the "Hello world" sketch does not activate the backlight LEDs.

Final Assembly

Now that you have completed the "heavy lifting" portion of the assembly, it is now time to bring it all together. The Arduino, display, and interface shields are mounted in an enclosure of your choice. We used an extruded Aluminum box from LMB; the same style as we used for the sequencer from Chapter 13. Refer back to Figure 14-3 for help identifying the various connectors and jumper cables.

We made two additional cables to connect the sensor to the interface shield. The first cable allows you to place the sensor remotely from the display. Use a cable length you feel appropriate to how you would like to place the sensor in relationship to the Arduino and display. Our cable is about six feet in length, uses four-conductor shielded wire, and has a 4-pin female mic connector at each end. The cable is wired pin for pin, or in other words, pin 1 is connected to pin 1 at the other end, pin 2 is connected to pin 2, and so on. The second cable connects a second chassis-mounted 4-pin mic connector to the interface shield inside the enclosure. The cable is shown in Figure 14-26 and shows the Molex connector, P2, pin positions.

The completed pieces, the sensor assembly, interconnecting cable, and display unit, are shown in Figure 14-27. All that is left is to test the build by loading up the test software, and begin calibrating the meter! The listing for the test software is provided in Listing 14-1.

302 Arduino Projects for Amateur Radio

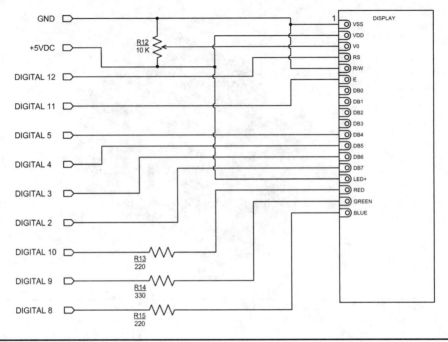

FIGURE 14-24   RGB LCD shield schematic.

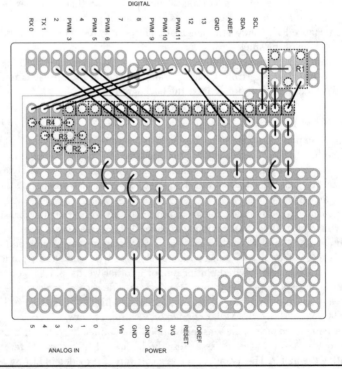

FIGURE 14-25   RGB LCD shield layout and wiring.

Chapter 14: A Directional Watt and SWR Meter 303

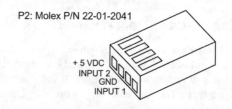

FIGURE 14-26 Display assembly jumper cable.

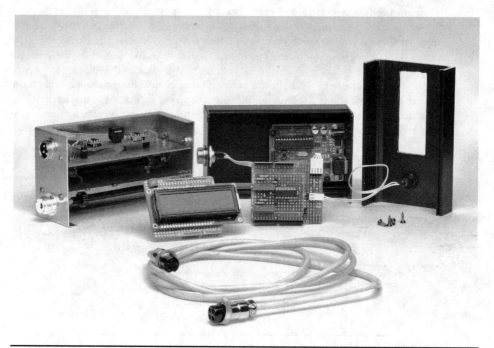

FIGURE 14-27 Completed components of the SWR/Wattmeter.

Testing the Directional Wattmeter/SWR Indicator

Testing and calibration requires a source RF between 25 and 50 W in the HF range. A modern HF transceiver with an adjustable output level is perfect. It is helpful if you have a known good Wattmeter as a reference. If you don't have a suitable Wattmeter, look to borrow one from another local ham. Also, local high schools and community colleges may also be a source for you. Ideally, a dummy load with a built-in Wattmeter would be perfect (see Chapter 6 for the Dummy Load/Wattmeter).

Connect the RF source, dummy load, and directional coupler as shown in Figure 14-28. The dummy load must be capable of dissipating the 25 to 50 W we need for calibration, well within the limits of the dummy load described in Chapter 6. We use 50 Ω RG-8X jumpers or the equivalent for testing. Before connecting the pieces as shown, it is always a good idea to make sure that the cables, radio, and dummy load are functioning correctly. Connect the radio to the dummy load using each of the coax jumpers, testing them to make sure that they are making good connections. Bad cables lead to bad results.

We used a simple program shown in Listing 14-1 to check that the sensors are working and that the Arduino is able to read the output of each of the log amplifiers, detect the switch, and change the LCD backlight colors. We also use this program to set the calibration points for use in the operating program shown in Listing 14-2. With a low power RF source attached as shown in Figure 14-28, apply power and observe the LCD display. As power increases, so do the values shown in the display. Pressing the pushbutton causes the LCD backlight to cycle through its colors. Once you have verified that things are working as they should, we calibrate the meter.

Calibrating the Directional Wattmeter

Using the RF source as in Figure 14-25, set the output power to 20 W. The Wattmeter included with the Dummy Load in Chapter 6 is perfect for measuring the power. After verifying that the power level is 20 W, adjust the potentiometer, R5, so that the upper line of the display indicates a value of 1000. This is our first calibration point for forward power sensor. We know that 20 W input produces a count of 1000. We need one additional point to complete the calibration of the forward sensor. To ensure the accuracy of the Directional Wattmeter, the greater the difference between our calibration points, the better. We chose a power level of 20 mW for the second calibration point. However, we had the ability to accurately measure the power at that level using our HP 438 Power Meter. Record the results of the measurement. Make the second power measurement at as low a power level as you are able that is still within the range of the Directional Wattmeter. In other

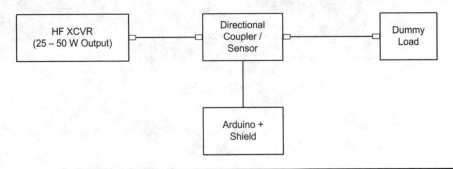

FIGURE 14-28 Test and calibration setup.

words, it should be greater than 100 µW. Again, record your results. Make sure that you record the power level you are using and the result displayed on the LCD.

The next step is to reverse the connections and calibrate the reflected power sensor. Repeat the two measurements, adjusting the reflected power calibration potentiometer, R9, for a reading of 1000 with 20 W input. Reduce the power and make a second measurement. The recorded values are used in the operating program for the Directional Wattmeter. You should now have eight values recorded: the high and low power settings in the forward direction and their corresponding analog output value, as well as those for the reflected direction. The next step is to edit the operating software with the measured calibration values and start using your new Directional Wattmeter/SWR Indicator.

```
/******
 *
 *   Test program to verify that sensors, switch and display are working
 *   by W6DQ D.Kidder 1 March 2014
 *
 ******/

#include <LiquidCrystal.h>

#define LCDCOLS              16
#define LCDROWS              2
#define FORWARDPOWERINPUT    1
#define REFLECTEDPOWERINPUT  2
#define SWITCH               7

#define DEBOUNCEDELAY        10

#define BLUELIGHT            8
#define GREENLIGHT           9
#define REDLIGHT             10

LiquidCrystal lcd(12, 11, 5, 4, 3, 2); // For LCD display from Chapter 3

void setup()
{

  analogReference(EXTERNAL);      //  AREF from LM324 voltage follower

  lcd.begin(LCDROWS, LCDCOLS);    // initialize the LCD
  lcd.clear();
  lcd.home();
  lcd.print("SWR  METER TEST");   //sappy stuff, name, version and date
  lcd.setCursor(0,1);
  lcd.print("ver0.1 01MAR14");
  delay(3000);                    // Delay probably okay since we're not doing
                                  // anything else

  lcd.clear();

  pinMode(REDLIGHT,OUTPUT);
```

LISTING 14-1   Test software for the Directional Wattmeter/SWR Indicator.

```
  pinMode(GREENLIGHT,OUTPUT);
  pinMode(BLUELIGHT,OUTPUT);

  pinMode(SWITCH,INPUT);
  digitalWrite(SWITCH,HIGH);

}

void loop()
{

  digitalWrite(REDLIGHT,LOW);
  digitalWrite(GREENLIGHT,LOW);
  digitalWrite(BLUELIGHT,LOW);

  lcd.home();
  lcd.print(analogRead(FORWARDPOWERINPUT));
  lcd.setCursor(0,1);
  lcd.print(analogRead(REFLECTEDPOWERINPUT));

  if (debounce(SWITCH) == true){

    digitalWrite(REDLIGHT,LOW);
    digitalWrite(GREENLIGHT,HIGH);
    digitalWrite(BLUELIGHT,HIGH);
    delay(1000);
    digitalWrite(REDLIGHT,HIGH);

    digitalWrite(GREENLIGHT,LOW);
    delay(1000);
    digitalWrite(GREENLIGHT,HIGH);

    digitalWrite(BLUELIGHT,LOW);
    delay(1000);
    digitalWrite(BLUELIGHT,HIGH);

  }

}
/*****
  This function is a pretty standard debounce function to determine
  that the switch has really been pressed.

  Parameter list:
    int pin              the pin that registers the switch state

  Return value:
    boolean              true if the switch was pressed, false otherwise

*****/
```

LISTING 14-1 Test software for the Directional Wattmeter/SWR Indicator. (*continued*)

```
boolean debounce(int pin)
{
  boolean currentState;
  boolean previousState;
  int i;

  previousState = digitalRead(pin);
  for (i = 0; i < DEBOUNCEDELAY; i++) {
    delay(1);     // small delay
    currentState = digitalRead(pin);  // Read it now
    if (currentState != previousState)
    {
      i = 0;
      previousState = currentState;
    }
  }
  if (currentState == LOW)
    return true;
  else
    return false;
}
```

LISTING 14-1   Test software for the Directional Wattmeter/SWR Indicator. (*continued*)

Software Walk-Through

The software for the SWR/Wattmeter is shown in Listing 14-2. For the most part, this program is straightforward and using the techniques we have been describing throughout the book. However, there are several new items that we discuss in more detail. What is important to know from this section are the definitions that are used to tailor the software to your build. Variations in component values result in different readings in the meter, hence the need for the calibration values that were recorded previously. The software uses the calibration data to set constants used to adjust the calculations so that the readings are correct.

```
/******
 *
 *  Directional Wattmeter / SWR Indicator by D. Kidder W6DQ
 *  Version 1.7 26 Aug 2014
 *
 *  Works with the hardware described in "Arduino Projects for Ham Radio"
 *  by Jack Purdum, W8TEE, and Dennis Kidder, W6DQ.
 *
 *  The hardware uses a Stockton Coupler with AD8307 Log-Lin Amplifier as
 *  a sensor. As designed, the Wattmeter is designed for 25 Watts max but
 *  will produce an alarm if the power exceeds 20 Watts. With a backlit
 *  RGB LCD, when in alarm, the backlight is red. An audible alert is also
 *  used.
 *
```

LISTING 14-2   Arduino-based Directional Wattmeter/SWR Indicator.

```
 *   The SWR indicator can be switched between two modes - one with a digital
 *   indication, and the second with a bar graph display. THe SWR indicator
 *   also uses the LCD backlight to display status of the SWR. With a low SWR
 *   the display is green. As the SWR begins to rise, the discplay is yellow.
 *   A high SWR condition is shown as red with an audible alert.
 *
 ******/

#include <LiquidCrystal.h>
#include <LcdBarGraph.h>
#include <math.h>
#include "TimerOne.h"

// ================= LCD configuration =====================================
#define LCDCOLS             16          // 16 columns
#define LCDROWS             2           // 2 rows

#define BARGRAPHROW         1           // Draw graph on second row
#define BARGRAPHCOL         0           // Start graph in column 0

#define DISPLAYFLASHRATE    300000      // Update rate in microseconds

#define POWERWIDTH          5           // Numeric field width, include '-', '.'
#define SWRWIDTH            4           // Numeric field width, include '-', '.'
#define DEFAULTPRECISION    1           // Number of decimal places to display
#define PRECISIONTWO        2           // Number of decimal places to display
#define PRECISIONZERO       0           // Number of decimal places to display

#define STARTUPDELAY        3000        // Display splash screen
#define DEBOUNCEDELAY       50

// ================= Define inputs and outputs =============================

#define FORWARDPOWERINPUT   1           // Forward power sensor: analog input 1
#define REFLECTEDPOWERINPUT 2           // Reflected power sensor: analog input 2
#define SWITCH              7           // Display mode switch: digital pin 7
#define BLUELIGHT           8           // Display backlight BLUE digital pin 8
#define GREENLIGHT          9           // Display backlight GREEN digital pin 9
#define REDLIGHT            10          // Display backlight RED digital pin 10

/******
 *
 *   The following eight values are modifiable by the user. They are used to
 *   derive constants used to calculate the input power. The values consist of
 *   two input power levels and the corresponding analog input value for both
 *   forward and reverse measurements. In the case of *POWERONE, we want the
 *   analog value to be the max before alarm. We have arbitrarily used "1000"
 *   so that anything great than 1000 (in other words 1000 to 1023) will cause
 *   an alarm.
 *
 ******/
```

LISTING 14-2 Arduino-based Directional Wattmeter/SWR Indicator. (*continued*)

```
#define FORWARDPOWERONE      20.0     // Forward power at max reading
#define FORWARDANALOGONE     1000     // Analog value at max forward
#define FORWARDPOWERTWO      .020     // Forward power at second reading
#define FORWARDANALOGTWO     706      // Analog value at second forward
#define REFLECTEDPOWERONE    20.0     // Reflected power at max reading
#define REFLECTEDANALOGONE   1000     // Analog value at max refelected
#define REFLECTEDPOWERTWO    .020     // Refelected power at second reading
#define REFLECTEDANALOGTWO   701      // Analog vale at second reflected

// ================= Alarm Levels ==========================================

#define POWERLEVELALARM      1000     // Analog value for power that
                                      // generates an alarm
                                      // (we fudge it a little bit)
#define POWERLEVELALARMBAD   1010     // Start flashing the display and
                                      // make noise!
#define SWRYELLOW            5.0      // Threshold for caution SWR
#define HIGHSWR              10.0     // Threshold for SWR that is OFF THE
                                      // CHARTS
#define SWRALARMTHRESHOLD    10.0     // Threshold to disable the alarm
                                      // functions (dBm)

boolean alarmLock = false;            // Used to suppress display during
                                      // hi power alarm states
boolean lowPowerAlarmLock = false;    // Used to suppres alarms when power
                                      // is very low
boolean hiSwrLock = false;            // Used to suppress display during
                                      // high SWR alarm states

#define ALARM1               1        // Power exceeds POWERLEVELALARM
#define ALARM2               2        // Power exceeds POWERLEVELALARMBAD
#define ALARM3               3        // Reflected power > Forward power
#define ALARM4               4        // SWR Rising
#define ALARM5               5        // High SWR -- off the charts!!


// ================= Define the different backlight modes for the LCD =======

#define GOWHITE   {digitalWrite(REDLIGHT, LOW);
                   digitalWrite(GREENLIGHT, LOW);digitalWrite(BLUELIGHT, LOW);}
#define GOBLACK   {digitalWrite(REDLIGHT, HIGH); digitalWrite(GREENLIGHT,
                   HIGH);digitalWrite(BLUELIGHT, HIGH);}
#define GORED     {digitalWrite(REDLIGHT, LOW); digitalWrite(GREENLIGHT,
                   HIGH);digitalWrite(BLUELIGHT, HIGH);}
#define GOYELLOW  {digitalWrite(REDLIGHT, LOW);digitalWrite(GREENLIGHT, LOW);
                   digitalWrite(BLUELIGHT, HIGH);}
#define GOGREEN   {digitalWrite(REDLIGHT, HIGH);digitalWrite(GREENLIGHT,
                   LOW); digitalWrite(BLUELIGHT, HIGH);}
```

LISTING 14-2   Arduino-based Directional Wattmeter/SWR Indicator. (*continued*)

```
//================= Analog Input Buffer setup =================================

#define BUFFERSIZE  10        // Depth of the circular buffer

int index;                    // Index into the buffer array

int forwardBufferTotal;       // The running sum of the forward power buffer
int reflectedBufferTotal;     // The running sum of the reflected power buffer

int forwardPowerCircularBuffer[BUFFERSIZE];    // Forward power buffer
int reflectedPowerCircularBuffer[BUFFERSIZE];  // Reflected power buffer

//================= Power expressed in different units =====================

#define Watts     1L          // Used to scale power to Watts
#define mW        1000L       // Used to scale power to milliatts
#define uW        1000000L    // Used to scale power to microwatts

int forwardPowerLog = 0;      // Log of forward power
int reflectedPowerLog = 0;    // Log of reflected power

float forwardPower = 0.0;     // Power (in Watts) with different scaling
float reflectedPower = 0.0;
float forwardPowermW = 0.0;   // milliwatts
float reflectedPowermW = 0.0;
float forwardPoweruW = 0.0;   // microwatts
float reflectedPoweruW = 0.0;

float forwardPowerDbm;        // dB milliwatts (dBm)
float reflectedPowerDbm;

// ================= Constants used to calculate power from log amp output ==
float forwardPowerSlope;
float forwardPowerIntercept;
float reflectedPowerSlope;
float reflectedPowerIntercept;

float swr;
int   swrgraph;

int buttonState;
int buttoncount = 0;

volatile int alarmType;
volatile boolean state;
volatile boolean OneShotOnly;

//================= Set the parameters for the LCD and the bar graph ========
```

LISTING 14-2 Arduino-based Directional Wattmeter/SWR Indicator. (*continued*)

```
LiquidCrystal lcd(12, 11, 5, 4, 3, 2);        // For LCD display from Chapter 3
LcdBarGraph    lbg(&lcd, LCDCOLS, BARGRAPHCOL, BARGRAPHROW);

//================= Setup buffers used for creating displays ================

char buffer[ LCDCOLS + 1 ];
char lineOneBufferout[ LCDCOLS + 1 ];

// ==========================================================================
// ================= setup() Portion of the program =========================
// ==========================================================================

void setup()
{
//================= Initialize IO Pins =====================================

Serial.begin(115200);

  pinMode(SWITCH,INPUT);         // Assign the display selector switch and
                                 // turn on the pullup
  digitalWrite(SWITCH,HIGH);

  pinMode(REDLIGHT,OUTPUT);      // set the mode of the pins to drive the LCD
                                 // backlight
  pinMode(GREENLIGHT,OUTPUT);
  pinMode(BLUELIGHT,OUTPUT);

  analogReference(EXTERNAL);     // use ext reference from LM324 voltage
                                 // follower (~3.7 VDC)

//================= Initialize the display plus add splash screen ===========

  GOWHITE;                       // Turn on the backlight!
  lcd.begin(LCDROWS, LCDCOLS);   // initialize the LCD
  lcd.clear();
  lcd.home();
  lcd.print("Watt/SWR Meter");   //sappy stuff, name, version and date
  lcd.setCursor(0,1);
  lcd.print("ver 1.7 26AUG14");
  delay(STARTUPDELAY);           // Delay probably okay since we're not doing
                                 // anything else
  lcd.clear();

//================= Initialize Timer#1 to flash the display in alarm states =

  Timer1.initialize(DISPLAYFLASHRATE);

/* ================= Initialize slope and intercept values ==================
*
```

LISTING 14-2 Arduino-based Directional Wattmeter/SWR Indicator. (*continued*)

```
 * Derive the slope and intercepts for the forward and reflected power curves.
 * Using the slope and intercept, we can derive the power with any given
 * analog input from the sensors.
 *
 *     Slope = (analog1 - analog2) / (power1 - power2)
 *
 *     Intercept = power1 - (analog1 / slope )
 *
 * power is expressed in dBm, analog is the analog input value
 *
 *****/

  forwardPowerSlope = slope(FORWARDANALOGONE, FORWARDPOWERONE,
              FORWARDANALOGTWO, FORWARDPOWERTWO);
  reflectedPowerSlope = slope(REFLECTEDANALOGONE, REFLECTEDPOWERONE,
              REFLECTEDANALOGTWO, REFLECTEDPOWERTWO);

  forwardPowerIntercept = intercept( FORWARDPOWERONE, FORWARDANALOGONE,
              forwardPowerSlope );
  reflectedPowerIntercept = intercept( REFLECTEDPOWERONE, REFLECTEDANALOGONE,
              reflectedPowerSlope );

//================= Initialize the input circular buffers to zero ===========

  memset(forwardPowerCircularBuffer, 0, BUFFERSIZE);
  memset(reflectedPowerCircularBuffer, 0, BUFFERSIZE);

/*
  for(int index = 0; index < BUFFERSIZE; index++) {   // initialize the
                                                       // buffers with zeros
    forwardPowerCircularBuffer[index] = 0;
    reflectedPowerCircularBuffer[index] = 0;

  }
*/}
// =========================================================================
// ================= loop() Portion of the program =========================
// =========================================================================

void loop()
{
/* ================ Circular Input Buffer and Averaging Code ==============
 *
 *   In order to stabilze the radings from the sensors, we use a circular
 *   buffer. The buffer is an array that is of size BUFFERSIZE. We read from
 *   the analog inputs into the buffer and then after each value is added, sum
 *   the values in the buffer and divide by BUFFERZSIZE, giving an average
 *   value.
 *
 *   In adddition, we monitor each reading to ensure that the maximum power
 *   levels are not exceeded. Should the instantaneous input power exceed
 *   preset limits, an alarm is raised and the display goes yellow,
```

LISTING 14-2   Arduino-based Directional Wattmeter/SWR Indicator. (*continued*)

```
 *    indicating that the power is rising to a dangerous level, or flashing
 *    red, indicating that the power is sufficiently high as to cause permanent
 *    damage to the sensor. Returning to normal levels removes the alarm.
 *
 *****/

     forwardBufferTotal = forwardBufferTotal -
                forwardPowerCircularBuffer[index];
     forwardPowerCircularBuffer[index] = analogRead(FORWARDPOWERINPUT);
     forwardBufferTotal = forwardBufferTotal +
                forwardPowerCircularBuffer[index];

     reflectedBufferTotal = reflectedBufferTotal -
                reflectedPowerCircularBuffer[index];
     reflectedPowerCircularBuffer[index] = analogRead(REFLECTEDPOWERINPUT);
     reflectedBufferTotal = reflectedBufferTotal +
                reflectedPowerCircularBuffer[index];

/* ============================================================================
= This section of code is used to detect alarm conditions ... excessive power
*/

     if ( forwardPowerCircularBuffer[index] > POWERLEVELALARMBAD ||
             reflectedPowerCircularBuffer[index] > POWERLEVELALARMBAD ){
       Timer1.attachInterrupt(FlashDisplay);
       alarmType = ALARM2;
       DisplayAlarm(alarmType);
       alarmLock = true;
     }
     else if ( forwardPowerCircularBuffer[index] > POWERLEVELALARM ||
             reflectedPowerCircularBuffer[index] > POWERLEVELALARM ){
       Timer1.attachInterrupt(FlashDisplay);
       alarmType = ALARM1;
       DisplayAlarm(alarmType);
       alarmLock = true;
     }
     else {
       alarmLock = false;
       Timer1.detachInterrupt();
     }

//================= Alarm cleared, back to normal ============================

     index++;

     if ( index >= BUFFERSIZE ) index = 0;

     forwardPowerLog = forwardBufferTotal / BUFFERSIZE;
     reflectedPowerLog = reflectedBufferTotal / BUFFERSIZE;

//================= Series of conversions from the log output ===============
```

LISTING 14-2 Arduino-based Directional Wattmeter/SWR Indicator. (*continued*)

```
/********
 *
 *  Convert the log output of the sensor AD8307 to the actual power in dBm.
 *
 ******/

    forwardPowerDbm = ( forwardPowerLog / forwardPowerSlope ) +
                      forwardPowerIntercept;
    reflectedPowerDbm = ( reflectedPowerLog / reflectedPowerSlope ) +
                      reflectedPowerIntercept;

/*****
 *
 *  Convert the output power (dBm) to Watts
 *
 *****/

    forwardPower = dBmToWatts(forwardPowerDbm, Watts);
    reflectedPower = dBmToWatts(reflectedPowerDbm, Watts);

/*****
 *
 *  Convert the output power (dBm) to milliwatts
 *
 *****/

    forwardPowermW= dBmToWatts(forwardPowerDbm, mW);
    reflectedPowermW  = dBmToWatts(reflectedPowerDbm, mW);

/*****
 *
 *  Convert the output power (dBm) to microwatts
 *
 *****/

    forwardPoweruW = dBmToWatts(forwardPowerDbm, uW);
    reflectedPoweruW = dBmToWatts(reflectedPowerDbm, uW);

/*****
 *
 * This section creates a "lockout" on certain alarms. When there is no
 * power applied or very low power levels, it is easy to put the code into an
 * alarm state. The lockout is used to suppres alarms until a minimum power
 * level is reached.
 *
 *****/

    if ( forwardPowerDbm > SWRALARMTHRESHOLD || reflectedPowerDbm >
            SWRALARMTHRESHOLD ) {
      lowPowerAlarmLock = false;
    }
```

LISTING 14-2 Arduino-based Directional Wattmeter/SWR Indicator. (*continued*)

```
    else {
      lowPowerAlarmLock = true;
    }

/*****
*
*  Calculate the SWR from the forward and reflected power. The relationship
*  between SWR and polwer is expressed as:
*
*     SWR = ( 1 + SQRT( Pref / Pfor)) / SQRT( 1 - ( Pref / Pfor))
*
*****/

    swr = ( 1 + sqrt( reflectedPowermW  / forwardPowermW)) / ( 1 - sqrt(
             reflectedPowermW  / forwardPowermW));
    swr = constrain(swr, 1.0, 20.0 );



//================= Pushbutton is used to select from four display modes ====

    buttonState = digitalRead(SWITCH);

    if (buttonState == LOW) {
      deBounce(SWITCH);
      buttoncount++;
      if ( OneShotOnly == false ) OneShotOnly = true;
      lcd.clear();
      GOWHITE;
    }

/******
*
*  This portion of the code is used to detect alarm states NOT detected
*  previously. This section deals with high SWR and/or reversed input and
*  output.
*    1. High SWR. Result: flashing red dsiplay and error message
*    2. Reflected power exceeds forward power. Result: Flashing red display
*  and error message
*
*  When the alarm condition is cleared, we continue to the normal display
*  modes. There are four display modes:
*
*    1. Forward and Reflected power displayed in Watts, milliwats and
*       microwatts.
*    2. Forward and Reflected power displayed in dBm.
*    3. SWR displayed on top line, bar graph on lower line. 3:1 is approx
*       mid-scale.
*    4. Abbreviated Forward and Reflected power with SWR display.
*
*  Alarms are disabled at very low power levels. Variations in the output of
```

LISTING 14-2 Arduino-based Directional Wattmeter/SWR Indicator. *(continued)*

```
 *  the log amps can cause an unwanted alarm with no RF input power at all.
 *  The value of ALARMTESTTHRESHOLD sets the minimum amount of power to be
 *  applied before the alarm tests are enabled.
 *
 ******/

    if (alarmLock != true ){

      if ( swr > HIGHSWR && lowPowerAlarmLock == false ){
        Timer1.attachInterrupt(FlashDisplay);
        alarmType = ALARM5;
        DisplayAlarm(alarmType);
        hiSwrLock = true;
      }
      else if ( swr > SWRYELLOW && lowPowerAlarmLock == false ){

        dtostrf(swr, SWRWIDTH, DEFAULTPRECISION, buffer);
        sprintf(lineOneBufferout,"   SWR: %s:1   ",buffer);
        Timer1.attachInterrupt(FlashDisplay);
        alarmType = ALARM4;
        DisplayAlarm(alarmType);
        hiSwrLock = true;
      }
      else if ( reflectedPowerDbm > forwardPowerDbm && lowPowerAlarmLock ==
          false ){
        Timer1.attachInterrupt(FlashDisplay);
        alarmType = ALARM3;
        DisplayAlarm(alarmType);
        hiSwrLock = true;
    }
      else {
        hiSwrLock = false;
        Timer1.detachInterrupt();
      }
    }

    if (hiSwrLock != true){

      GOWHITE;

      switch (buttoncount){

        case 0:       // Display No. 1: Forward and Reflected power in Watts

          if ( forwardPowerDbm > 40.0 ){
            dtostrf(forwardPower, POWERWIDTH, DEFAULTPRECISION, buffer);
            sprintf(lineOneBufferout,"Forward:%s W ",buffer);
          }
          else if ( forwardPowerDbm > 30.0 ){
            dtostrf(forwardPower, POWERWIDTH, PRECISIONTWO, buffer);
            sprintf(lineOneBufferout,"Forward:%s W ",buffer);
          }
```

LISTING 14-2 Arduino-based Directional Wattmeter/SWR Indicator. (*continued*)

```
    else if ( forwardPowerDbm > 20.0 ){
      dtostrf(forwardPowermW, POWERWIDTH, PRECISIONZERO, buffer);
      sprintf(lineOneBufferout,"Forward:%s mW",buffer);
    }
    else if ( forwardPowerDbm > 10.0 ){
      dtostrf(forwardPowermW, POWERWIDTH, DEFAULTPRECISION, buffer);
      sprintf(lineOneBufferout,"Forward:%s mW",buffer);
    }
    else if ( forwardPowerDbm > 0.0 ){
      dtostrf(forwardPowermW, POWERWIDTH, PRECISIONTWO, buffer);
      sprintf(lineOneBufferout,"Forward:%s mW",buffer);
    }
    else {
      dtostrf(forwardPoweruW, POWERWIDTH, PRECISIONZERO, buffer);
      sprintf(lineOneBufferout,"Forward:%s uW",buffer);
    }

    lcd.home();
    lcd.print(lineOneBufferout);

    if ( reflectedPowerDbm > 40.0 ){
      dtostrf(reflectedPower, POWERWIDTH, DEFAULTPRECISION, buffer);
      sprintf(lineOneBufferout,"Reflctd:%s W ",buffer);
    }
    else if ( reflectedPowerDbm > 30.0 ){
      dtostrf(reflectedPower, POWERWIDTH, PRECISIONTWO, buffer);
      sprintf(lineOneBufferout,"Reflctd:%s W ",buffer);
    }
    else if ( reflectedPowerDbm > 20.0 ){
      dtostrf(reflectedPowermW , POWERWIDTH, PRECISIONZERO, buffer);
      sprintf(lineOneBufferout,"Reflctd:%s mW",buffer);
    }
    else if ( reflectedPowerDbm > 10.0 ){
      dtostrf(reflectedPowermW , POWERWIDTH, DEFAULTPRECISION, buffer);
      sprintf(lineOneBufferout,"Reflctd:%s mW",buffer);
    }
    else if ( reflectedPowerDbm > 0.0 ){
      dtostrf(reflectedPowermW , POWERWIDTH, PRECISIONTWO, buffer);
      sprintf(lineOneBufferout,"Reflctd:%s mW",buffer);
    }
    else {
      dtostrf(reflectedPoweruW, POWERWIDTH, PRECISIONZERO, buffer);
      sprintf(lineOneBufferout,"Reflctd:%s uW",buffer);
    }

    lcd.setCursor(0,1);
    lcd.print(lineOneBufferout);

    break;
```

LISTING 14-2 Arduino-based Directional Wattmeter/SWR Indicator. *(continued)*

```
      case 1:        // Display No. 2: Forward and Reflected power in dBm

        if ( forwardPowerDbm < -10.0 ){
          dtostrf(forwardPowerDbm, POWERWIDTH - 1, PRECISIONZERO, buffer);
        }

        else{

          dtostrf(forwardPowerDbm, POWERWIDTH - 1, DEFAULTPRECISION,
               buffer);
        }

        sprintf(lineOneBufferout,"  FWD: %s dBm ",buffer);

        lcd.home();
        lcd.print(lineOneBufferout);

        dtostrf(reflectedPowerDbm, POWERWIDTH - 1, DEFAULTPRECISION,
 buffer);
        sprintf(lineOneBufferout,"  REF: %s dBm ",buffer);

        lcd.setCursor(0,1);
        lcd.print(lineOneBufferout);

        break;

      case 2:                           // Display No. 3: SWR with bar graph

        GOWHITE;
        dtostrf(swr, SWRWIDTH, DEFAULTPRECISION, buffer);
        sprintf(lineOneBufferout,"  SWR: %s:1   ",buffer);
        swrgraph = constrain( (int) (( swr ) * 100), 100, 580);
        swrgraph = map(swrgraph, 100, 580, 0, 490) + 10;      // Added the "
                             // +10" so that something is always visible

        lcd.home();
        lcd.print(lineOneBufferout);
        ClearBarGraph();
        lbg.drawValue(swrgraph, 500); // display bar graph of SWR
        break;

      case 3:                           // Display No. 4:  Forward and
                                        // Reflected power in Watts plus SWR

 if ( forwardPowerDbm > 40.0 ){
         dtostrf(forwardPower, POWERWIDTH, DEFAULTPRECISION, buffer);
         sprintf(lineOneBufferout,"FWD%s W   SWR:",buffer);
        }
        else if ( forwardPowerDbm > 30.0 ){
          dtostrf(forwardPower, POWERWIDTH, PRECISIONTWO, buffer);
          sprintf(lineOneBufferout,"FWD%s W   SWR:",buffer);
        }
```

LISTING 14-2 Arduino-based Directional Wattmeter/SWR Indicator. (*continued*)

```
    else if ( forwardPowerDbm > 20.0 ){
      dtostrf(forwardPowermW, POWERWIDTH, PRECISIONZERO, buffer);
      sprintf(lineOneBufferout,"FWD%s mW SWR:",buffer);
    }
    else if ( forwardPowerDbm > 10.0 ){
      dtostrf(forwardPowermW, POWERWIDTH, DEFAULTPRECISION, buffer);
      sprintf(lineOneBufferout,"FWD%s mW SWR:",buffer);
    }
    else if ( forwardPowerDbm > 0.0 ){
      dtostrf(forwardPowermW, POWERWIDTH, PRECISIONTWO, buffer);
      sprintf(lineOneBufferout,"FWD%s mW SWR:",buffer);
    }
    else {
      dtostrf(forwardPoweruW, POWERWIDTH, PRECISIONZERO, buffer);
      sprintf(lineOneBufferout,"FWD%s uW SWR:",buffer);
    }

    lcd.home();
    lcd.print(lineOneBufferout);

    if ( reflectedPowerDbm > 40.0 ){
      dtostrf(reflectedPower, POWERWIDTH, DEFAULTPRECISION, buffer);
      sprintf(lineOneBufferout,"REF%s W  ",buffer);
    }
    else if ( reflectedPowerDbm > 30.0 ){
      dtostrf(reflectedPower, POWERWIDTH, PRECISIONTWO, buffer);
      sprintf(lineOneBufferout,"REF%s W  ",buffer);
    }
    else if ( reflectedPowerDbm > 20.0 ){
      dtostrf(reflectedPowermW , POWERWIDTH, PRECISIONZERO, buffer);
      sprintf(lineOneBufferout,"REF%s mW ",buffer);
    }
    else if ( reflectedPowerDbm > 10.0 ){
      dtostrf(reflectedPowermW , POWERWIDTH, DEFAULTPRECISION, buffer);
      sprintf(lineOneBufferout,"REF%s mW ",buffer);
    }
    else if ( reflectedPowerDbm > 0.0 ){
      dtostrf(reflectedPowermW , POWERWIDTH, PRECISIONTWO, buffer);
      sprintf(lineOneBufferout,"REF%s mW ",buffer);
    }
    else {
      dtostrf(reflectedPoweruW, POWERWIDTH, PRECISIONZERO, buffer);
      sprintf(lineOneBufferout,"REF%s uW ",buffer);
    }

    lcd.setCursor(0,1);
    lcd.print(lineOneBufferout);

    dtostrf(swr, SWRWIDTH - 1, DEFAULTPRECISION, buffer);
```

LISTING 14-2 Arduino-based Directional Wattmeter/SWR Indicator. (*continued*)

```
            lcd.setCursor(13,1);
            lcd.print(buffer);
            
            break;
        }
      }
    }
    if (buttoncount == 4) {               // reset BUTTONCOUNT so we can
                                          // start from the beginning
      buttoncount = 0;
    }
}
// ==========================================================================
//================= End of loop() ==========================================
// ==========================================================================


/* ==========================================================================
 *
 * Calculate the slope of the log amp output curve
 *
 * Input parameters:
 *    int X1    - analog value for sample #1
 *    double Y1 - power for sample #1
 *    int X2    - analog value for sample #2
 *    double Y1 - power for sample #2
 *
 * Return parameters:
 *    float val - slope of the two sample points
 *
 ******/

float slope(int X1, double Y1, int X2, double Y2){
    
    float  val = ( X1 - X2 ) / (( 10.0 * log10( 1000.0 * Y1)) -
                 (10.0 * log10( 1000.0 * Y2)));
                 
    return val;
    
}

/* ==========================================================================
 *
 * Calculate the intercept of the log amp output curve
 *
 * Input parameters:
 *     double Y1  - power
 *     int X1     - analog value
 *     slope      - slope
 *
```

LISTING 14-2   Arduino-based Directional Wattmeter/SWR Indicator. (*continued*)

```
 *   Return parameters:
 *       float val - slope of the two sample points
 *
 ******/

float intercept(float Y1, int X1, float slope){

  float  val = (10.0 * log10(1000.0 * Y1)) - ( X1 / slope);

  return val;

}

/* ==========================================================================
 *
 *   Convert dBm to Watts
 *
 *   Input parameters:
 *       float dBm     power in dBm to be converted
 *       int   order   order of magnitude ... Watts = 1,
 *                                            mW = 1000 or
 *                                            uW = 1000000
 *
 *   Return parameters
 *        float power
 *
 *****/

float dBmToWatts(float dBm, long order){

  float power = (float) pow(10.0, (( dBm - 30.0) / 10.0 )) * order;

  return power;

}

/* ==========================================================================
 *
 *   Simple button debounce routine.
 *
 *   Input parameters:
 *     int buttonPin                          // where the switch is connected!
 *
 *   Return parameters:
 *     none
 *
 ******/
```

LISTING 14-2   Arduino-based Directional Wattmeter/SWR Indicator. (*continued*)

```
void deBounce(int buttonPin)
{
  unsigned long now = millis ();
  do
  {
    if (digitalRead(buttonPin) == LOW)   // on bounce, reset time-out
      now = millis ();
  } while (digitalRead(buttonPin) == LOW || (millis() - now) <=
          DEBOUNCEDELAY);
}
/* ========================================================================
 *
 * LCD Bar Graph does not clear any data from the row prior to writing to it.
 * Under normal circumstances an lcd.clear() would be sufficient. However,
 * we don't want to use lcd.clear on every write. This function executes an
 * lcd.clear once following an alarm cleared or after a button push and
 * branching to display case 3 ... SWR plus SWR bar graph.
 *
 * Input parameters:
 *    void
 *
 * Return parameters:
 *    none
 *
 ******/

void ClearBarGraph()
{
  if ( OneShotOnly == true ){
    lcd.clear();
    OneShotOnly = false;
  }
}
/* ========================================================================
 *
 * Alarm function -- updates deiplay to reflect alarm state
 *
 * Input parameters
 *    int alarmtype
 *
 * Return values
 *    none
 *
 *
 *****/

void DisplayAlarm(int alarmtype)
{
    switch(alarmtype){
      case ALARM1:                       // Power exceeds 20 Watts
        lcd.home();
```

LISTING 14-2 Arduino-based Directional Wattmeter/SWR Indicator. *(continued)*

```
                lcd.print("  INPUT > 20 W   ");
                lcd.setCursor(0,1);
                lcd.print("  REDUCE POWER  ");
                OneShotOnly = true;
                break;

        case ALARM2:                    // Power greatly exceeds 20 Watts DANGER!!
                lcd.home();
                lcd.print("  INPUT >> 25 W  ");
                lcd.setCursor(0,1);
                lcd.print("REDUCE PWR NOW!!");
                OneShotOnly = true;
                break;
        case ALARM3:                    // Reflected power exceeds forward power
                lcd.home();
                lcd.print("     ERROR!     ");
                lcd.setCursor(0,1);
                lcd.print("REF EXCEEDS FOR ");
                break;
        case ALARM4:                    // SWR is increasing to a dangerous level
                lcd.home();
                lcd.print(lineOneBufferout);
                lcd.setCursor(0,1);
                lcd.print("   HIGH SWR!!!  ");
                break;
        case ALARM5:                    // SWR is increasing
                lcd.home();
                lcd.print("     HIGH SWR   ");
                lcd.setCursor(0,1);
                lcd.print("NEEDS ATTENTION!");
                break;
    }
}
/* ==========================================================================
 *
 * Function used to flash backlight with timer1 interrupt. Enabled by
 * attaching interrupt:
 *
 *    timer1.attachInterrupt(FlashDisplay);
 *
 * and it is disabled by detaching it:
 *
 *    timer1.detachInterrupt();
 *
 * Input parameters
 *    void
 *
 * Return values
 *    none
 *
 ******/
```

LISTING 14-2 Arduino-based Directional Wattmeter/SWR Indicator. (*continued*)

```
void FlashDisplay()
{

  if (state != true) state = true;
  else state = false;

  switch(alarmType){
    case ALARM1:          // Power exceeds 20 Watts
    case ALARM3:          // Reflected power is greater than forward power
    case ALARM4:          // Rising SWR Caution ...
    {
      GOYELLOW;
      break;
    }
    case ALARM2:          // Power greatly exceeds 20 Watts DANGER!!
    case ALARM5:          // High SWR
    {
    if (state == true){
      GORED;
    }
    else GOBLACK;
      break;
    }
  }
}
```

LISTING 14-2 Arduino-based Directional Wattmeter/SWR Indicator. (*continued*)

Definitions and Variables

We begin with several *#include* compiler directives; *Liquid.Crystal.h* for LCD support, *LcdBarGraph.h* to generate a bar graph display as we did for the panel meter in Chapter 5, and two new libraries, *math.h* and *timerOne.h*. The *math.h* library is part of the Arduino distribution and provides a logarithm function needed for calculating RF power using a log amp. The *timerOne.h* library provides a timer interrupt function that we use. The *timerOne.h* library is downloaded from http://code.google.com/p/arduino-timerone/downloads/list.

The set of definitions starting with *FORWARDPOWERONE* are modified by the user to calibrate the power meter. As you know, the log amp measures power over a logarithmic scale and produces an output voltage that is the log of the input power. In order to convert to the equivalent power read from the input power read, we use a little bit of geometry, creating a line (representing the input power versus output voltage) with a slope and an intercept. Knowing the slope and intercept of the line, we are able to calculate the input power for any output voltage.

During the checkout and calibration procedure you captured two data points for each of the two sensors, forward and reflected. The calibration values are substituted into the eight *#define* statements as shown. The comments indicate which value goes where. The power values are the input RF power and the analog values are what was recorded form the LCD screen. The values included in the listing are for the sensor we built. The program uses these values to determine the slope and intercept for the forward and reflected power conversion.

The Wattmeter/SWR Indicator also includes several useful alarm functions. The alarm functions warm the operator of potentially harmful conditions such as high SWR or excessive power. High SWR can damage transmitting equipment so it is a good idea to avoid this as often as possible. Excessive power, on the other hand, can damage the power sensors in the directional coupler. The Alarm level section contains the threshold values for excessive power and SWR. The power alarm is of two stages, meaning that there is a "nudge" warning as you approach the dangerous power level, changing the color of the display backlight to yellow. The second stage is more insistent by flashing the display with a red backlight. If you should desire to do so, it is a simple matter to add an audible warning as well by adding another digital pin as an output to drive a buzzer.

We use a series of macro definitions provide a simple way of executing repetitive sequences of commands with a single name. What we have done is defined a series of *digitalWrite()* commands to set the different backlight colors. This macro definition:

```
#define GOWHITE   {digitalWrite(REDLIGHT, LOW);
digitalWrite(GREENLIGHT, LOW);digitalWrite(BLUELIGHT, LOW);}
```

creates *GOWHITE* that may be used anywhere in the program where we wish to set the backlight to white. Note the braces that surround the macro definition. In order to flash the red backlight, we use a timer interrupt and set the timer delay to the value defined by *DISPLAYFLASHRATE*.

One trait of the Arduino analog input functions is that, with any given analog input, there is more than likely going to be some jitter in the output values. This is not unusual for an analog to digital converter. In order to provide a more stable display, we decided to average a number of consecutive readings, thus smoothing the values being displayed. This is also called "smoothing." To average the values we set up two arrays: `forwardPowerCircularBuffer` and `reflectedPowerCircularBuffer`. As the name implies, these are "circular buffers," a fixed length buffer that is connected end-to-end, giving the appearance of an endless buffer. The key here is that with a buffer length of "n," the "n+1th" entry overwrites the "0th" entry. We then take the sum of the entries and divide by "n," the result being the average of the last "n" values. `BUFFERSIZE` sets the value of n.

setup()

Within `setup()` we initialize the input and output pins, the LCD and set the contents of the circular buffers to "0." To initialize the buffers, we use the `memset()` function:

```
memset(forwardPowerCircularBuffer, 0, BUFFERSIZE);
```

to set the contents of the arrays to zero. The *memset()* function is a standard library function that is tweaked for efficient setting of memory blocks to a given value. In addition, we calculate the calibration constants for slope and intercept from the calibration values edited from the test results.

Recall that we are using a circuit similar to that of the panel meter from Chapter 5. In this case as in Chapter 5, the op amp is used as a DC amplifier to scale the voltage from the log amp to a range more compatible with the analog input pins of the Arduino. In Chapter 5, we discussed the need to use an external voltage reference for the analog to digital converter (ADC). We have used the same design here, using one section of the quad op amp to provide the external reference voltage for the ADC. Remember that you must have executed the `analogReference(EXTERNAL)` command prior to connecting your interface shield to the Arduino. Just as in Chapter 5, there is a risk to the Arduino processor when the internal and external references are connected at the same time. To make things easier, we include a jumper pin on our shield to disconnect AREF if it is needed. It's much easier to pull the jumper than disassemble the stack of cards.

To determine the slope and intercept, we use two values of power and voltage for each of the two sensors. The slope is derived by the following equation, taken from the AD8307 datasheet:

$$slope = (analog\ 1 - analog\ 2)/(power\ 1 - power\ 2)$$

where analog 1 and analog 2 are the output of the ADC for the corresponding input power levels, power 1 and power 2. The equation for determining the intercept also comes from the AD8307 datasheet and thus:

$$intercept = power\ 1 - (analog\ 1/slope)$$

where power 1 and analog 1 are as used by the slope equation and slope being the result of the slope calculation.

loop()

The `loop()` section of the program is where the input and output processing takes place. We break `loop()` down in to a series of functions that, in sequence, read the raw data from the sensor, perform conversions between the value read and the values to be presented as output, and lastly, format the output data and send it to the display.

The first thing we do in `loop()` is read the analog inputs from the two sensors and average them using the circular buffer. The following snippet of code illustrates the code for the forward power sensor:

```
forwardBufferTotal = forwardBufferTotal - forwardPowerCircularBuffer[index];
forwardPowerCircularBuffer[index] = analogRead(FORWARDPOWERINPUT);
forwardBufferTotal = forwardBufferTotal + forwardPowerCircularBuffer[index];

...

index++;
if ( index >= BUFFERSIZE )
   index = 0;
forwardPowerLog = forwardBufferTotal / BUFFERSIZE;
```

The variable, `forwardBufferTotal`, is the sum of the values in the buffer. The first thing we do is subtract the `forwardPowerCircularBuffer[index]` value from the total. We then read the analog input and load the result into the array at `index`, and then add the value to `forwardBufferTotal`. We then increment `index`, and if it is greater than or equal to `BUFFERSIZE` it is reset to zero. The total is then averaged by diving by `BUFFERSIZE`. The remainder of `loop()` is processed and the sequence is repeated.

The circular buffer code also contains tests for the high power alarms. As the buffering and averaging produce some delay, we decided that it would be a good idea to detect high power and raise the alarm as quickly as possible. Rather than waiting for the average value, we alarm on the instantaneous value of each sensor. In other words, we test each reading to see that they are in the safe range. The following is the power alarm detection:

```
if ( forwardPowerCircularBuffer[index] > POWERLEVELALARMBAD ||
             reflectedPowerCircularBuffer[index] > POWERLEVELALARMBAD ){
      Timer1.attachInterrupt(FlashDisplay);
      alarmType = ALARM2;
      DisplayAlarm(alarmType);
      alarmLock = true;
}
```

```
    else if ( forwardPowerCircularBuffer[index] > POWERLEVELALARM ||
                   reflectedPowerCircularBuffer[index] > POWERLEVELALARM ){
        Timer1.attachInterrupt(FlashDisplay);
        alarmType = ALARM1;
        DisplayAlarm(alarmType);
alarmLock = true;
}else {
        alarmLock = false;
        Timer1.detachInterrupt();
}
```

The tests for high power are made against the raw, analog value, rather than going through the conversion process. Since 20 W is represented by an analog value of 1000, anything over 1000 is considered an alarm. An analog reading of 1010 is roughly 25 W and that is our second threshold for alarming. Above 20 W, the display changes to a yellow backlight. Above 25 W the display changes to a flashing red backlight. We use an interrupt from timer1 to flash the red backlight on and off.

One way to flash the display backlight is to use a *delay()* statement, but *delay()* has some negative effects on programs, primarily, when the *delay()* is executing the processor is not able to do any other processing, including interrupts. Rather than using *delay()* statements, we opted to use an interrupt set by timer1 and use the interrupt to turn the red backlight on and off. We set timer1 to generate an interrupt after 300,000 µS have elapsed. The statement:

```
Timer1.attachInterrupt(FlashDisplay);
```

is used to "attach" the timer to the function *FlashDisplay()*, in a sense, enabling the interrupt. The timer is first initialized in *setup()* with:

```
Timer1.initialize(DISPLAYFLASHRATE);
```

setting the timer's overflow value at *DISPLAYFLASHRATE*, or 300,000 µS. Once the timer interrupt is enabled, the overflow causes an interrupt to occur every 0.3 seconds, invoking the function, *FlashDisplay()*. The function sets the backlight color and depending on the alarm, flashes the backlight.

The boolean, *alarmLock*, is used to lockout additional display processing during an alarm condition. When there is an alarm condition, *alarmLock* is set to *true*. When the alarm condition is cleared, *alarmLock* is set back to *false*.

All of our power measurements from the AD8307 are a voltage representing the power expressed in dBm or "decibels milliwatt." Since power is a logarithmic function, dBm provides a linear progression through decades of power levels. The value "0 dBm" is equal to 1 mW. Therefore, 10 mW would be 10 dBm, 100 mW would be 20 dBm, and 100 µW would be –10 dBm. By converting the analog input value to dBm, the measured power may be displayed in Watts, milliwatts, microwatts, and of course, dBm. We again use a little geometry to calculate the power in dBm from the input analog value. Recall that in *setup()* we derived a slope and intercept for each sensor? Now it's time to use those values to calculate the power. We use the equation:

Power (in dBm) = (analog input/slope) + intercept

where slope and intercept are the values calculated in setup() for the sensors. The result is expressed in dBm.

It is now a simple matter to convert from dBm to power in Watts. The equation:

Power (in Watts) = 10 to the power (power in dBm – 30) / 10

Is used for the conversion. The code looks like this:

```
forwardPower = (float) pow(10.0,((forwardPowerDbm - 30.0) / 10.0));
reflectedPower = (float) pow(10.0,((reflectedPowerDbm - 30.0) / 10.0));
```

where we are using a new function *pow()*. The function *pow()* is used to raise a number to an exponent, in this case power in dBm minus 30 and then divided by 10 is an exponent of the base, 10.0. The resulting value is in Watts. Since 1 W is 1000 times 1 mW and 1 mW is 1000 times a μW, it becomes simple math to convert between these different units of measure.

One of the major functions of the Wattmeter project is to provide an SWR reading. The next steps calculate SWR, using the power measurements as input. Recall from earlier that the formula for calculating SWR is:

$$SWR = \frac{1 + \sqrt{\frac{P_R}{P_F}}}{1 - \sqrt{\frac{P_R}{P_F}}}$$

where P_R is the reflected power and P_F is the forward power.

Just as was done for excessive power levels, we test for high SWR and set alarms. As in the power alarms, the SWR alarms are set in two steps, the first being an SWR above 5:1, and the second, 10:1. Above 5:1, the backlight is set to yellow, and above 10:1, the backlight is set to red and flashed on and off using *FlashDisplay()*.

The SWR alarms use two additional locks: *highSwrLock* and *lowPowerAlarmLock*. The first lock, *highSwrLock*, is used to stop additional display processing while the alarm is on the screen, just as we did with the high power alarms. The second lock, *lowPowerAlarmLock*, serves a slightly different purpose. If the reflected power is greater than the forward power, the resulting SWR is a negative value. Applying power with the coupler connected in reverse is one way that this happens. There are also situations where the apparent reflected power is greater than the forward power. Even with no power applied, it is possible to produce a negative SWR. With no power applied, the analog input values are the quiescent output voltage of the log amps.

Because of variations between parts, it is possible to have the reflected power log amp having a larger quiescent voltage output then the forward log amp, hence a negative SWR reading. Since we really don't want to have an alarm when there is no power applied, *lowPowerAlarmLock*, prevents the SWR alarms from occurring below a preset power level, in this case 10 mW. The quiescent levels of the log amps with no input are typically in the 100 μW or −10 dBm range.

Last but not least, we format the power readings and SWR values to be sent to the display. Placing the decimal point becomes important when displaying data over many orders of magnitude range. Our meter directly measures power levels from less than 100 μW up to 20 W, over five orders of magnitude. In testing our meter in Dennis's lab, we found that the measured values were typically within 5% of readings made with his Hewlett Packard Model 438 Power Meter. More often than not, they were in the 1 to 2% range of error. Understanding the accuracy of an instrument is important to formatting how the results are displayed. For instance, given an accuracy of 5%, we would not provide a display of 10.00 W because that last fractional digit would be ambiguous as it is significantly smaller than 5% of the total value. We would choose to display the results as 10.0 because 5% of 10.0 is 0.5 corresponding to one fractional decimal place in the display.

Thus we adjust the display format based on the order of magnitude of the reading. The first step prior to any display formatting is to test for excessive input power, excessive SWR of reversed forward and reflected connections. If there are no alarm conditions, we proceed to the display selection based on the number of times the input switch is pressed.

There are four possible display formats:

1. Forward and Reflected power in Watts, mW and uW.
2. Forward and Reflected power in dBm.
3. SWR displayed in addition to a bar graph.
4. Forward and Reflected power in Watts with SWR.

There is one known bug in the program. Repeatedly pressing the pushbutton cycles through the four display modes of the Directional Wattmeter. The first time through the displays, the SWR bar graph displays as expected. However, subsequent cycles bringing up the SWR bar graph do not display the bar graph unless the input values for either forward or reflected power change. In a practical sense, there is enough jitter and noise in the system such that this is not a big issue. A little bit of noise on the input and the display pops right up. It's one of those items that just nags at a developer to make it right.

Further Enhancements to the Directional Wattmeter/SWR Indicator

The Directional Wattmeter was designed for QRP operation and has a limit on the amount of power that can be applied without damaging the sensor's log amps. To measure higher power levels, it is a simple of matter of inserting an attenuator in front of the inputs to the log amps. The attenuator must be designed for the characteristic impedance of the line section, or in our case an input impedance of 50 Ω. The log amp itself represents a load of about 1000 Ω. Another consideration for high-power operation is the choice of ferrite cores in the sensor. The cores we have used become saturated at higher power levels, the result being that the higher power readings become nonlinear. The design for a higher power sensor requires larger ferrite cores, such as the Amidon FT-82-67, and rather than RG-8X, uses RG-8/U or the equivalent. The turns ratios remain the same; the higher-power cores have 32 turns just as the ones we made for this Wattmeter.

Conclusion

There is always room for improvement in any endeavor. This project is no exception. We leave it up to the user to create new display formats for the project. Maybe you would like to see two bar graphs indicating power or possibly even a four-line display. Feel free to exercise your creativity, but be sure and share your ideas and changes with us on our web site (http://www.arduinoforhamradio.com).

CHAPTER **15**

A Simple Frequency Counter

In this chapter we construct a simple device for measuring frequency called a frequency counter. It uses the timer function in the Arduino's Atmel processor and measures frequencies from the audio range to a little over 6 MHz. It is constructed on a single Arduino R3 prototyping shield and includes an LCD to readout the frequency. The counter is surprisingly accurate and can easily be calibrated with a known accurate frequency source. The chapter ends with a discussion of how to use the frequency counter with a typical QRP radio that lacks a frequency readout. We use the Arduino Frequency Counter to measure the Variable Frequency Oscillator or "VFO" frequency and then display the corresponding receive or transmit frequency on an LCD. The frequency counter is a nice addition to many low-cost QRP rigs. Figure 15-1 shows the

FIGURE 15-1 The Arduino frequency counter connected to an MFJ Cub QRP transceiver.

frequency counter being used with an MFJ 40 Meter Cub QRP transceiver. In this case we use the Arduino to do a little bit of math to convert the VFO frequency to the actual frequency of the radio. We will discuss this in greater detail later in this chapter.

There are some indispensable tools that we use when working with radio gear. There are the obvious ones: side cutters, pliers, screwdrivers, soldering iron, and so on. A good multimeter is a useful tool as well. One tool that doesn't get much mention but is equally useful to the multimeter is a frequency counter. The name is a bit of a misnomer because you don't really count frequencies. What you are doing instead is counting the number of transitions, or "cycles," a signal makes over a known period of time. From the count we determine the frequency.

Frequency is nothing more than expressing the "cycles per second" or "Hertz" of a signal. So, if we measure 3000 transitions of a waveform over a period of one second, the result is 3000 cycles per second. Or is it? Take a look at Figure 15-2 depicting a sine wave. We see that there are actually TWO transitions in each "cycle," one "rising" and one "falling." So, what do we actually measure? We don't measure ALL of the transitions. Instead we count the cycles using *only* the rising or falling edge of the signal. If we count only the rising edges of the 3000 cycles per second signal, we actually count 3000 transitions.

But wait, there's more! Counting and timing are two things that a processor can do well. Arduino is no exception. It turns out that the Atmel processors used in the Arduino have a clever little added bonus; the controllers have internal programming that allows the hardware to count transitions on an input without the need for external software via a dedicated hardware function! All we have to do is turn it on and turn it off to perform the count and then read back the results.

Of course, the Atmel processor is a digital device and the counter hardware function is implemented on a digital IO pin. To be more specific, digital pin 5 implements the hardware counting function. However, the sine wave depicted in Figure 15-2 is an analog signal and may have varying amplitude; therefore, we must do something to convert the analog signal into something more compatible with the digital input. One way to convert the sine wave to a digital signal is with a comparator (a device that when the input exceeds a set threshold, the output, normally zero or low, goes high). Another method is to use a high gain amplifier and have the input signal drive the amplifier into "saturation," meaning that as soon as the input rises above zero, the output goes full "on" or is saturated. It is the saturation method we have chosen to "condition" our input signal so that the Arduino may count the transitions. Figure 15-3 shows how the sine wave is "transformed" into a digital square wave for the counter.

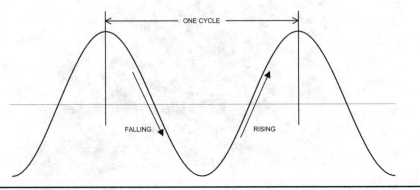

FIGURE 15-2 Transitions in a sine wave.

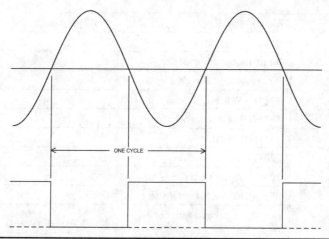

FIGURE 15-3   Deriving a digital signal from the sine wave.

Circuit Description

The input conditioning circuit for the frequency counter is shown in Figure 15-4. Two BC547B NPN bipolar transistors, Q1 and Q2, are cascaded to provide sufficient gain that a relatively low level signal produces adequate signal level to pin 5 of the Arduino for the counter to function. Table 15-1 is the parts list for the frequency counter.

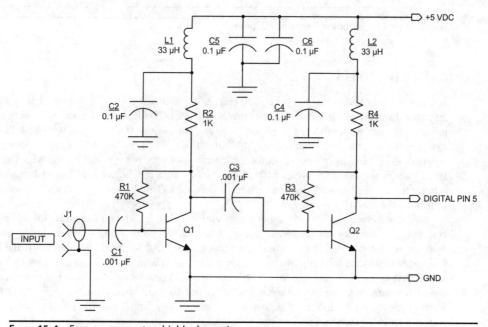

FIGURE 15-4   Frequency counter shield schematic.

Ref	Description	Part Number	Source / Mfg
C1, C3	0.001 µF monolithic capacitor		
C2, C4, C5, C5	0.1 µF monolithic capacitor		
J1	2-pin header, 0.1-in. centers		
L1, L2	33 µH ¼ W fixed inductor		
R1, R3	470 kΩ ¼ W resistor 5%		
R2, R4	1 kΩ ¼ W resistor 5%		
R5	220 Ω ¼ W resistor 5%		
R6	10 kΩ potentiometer		
Q1, Q2	BC547B NPN bipolar transistor		
misc	prototyping shield with header pins	ProtoPro-B, or equiv.	Omega MCU Systems
P1	2-pin plug and pins, 0.1-in. center	shell: 22-01-2021 pin: 08-50-0114	Molex
	Coax cable. 2 feet in length	RG-174 or similar	
	Test lead "grabbers"	XM25S	E-Z-hook, eBay
Optional prescaler:			
U1	Dual D-type Flip-flop	SN7474 or equiv	Jameco, Mouser, eBay, etc
JP1	3-pin header, 0.1-in. centers		
	Shorting plug to use with JP1		

TABLE 15-1 Parts List for the Frequency Counter

Constructing the Shield

You may have noticed from the schematic (Figure 15-4) that the frequency counter uses digital input pin 5 and that we have already used this pin for the LCD display in Chapter 3. True, but pins in conflict is something that we encounter time and time again while working with Arduino and Arduino-related projects. What can we do? As designers, we do the best job we can of optimizing re-use of the preceding projects in constructing new projects, but we can't always win. You may recall that we discussed the issues with "deconflicting" pins in Chapter 10, "Project Integration." But, for this project, we have decided to use a dedicated LCD on the frequency counter shield, thus mitigating the conflict between pins.

The component layout for the frequency counter shield is shown in Figure 15-5. We have constructed the frequency counter shield on a ProtoPro-B prototyping shield. In laying out the components for the shield, we encounter several issues. First, we must allow space for the large amount of real estate taken up by the LCD on the shield. Having the LCD mounted on the shield means that we need to use some low-profile construction techniques. You could, of course, make the LCD removable. The second issue is with regards to good high-frequency design. We bring high-frequency signals into the frequency counter shield and it is important that we keep these signals from affecting other parts of the Arduino. Good construction practice is that we keep leads short and that we bypass all power leads. These practices do not completely eliminate RF signals from getting into places we don't want them; however, they reduce the chances of our getting into

Chapter 15: A Simple Frequency Counter

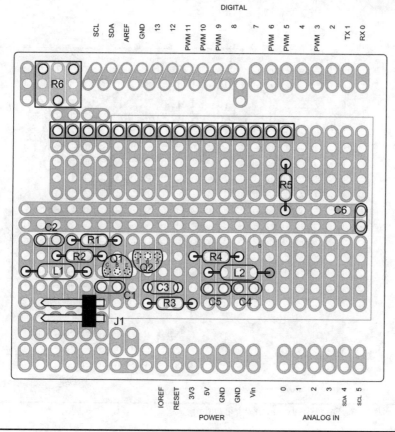

FIGURE 15-5 Frequency counter shield component layout.

trouble with stray RF. Of course, you can also take this project and mount it in a nice shielded enclosure.

The input cable for the counter is made from a piece of coax such as RG-174 or similar flexible, lightweight coax, that is terminated on a 2-pin Molex socket that connects to J1. Our test cable is about 2 ft in length. The Molex shell is a part number 22-01-2021 and the pins are part number 08-50-0114. The pins are a crimp-style part; however, if you don't have the correct crimping tool, you can solder them or crimp them carefully using a pair of needle nose pliers. The shell and pins are available from Mouser Electronics (www.mouser.com) and other places such as eBay. They are inexpensive and we use similar Molex parts in other projects in this book. This series of Molex connectors mates nicely with the 0.10-in. header pins that we use. You can terminate the other end of the coax with alligator clips.

Looking at Figure 15-6, the wiring diagram for the frequency counter shield, we see that pin 5 has been reused as the input for the frequency counter and that LCD data pins, DB4 through DB7, have been relocated to digital pin 6 through 9. As you might expect, the software (Listing 15-1) must also reflect this change in defining the data pins assigned to the LCD. We found that creating an Excel worksheet really helps in solving pin conflicts, but sometimes, there are no solutions. You can see our worksheet in Appendix C. While many pins on the Arduino serve multiple functions, some, like the hardware counting functions, are only available on dedicated pins, in this case, digital pin 5. The completed shield is shown in Figure 15-7.

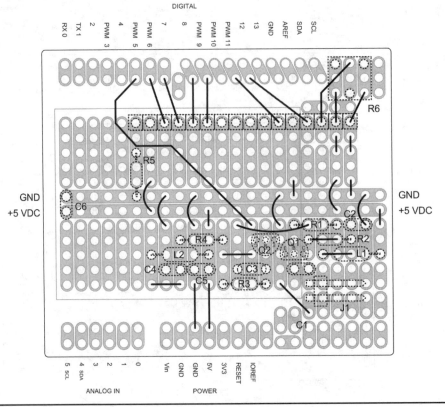

FIGURE 15-6 Frequency counter shield wiring.

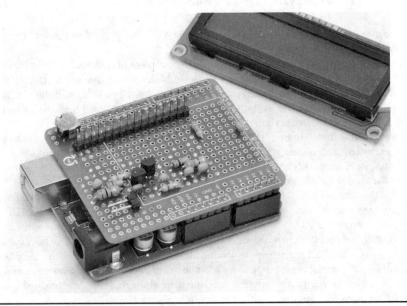

FIGURE 15-7 The completed frequency counter shield.

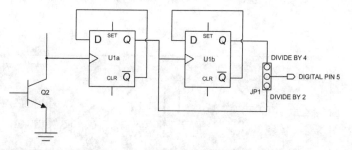

FIGURE 15-8   Divide by four prescaler using two D-Type Flip-Flops.

An Alternate Design for Higher Frequencies

One of the limitations of the Arduino and in particular the Atmel ATmega328 processor family is the maximum rate at which the timer can count. Using a 16 MHz clock rate (normal for the Arduino Duemilanove and Uno), the maximum frequency that can be expected to be read is a little above 6 MHz. So, if we want to measure higher frequencies, what do we do?

One alternative is to add a prescaler to the counter. The prescaler goes between the input signal conditioning transistor and digital pin 5. We use a binary counter to *divide* the frequency down to a range that can be measured. In this case, we are adding a TTL Dual D-Type Flip-Flop, an SN7474 14 Pin DIP part. A single D-Type Flip-Flop is configured as a "divide by two" counter by connecting the Q "NOT," or Q "bar" output, to the "D" input of the Flip-Flop and applying the input pulses to the clock input. The "Q" output is half the rate of the input pulse rate. Figure 15-8 shows both halves of an SN7474 Dual Flip-Flop configured as a divide by

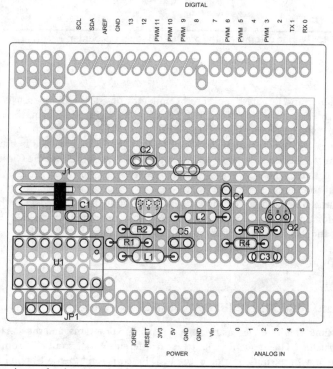

FIGURE 15-9   Parts layout for the prescaled frequency counter.

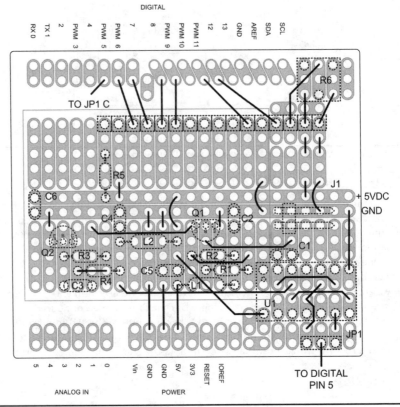

FIGURE 15-10 Wiring diagram for the prescaled frequency counter.

four counter and inserted between the collector of Q1, the input transistor, and digital pin 5 of the Arduino. Thus, for every four pulses of the input waveform, the Arduino counts only one. So, instead of being limited to a little over 6 MHz on our counter, we can now measure frequencies to approximately 24 MHz. Jumper JP1 in Figure 15-8 is used to select between divide by two or divide by four. A CMOS Dual D-Type Flip-Flop, the CD4013B, can be used as an alternative to the TTL SN7474. Either part works; however, the CD4013B has a different pinout than the SN7474.

Figures 15-9 and 15-10 show the way we built the "prescaler" with a parts layout and wiring for this add-on design. To read the correct frequency on the LCD, we also made a modification to the software to multiply the count by one, two, or four by uncommenting the appropriate *#define* statement. This means that the prescaler is selectable when you compile the software.

Code Walk-Through for Frequency Counter

By this time, you have seen many different Arduino programs, and they all follow much the same basic structure. The frequency counter is no different. We start out with some definitions and then follow that with the initialization of the LCD in *setup()* and then the main body of the program is *loop()*. The only new item in this program is a new library, FreqCounter.h. Download the library using this link:

http://interface.khm.de/wp-content/uploads/2009/01/FreqCounter_1_12.zip

and unzip the file into your library directory in the same manner as with other new libraries (e.g., Chapter 4).

The source code for the frequency counter is provided in Listing 15-1. The design is based on work done by Martin Nawrath from the Academy of Media Arts and utilizes the FreqCounter.h library created by Martin. The library uses the ATmega328 programmable timers, TIMER/COUNTER1 to count the transitions on digital pin 5, and TIMER2 to set the gate time.

Essentially, TIMER2 is set up to count clock ticks to equal the preset gate time. During the period of time that TIMER2 is counting, TIMER/COUNTER1 is counting the rising edge transitions on digital pin 5. Once the count is completed, the value is passed back to the program.

The returned value is converted to an actual frequency (cycles per second, or Hertz) by multiplying the number of gate times in one second, in this case 1000 milliseconds divided by 100 milliseconds gate time or 10, and then the result is divided by 10E6 to obtain the result in megahertz (MHz). The resulting frequency is then cast as a floating point number.

We use the stdlib.h functon, *dtostrf()* to convert from the *float* value to an ASCII text string and stuff the result into *buffer*. We then concatenate the contents of *buffer* with the scale units string (MHz) and then output the contents of *buffer* to the display.

One thing to note: The resolution of the counter is determined by the gate time. With a gate time of 100 milliseconds, the counter is limited to 10 Hz resolution. To increase the resolution to 1 Hz, the gate time is changed to 1 second by setting the value of GATETIME to 1000.

```
/************
 *    A Simple Frequency Counter   ver 1.0  27 March 2014 D.Kidder W6DQ and
 J.Purdum, W8TEE
 *
 *    This frequency counter is based on the work of:
 *
 *    Martin Nawrath KHM LAB3
 *    Kunsthochschule f,r Medien K^ln
 *    Academy of Media Arts
 *    http://www.khm.de
 *    http://interface.khm.de/index.php/labor/experimente/
 *
 *    and uses the FreqCounter.h library from the same source.
 *
 *    Pulses are counted on Digital Pin 5 during the gate time and then converted
 *    into the frequency in Hz, KHz, or MHz, depending on settings in the code used
 *    during compile. The counter has a maximum frequency range of a little over
 *    6 MHz without prescaling. By adding a simple divider consisting of one or two
 *    D-Type Flip-Flops arranged as divide by 2 or by 4 respectively.  Using divide
 *    by 4 raises the maximum frequency to roughly 24 MHz.
 *
 *************/
```

LISTING 15-1 Frequency counter source listing.

```
#include <FreqCounter.h>
#include <LiquidCrystal.h>
#include <stdlib.h>

#define DEFAULTPRECISION 6          // Default decimal places
#define VALUEWIDTH       10         // Numeric field width

#define LCDNUMCOLS       16         // number of columns on LCD
#define LCDNUMROWS       2          // number of rows

#define STARTUPDELAY     3000       // Delay used for splash screen at startup
#define COUNTDELAY       20         // Measurement delay

#define CALIBRATION      0          // Adjusts frequency for variation in Arduino
                                    // clock rate
#define GATETIME         100        // Gate time in milliseconds

#define MEGAHERTZ        1000000    // Let's define a couple of constants
#define MILLISECONDS     1000       // Number of milliseconds in a second

/******
 *
 * The following three definitions are for using different prescaling divider
 * circuits.  Only one line should ever be uncommented.  The prescaler circuit
 * divides the input by 2 or 4 or not at all.  To adjust the displayed frequency,
 * we multiply the result by the same amount.
 *
 ******/

#define PRESCALER        1          // Uncomment when no divider is used
//#define PRESCALER      2          // Uncomment when divide by two is used
//#define PRESCALER      4          // Uncomment when divide by four is used


#define FIRSTIF          12.0       // First IF frequency of the radio
#define CALOFFSET        -.004988   // Determined through testing ...
                                    // offset from 1st IF

char  scale[] = " MHz";             // Scale string for display units
float scaleFactor = 1.0;            // Scale factor used to adjust units.
                                    // Default is 1.0 for Hz, 0.001 for KHz and
                                    // .000001 for MHz

LiquidCrystal lcd(12, 11, 9, 8, 7, 6); // Setup the LCD hardware interface pins

void setup()
{
```

LISTING 15-1 Frequency counter source listing. (*continued*)

```
    lcd.begin(LCDNUMCOLS,LCDNUMROWS);    // initialize the LCD

    lcd.print("Freq Counter ");          //sappy stuff, name, version and date
    lcd.setCursor(0,1);
    lcd.print("ver 1.0 27Mar14");
    delay(STARTUPDELAY);
    lcd.clear();
    lcd.setCursor(0,0);
    lcd.print("Frequency:");

}

void loop() {

  char buffer[LCDNUMCOLS + 1];           // Make it big enough for a 2x16 display
  unsigned long dataIn;
  float val;
  int i;

  FreqCounter::f_comp=CALIBRATION;       // Calibrate with known source
  FreqCounter::start(GATETIME);          // Count pulses for specified gating period

  while (FreqCounter::f_ready == 0)

     dataIn = FreqCounter::f_freq;;

  delay(COUNTDELAY);

  val = (float) dataIn * PRESCALER * scaleFactor * (MILLISECONDS / GATETIME) /
MEGAHERTZ ; // Scale the input reading...


  val = FIRSTIF + CALOFFSET - val; // Add the frequencies from the stages:
                                   // VFO plus IF.  CALOFFSET is used to
                                   // calibrate the display using a known source


  dtostrf(val, VALUEWIDTH, DEFAULTPRECISION, buffer); // Convert and format value

  strcat(buffer, scale);    // Concatenate contents of buffer and the scale units
                            // into buffer

  lcd.setCursor(0,1);       // Set up the display...
  lcd.print(buffer);

}
```

LISTING 15-1   Frequency counter source listing. (*continued*)

Displaying the Tuned Frequency of Your Display-less QRP Rig

There are a number of QRP radios that take a minimalist approach in their design and as a result have no frequency display. For many of these rigs, it is a fairly simple matter to connect a frequency counter, such as the one described in this project, to obtain a direct frequency readout. We describe how to connect the Arduino Frequency Counter to a typical QRP radio and then give several examples.

There are several different approaches to the design of QRP radios. The simplest is the "direct conversion" or "zero IF" receiver/transmitter, but more and more dual-conversion transceivers are being built because availability of low-cost circuits like the SA602/SA612 Double Balanced Mixer/Oscillator chip. We discuss the dual-conversion applications first and then the direct conversion sets. We show how the Arduino Frequency Counter is connected to several popular QRP radios available today. We have selected the MFJ Cub QRP Transceiver as an example. The Cub has six different models, each a single band transceiver.

Double Conversion Applications

The majority of QRP radios these days use a double conversion scheme. The availability of low-cost mixer/oscillator combination chips, such as the SA602, has made it very economical and provides a high degree of stability and performance. But, many times what comes with the low cost is no frequency display. You get a radio with a tuning knob. And with that, you have no idea where you are on the band. What these radios need is a frequency display.

The double conversion receiving scheme is shown in Figure 15-11. A double conversion transmitter would be very similar, with the signal flow in the opposite direction. The first mixer combines the incoming signal with the VFO to produce a signal in the intermediate frequency or "IF" passband. The second mixer combines the output of the IF with the BFO (Beat Frequency Oscillator) to produce the recovered audio signal. The key elements of this project are determining the frequency of the VFO and the IF. The IF is a fixed frequency and should be known from the radio design. The VFO is variable and is what we want to measure with the frequency counter. In many cases, the transmitter is also controlled by the same VFO as the receiver, but the second mixer may be separate and operating at a slightly different frequency than the receiver. Why is the transmitter on a different frequency? Often we use the transmitted signal as a sidetone to monitor our sending by leaving the receiver operating but with greatly attenuated output. If there was no

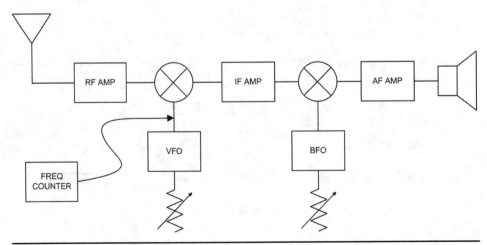

FIGURE 15-11 Double conversion scheme.

offset between the transmitted and received frequencies, we would hear a tone of zero Hz. A better way to state this is that the resulting tone would be zero Hz and we can't hear that.

It may actually be more important to know the transmitted frequency rather than the received frequency. Many times, this choice really depends on the design of the radio and how accessible the transmit frequency is. In either case, the methodology remains the same.

As an example, let's use a radio operating at 14 MHz with a 5 MHz IF. Assume the radio tunes 14.000 through 14.070 MHz. With these numbers in mind, the VFO operates at the difference of the two frequencies; therefore when the radio is tuned to 14.000, the VFO is 14.000 minus 5.000, yielding a difference of 9 MHz. We use the frequency counter to measure the VFO frequency (i.e., 9 MHz) and that result is added to the IF (i.e., 5 MHz) and the sum of these two values is the radio's frequency (14 MHz). Again, there may be small offsets introduced between transmit and receive to provide a sidetone frequency on transmit, but these can be determined and taken into account when the display frequency is calculated. Because the IF is fixed, if we know the result at one VFO setting, we use the same calculation at any VFO setting for this radio on this band.

Adding a Frequency Display to the MFJ Cub QRP Transceiver

MFJ has a series of single-band QRP radios. These offer some good performance for the cost, but they do lack a frequency display. We chose to add the frequency display to the 40 Meter Cub, the MFJ 9340 transceiver. From the information provided by MFJ in the Cub's manual, the VFO operates in the 5 MHz range and the IF is 12 MHz. Different bands use different combinations of VFO and IF frequencies as illustrated in Table 15-2.

When the VFO frequency is less than 6 MHz, the counter is used without a divider stage; hence the NA entries in Table 15-2. For the Cub, only two models require a divider: 9317 and 9315. A divide by 2 circuit is sufficient in these two cases.

As it turns out, the Cub makes it very easy to add the frequency display. In this case, we monitor the transmit frequency. The Cub uses an SA602 Mixer/Oscillator for the first mixer and VFO as shown in the partial schematic in Figure 15-12. The output of the oscillator not only is used in the receiver's first mixer, it is passed to the transmitter through a buffer transistor, Q4. The buffer is always on so we can monitor the transmit frequency all the time. Transmit power level is controlled by the divider network of R18 and R19. We tap into the transmit VFO across the pot, R19. Figure 15-13 shows how we installed a small length of RG-174 coax from our tie point on R19 to an RCA jack to access the VFO buffer output. We drilled a new hole and mounted the RCA jack on the rear panel. Figure 15-14 is a close-up, showing the center conductor of the coax (the white lead) connected to the "hot" side of the pot, R19. The coax shield is connected to the ground side of the pot and is the black wire

One thing to note: When the frequency counter is attached to the Cub, the VFO frequency is "pulled" off slightly. This is to be expected, as even with the buffer stage our circuit loads down the oscillator in the SA602. This is not a big deal, but we have to account for it in our calibration adjustments.

MODEL	BAND	VFO	IF	DIVIDER
9380	80	6	10	NA
9340	40	5	12	NA
9330	30	4.1	6	NA
9320	20	4	10	NA
9317	17	8.06	10	÷ 2
9315	15	9	12	÷ 2

TABLE 15-2 MFJ Cub Transceiver VFO and IF Frequencies

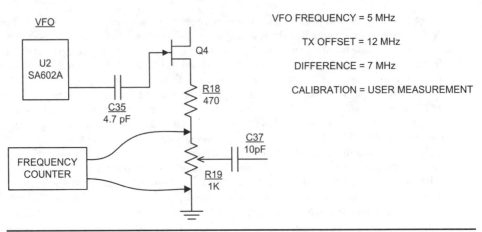

FIGURE 15-12   Attaching the frequency counter to the MFJ Cub transceiver.

FIGURE 15-13   The "pickoff point" from the frequency counter in the MFJ Cub. *(Cub courtesy of MFJ Enterprises)*

Calibration of the frequency counter with the Cub is accomplished using a known reference source. The most readily available known reference is a good communications receiver. With the 40 Meter Cub, we set the receiver to 7.000 MHz. Depending on which model Cub is used, set the receiver to the appropriate frequency. The Cub is connected to a 50 Ohm dummy load (e.g., Chapter 6) and the frequency counter. Tune the Cub to zero beat on the receiver while transmitting a signal. A frequency close to the expected 7.000 MHz frequency should appear on the LCD. Once

Chapter 15: A Simple Frequency Counter   345

FIGURE 15-14   Close-up of the pickoff installed in the 40 Meter MFJ Cub QRP transceiver.

the signal is zero beat on the receiver, note the displayed frequency on the Arduino LCD. The difference between the displayed frequency and the receiver's 7.000 is the calibration offset that is used in the frequency counter. It is important to note whether the displayed value is larger or smaller than the receiver's frequency as this difference determines whether the adjustment, or "offset," is added to or subtracted from the frequency counter's displayed value. The offset is used for the *CALOFFSET* definition in the program.

Adding a Frequency Display to a NorCal 40

The NorCal 40, developed by Wayne Burdick (N6KR) of Elecraft fame, was originally distributed by the Northern California QRP Club. The NorCal 40 is now available through Wilderness Radio (http://www.fix.net/~jparker/wild.html). This is a very popular radio and many of them have been sold over the years. The NorCal 40 does lack a frequency display, and adding one is fairly simple. Figure 15-15 shows the frequency plan of the radio.

The NorCal 40A uses a VFO covering the range of 2.085 to 2.125 MHz giving a tuning range of 40 kHz. The VFO is mixed with a 7.000 MHz incoming signal, the difference producing the IF of 4.9150 MHz. The IF is mixed with the BFO at 4.9157 MHz, the difference being 700 Hz, which is a pleasant pitch for copying Morse code on CW. The transmit scheme is the reverse, but with a slightly different mixer frequency. For 7.000 MHz out, the 2.085 MHz VFO is mixed with 4.9150, the sum being 7.000 MHz. By using a separate TX LO and RX BFO that differ by 700 Hz, the receive frequency is offset to provide sidetone and a pleasant tone for reception.

Figure 15-16 is a partial schematic of the NorCal 40A showing the best points to connect the frequency counter. JFET Q8 is the common VFO and the frequency counter is connected across the RF choke (RFC2) connected to Q8's source lead. Just as we did with the Cub, we use a short length of RG174 coax, connecting the center conductor to the Q8 source side and the shield braid

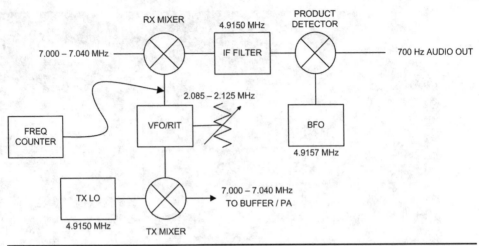

FIGURE 15-15 NorCal 40A partial block diagram showing frequency plan.

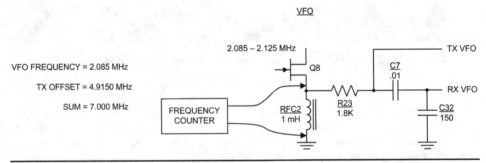

FIGURE 15-16 NorCal 40A partial schematic showing connection for frequency counter.

to the ground side of RFC2. The value in the program for *PRESCALER* is set to 1 and for *FIRSTIF* is set to 4.9150. Again, the value for *CALOFFSET* must be determined by testing.

Direct Conversion Applications

There are a number of Direct Conversion (DC) radio designs available. Figure 15-17 shows how a direct conversion receiver is configured. For the sake of clarity, only the receiver is shown. The receiver consists of an RF amplifier stage followed by a mixer. The VFO is operating at close to the same frequency as the incoming signal. The output is the difference between the frequency of the incoming signal and the VFO frequency. What we want in the output is in the audio range, so if the VFO is close to the received frequency (in other words, within a kHz or so), the difference is an audio signal. We can measure the frequency of the VFO and have a pretty good idea of the received frequency. It won't be exact because we designed the receiver to be slightly off frequency so that we would recover an audio output. If the VFO were exactly the same as the incoming frequency, the difference would be zero and there would be no recovered signal.

Because a DC radio uses a common VFO for transmit and receive, the transmitted frequency is the same as the receive frequency. Generally this causes a problem in that when the received and transmitted signals are on the same frequency, there is no difference signal to be heard as audio. The recovered signal is zero Hz. Many DC radios use a TX offset to move the transmitted

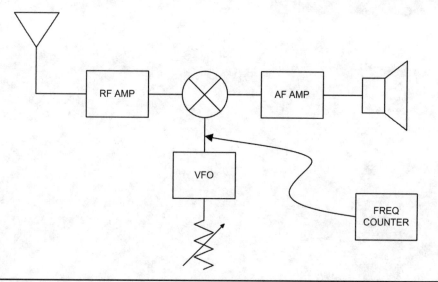

FIGURE 15-17   Direct conversion scheme.

frequency slightly away from the received frequency in order to have enough difference to produce an audio tone on the output. Typical offsets are in the range of 500 to 1000 Hz.

To connect the frequency counter to a DC radio, it is only necessary to identify the radio's VFO circuitry as we have done in the previous examples and connect the frequency counter's input to the output of the VFO.

Other Radio Applications

The Arduino Frequency Counter is usable with a wide variety of radios; not just QRP sets. Any radio can have a digital frequency display added. It is simply a matter of determining where to "pickoff" the signal to be measured and adding a connection. There are some complexity issues when dealing with a multiband radio, however. The QRP rigs we have discussed are kit-built, single-band rigs. The VFO and, in the case of dual-conversion schemes, the LO are operating over fixed ranges. However, in a multiband radio, while the VFO generally remains in a fixed range and the IF remains fixed, the LO changes to accommodate the different bands. This implies that, while the math may stay the same for calculating the actual frequency, the LO values change for different bands.

One possible solution is to provide a signal from the radio to the Arduino Frequency Counter to indicate what band is in use and the software then detects this signal and switches to the appropriate LO frequency. The radio's band switch is used to provide an output to a digital input on the Arduino to indicate the current band.

Conclusion

The possibilities are endless. We have provided a basic platform that can be used in many ways: as a piece of simple test gear for your workbench, to a frequency display for many QRP rigs. We hope to hear from you as to how you have used the frequency counter in your projects. Post your ideas and share your stories on our web site at www.arduinoforhamradio.com.

CHAPTER 16
A DDS VFO

There was a time when a transmitter consisted of a single vacuum tube operating as a keyed oscillator. Well, you can imagine what that sounds like on the air! "CW" was not called "chirpy warble" for nothing. Crystal oscillators improved on the stability of the early transmitters a great deal, but with a crystal, you are, for the most part, stuck on one frequency, which meant you needed a lot of crystals if you wanted to move around the bands. Along comes the Variable Frequency Oscillator or "VFO" and suddenly we were free to move about the bands. But, VFOs still tended to drift a lot during warm-up and were sensitive to many variables, not the least of which might be the operator's body capacitance.

Today, it seems, we can't get by without having a digital frequency readout down to the millihertz and stability that rivals the national frequency standards at NIST. Still, we do strive to know accurately our operating frequency and it is not all that hard today, with some readily accessible technologies. Enter, the Direct Digital Synthesis VFO or DDS VFO.

Synthesized frequency generators have been around for some time. Most modern amateur radios use one form of synthesizer or another. One common type is the Phase-Locked-Loop, or PLL, synthesizer. We have already played with a PLL circuit in the Morse code decoder in Chapter 8. Shown in Figure 16-1, a PLL synthesizer uses a Voltage Controlled Oscillator, or VCO, to generate the output frequency. A small portion of the output is fed back through a divider and then compared to a stable reference oscillator, generally a crystal oscillator. The difference between the reference and the divided down VCO output is then negatively fed back to the VCO so that any variation in the output frequency causes a small error voltage to be generated, pulling the VCO

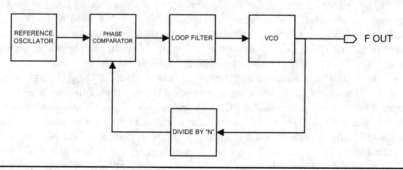

FIGURE 16-1 The Phase-Locked-Loop synthesizer.

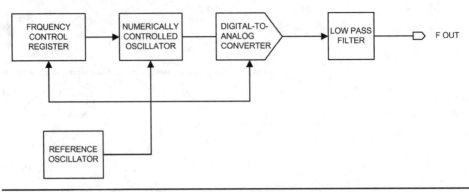

FIGURE 16-2   The direct digital synthesizer.

back in the opposite direction to the original variation. The result is a stable output frequency. Changing the divider value allows the VCO to move to other frequencies.

While PLL synthesizers are available, we chose not to use one for this project. We use a different type of synthesizer, the Direct Digital Synthesizer (DDS), for this Arduino Project.

Direct Digital Synthesis

Direct Digital Synthesis or DDS is a fancy name for a type of frequency source that digitally generates the desired frequency. As you can see in Figure 16-2, the key elements of the DDS are the Numerically Controlled Oscillator (NCO) and the Digital to Analog Converter (DAC). The NCO produces a numerical representation of the output of signal, generally a sine wave, that is then turned into an analog output by the DAC. The NCO is controlled by the "Frequency Control Register," a fancy name for a register that contains the digital value of the desired output frequency. The output is "cleaned up" with a low pass filter, removing any harmonic content. The entire device uses a stable reference clock to maintain accuracy. Of course, this architecture lends itself to being controlled by a digital processor such as an Arduino. Arduinos can be very good at stuffing a digital value into a register on some external device (as we have done in projects such as the Station Timer/Real Time Clock in Chapter 4).

The DDS VFO Project

Because of the complexity of the software, this project requires an ATmega328 or higher µC. The software requirements force this choice not because of the Flash memory limitations of the smaller chips, but rather because of the SRAM limitations. As you recall, SRAM is used for temporary storage for variables, plus the compiler copies string literals into SRAM unless you use the *F()* macro to prevent it. Finally, the software uses interrupt service routines (ISR) to react to signals sent from a rotary encoder. Because the 328 family of µC chips only has two interrupt pins (i.e., pins 2 and 3) for external interrupts, the LCD display is managed somewhat differently, too.

Our DDS VFO uses an Analog Devices AD9850 DDS chip. The AD9850 is a DDS synthesizer capable of generating a clean sine wave up to 62.5 MHz (spurious emissions at least 43 dB below the output frequency), operates on 5 VDC, and is housed in a 28-pin SSOP surface mount package. Now, before you get all excited about having to deal with a surface mount part again, even though we showed you the sure-fire method of soldering SMD parts, we use a preassembled breakout module that uses the AD9850. All the "dirty work" is already done! The module consists of the AD9850 DDS chip, a 125 MHz reference oscillator and the output low pass filter. The whole thing is built on a small circuit board that can be mounted on a prototype shield with header pins and sockets (so that the module can be removed) or soldered directly to the proto shield with header pins alone. A sampling

FIGURE 16-3 DDS breakout module.

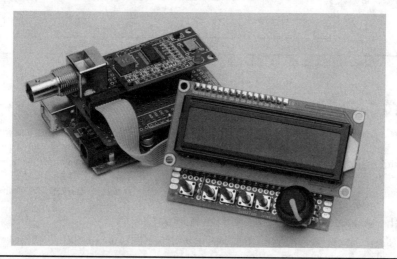

FIGURE 16-4 The completed DDS VFO with BNC output.

of eBay DDS modules using the AD9850 is shown in Figure 16-3. The DDS module we used is the one to the right in Figure 16-3. As shown in the same photograph, the underside of the module has the pinouts clearly marked. These modules are available on eBay for under $5!

As we stated earlier, a DDS module lends itself well to digital control. We take advantage of that by creating a VFO that employs many features that are provided through software. The completed DDS VFO is shown in Figure 16-4. The project comprises a DDS shield, the DDS VFO software, and a control panel board to operate the VFO. The control panel is the control panel board we used in Chapter 13 for the rotator controller. Here is a list of features we include in this design:

- Frequency coverage of HF ham bands (160 through 10 meters)
- Frequency readout to 10 Hz
- Marked band edges
- One memory per amateur band

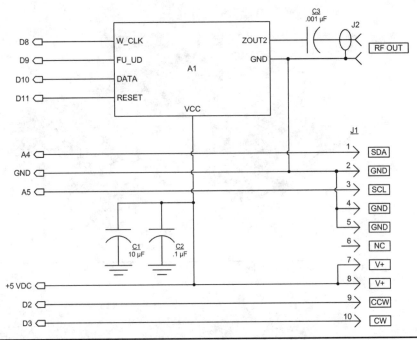

FIGURE 16-5   DDS VFO shield schematic.

DDS VFO Circuit Description

The schematic diagram for the DDS VFO shield is shown in Figure 16-5. As with the Rotator Controller in Chapter 13, we use the I2C bus interface (analog pins 4 and 5) and the external interrupt pins (digital pins 2 and 3), which are "passed through" the shield. Device A1 is the DDS module. The module has four inputs and one output that are used. Other than V_{cc} and ground, all remaining pins are not used. Table 16-1 lists the parts needed for the DDS VFO shield.

The Analog Devices AD9850 Breakout Module

The AD9850 breakout module that we obtained from eBay is a small circuit board with two 10-pin headers already installed as shown in Figure 16-3.

Ref	Description	Part Number	Source / Mfg
C1	10 µF Electrolytic, 16 VDC		Jameco, eBay
C2	0.1 µF monolithic capacitor		Jameco, eBay
C3	0.001 µF monolithic capacitor		Jameco, eBay
J1	10-pin header, 0.1-in. centers		Jameco, eBay
J2	BNC connector, PC mount		
A1	Analog Devices AD9850 mounted on breakout board		eBay
misc	prototyping shield with header pins	ProtoPro-B	Omega MCU Systems, eBay

TABLE 16-1   DDS VFO Shield Parts List

Constructing the DDS VFO Shield

We again chose to use the ProtoPro-B prototyping shield for assembling the DDS VFO. Header sockets are used to connect to the DDS breakout module and two 5-pin headers are used to provide a connection to the control panel board. The layout shown is for the DDS breakout module that we used. There are several different varieties available, so make sure that if yours is a different type, check the pinouts by the signal name as they are probably different.

We built several versions of the DDS VFO shield. One version uses a BNC connector for the output (see Figure 16-4) while another uses a 2-pin 90-degree header (J2) as shown by the layout in Figure 16-6. The connector you use is a matter of personal choice and depends upon how you intend to use the project. For the version with the 2-pin header, we assembled a short jumper cable that terminates with a BNC connector.

The wiring of the shield is shown in Figure 16-7. We used our standard wiring method of using bare tinned wire and Teflon tubing. Using a drill, we had to slightly enlarge the two holes for the mounting pins to go through the proto board in order to use the PC-mount BNC connector. Figure 16-8 is a photograph of the completed DDS VFO shield showing the BNC connector we used.

The wiring side of the completed DDS VFO shield is shown in Figure 16-9.

Adding an Output Buffer Amplifier for the DDS VFO

For some applications, the output level from the DDS module alone may be too low (about 500 mV). Such is the case if you drive vacuum tube gear, which typically needs a 5 to 7 V signal level. We added a simple buffer/gain stage to the output to kick the voltage level up a bit. We show two additional drawings for the buffered version of the DDS VFO. Figure 16-10 is the schematic for the

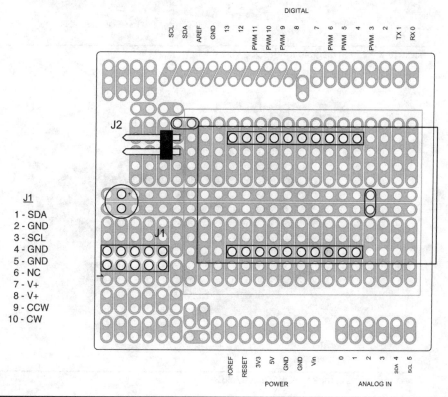

FIGURE 16-6   Parts layout for the DDS VFO shield.

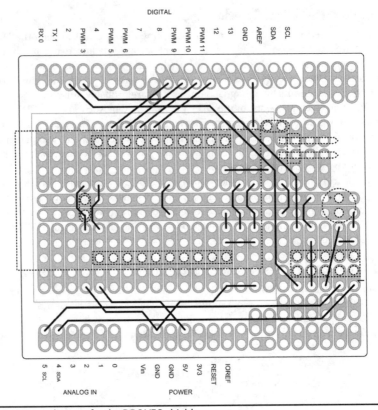

FIGURE 16-7 Wiring diagram for the DDS VFO shield.

FIGURE 16-8 The completed DDS VFO shield.

Chapter 16: A DDS VFO   355

FIGURE 16-9   Wiring side of the completed DDS VFO shield.

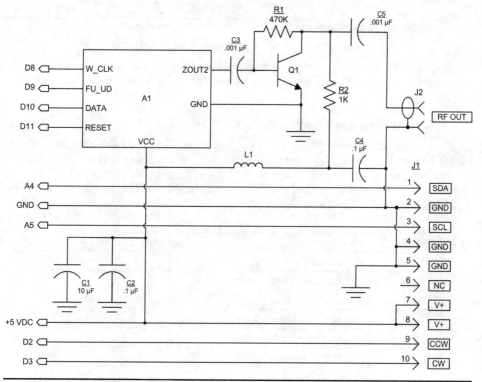

FIGURE 16-10   DDS VFO shield schematic with added output buffer.

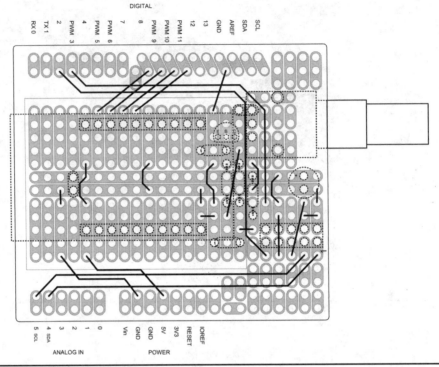

FIGURE 16-11   Wiring diagram for DDS VFO shield with buffered output.

shield and Figure 16-11 is the wiring diagram. The wiring diagram also shows our version that uses the PC-mount BNC connector. The additional parts needed for the output buffer are listed in Table 16-2.

The Front Panel and Interconnection

We use the rotator controller control panel board from Chapter 13 with no changes. Refer back to Chapter 13 for the details on constructing the control panel board. We made a short jumper cable to connect the control panel to the DDS VFO shield using ribbon cable and FC-10P connectors just as we did in Chapter 13. If you are mounting the DDS VFO in its own enclosure, make sure that you make the cable long enough.

Ref	Description	Part Number	Source / Mfg
C4	0.1 µF monolithic capacitor		
C5	0.001 µF monolithic capacitor		
L1	33 µH ¼ W fixed inductor		
R1	470 kΩ ¼ W resistor 5%		
R2	1 kΩ ¼ W resistor 5%		
Q1	BC547B NPN bipolar transistor		

TABLE 16-2   Additional Parts Needed for the DDS VFO Buffer

You are now ready to connect it all together and begin testing. First, as always, make sure that you double-check all of your wiring. Remember, you can use a multimeter to measure the continuity between the points on the schematic. If you don't have a multimeter, then perform a careful visual inspection. Once you are satisfied with your work, you can connect the pieces together and start loading the software.

However, before we jump into a discussion of the software, it is helpful to first understand the functionality of each of the controls depicted in Figure 16-4. Understanding what those controls do makes understanding the software that much easier.

DDS VFO Functional Description

The user interacts with the VFO via a program sketch (VFOControlProgram.ino) that manages the hardware. We refer to this program as the User Interface Program (UIP). There are five pushbutton switches (SW1 through SW5) that control most of the functionality of the VFO. A sixth switch (SW6), also a pushbutton switch, is part of the rotary encoder. An explanation of what each of these switches do provides a good description of the functionality of the VFO. We discuss these switches in a left-to-right manner, which is the order in which they appear on the control panel.

Overview

When the DDS VFO is initially turned on, the frequency stored when you last used the DDS VFO is read from EEPROM memory. That frequency is then displayed on line 1 of the LCD display. Line 2 of the display shows the ham band associated with the displayed frequency. Figure 16-12 shows what the display might look like when you power it up. In Figure 16-4, you can see the DDS shield with the breakout module that holds the AD9850 chip piggybacked onto an Arduino Duemilanove. The ribbon cable connects the DDS VFO shield to the control panel board, with the LCD, pushbutton switches, and rotary encoder. The display in Figure 16-12 shows the DDS VFO powered up, displaying the lower band edge for 80 meters.

EEPROM Memory Map

The ATmega328 µC family has 1024 bytes of EEPROM memory available for use. The software that controls the UIP views the EEPROM memory space as a sequence of 4-byte chunks, which we call a record. A *record* in our EEPROM memory space is defined as one *unsigned long*. The memory map for the EEPROM is shown in Figure 16-13. Table 16-1 presents the interpretation for each of the records shown in Figure 16-13. Because each record is a 4-byte memory space, record 0 is actually EEPROM address space 0-3. Record 1 occupies EEPROM addresses 4-7 and so on. While you could write all of the EEPROM memory addresses using the features of the VFO, we have written a short program (discussed later in this chapter) that writes the default frequencies mentioned in Table 16-3.

As mentioned earlier, on power-up, record 0 is fetched from EEPROM and displayed on the first line of the LCD display. The second line of the LCD display is used to display the band in

FIGURE 16-12 Initial VFO state on power up.

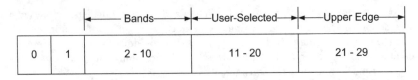

FIGURE 16-13 Record mapping for EEPROM memory.

Record Number	Description
0	The most recently recorded frequency. As you tune the VFO, if you pause on a frequency for more than 10 seconds and that frequency has not been previously stored, that frequency is written to this record.
1	This record is reserved, and currently unused.
2 through 10	These nine records store the frequencies of the lower band edges, starting with 10 meters through 160 meters. These should be viewed as READ-ONLY frequencies.
11 through 20	These 10 records hold the user-selected frequencies. We initialize these frequencies to a common-used QRP frequency for each band. Because there are only 9 bands but 10 records, the last record is initialized to the upper band edge for 10 meters.
21 through 29	These nine records hold the frequencies that represent the upper-edge frequencies for each band. These should be thought of as READ-ONLY frequencies.
30 through 255	Unused.

TABLE 16-3 EEPROM Memory Map for Records

which the frequency falls (e.g., "80m"). The output frequency of the VFO matches the frequency displayed on line 1 of the LCD display.

What happens next depends upon which switch you push. We'll assume you push them in left-to-right order (SW1 through SW5). The switch placement is illustrated in Figure 16-14.

SW1, the User Frequency Selection Switch (UFSS)

We named switch SW1 *User Frequency Selection Switch* because it is used to store and review user-selected frequencies. (It is the left-most pushbutton switch seen near the bottom of Figure 16-14.) In terms of the memory map shown in Figure 16-13, you are working with EEPROM records 11 through 20.

SW1 REVIEW Mode

When you press SW1, line 2 displays the word REVIEW on the LCD display to indicate you are in the REVIEW mode. The first time you run the UIP program, and assuming you have first run the VFO Initialization Program (discussed later), you should see the 160 meter QRP default frequency, 1818 kHz. This frequency is the first EEPROM record (record 11) in the user-selected memory space shown in Figure 16-13. You can now rotate the rotary encoder control (SW6 in Figure 16-14) to scroll through the 10 user-selected frequencies. As mentioned earlier, the first time through you are presented with the default (North America) QRP CW frequencies for each band. The tenth

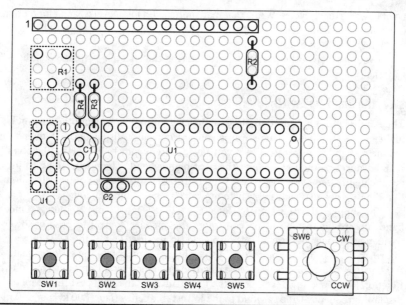

FIGURE 16-14   Component placement on control panel board.

address (i.e., record 20) is simply the upper-edge frequency for 10 meters. If you rotate past the last user-selected frequency, the display "wraps around" to the first frequency (i.e., moves from record 20 back to record 11).

SW1 STORE Mode
For example, suppose you want to change the upper-edge frequency for 10 meters (record 20) from 29,700,000 Hz to 29,699,990 Hz. To do that, first scroll to the frequency to change (i.e., 29,700,000 Hz) by rotating the encoder control while in the REVIEW mode. Once you see the target frequency displayed (i.e., 29,700.000 kHz), touch the UFSS switch, SW1, a second time. You immediately see the second line display change from REVIEW mode to STORE mode. This means you are going to store a new frequency in record 20 and replace the 29,700,000 Hz frequency with a new frequency.

Rotate the encoder control counter clockwise one "click," or one position. The frequency display should now show 29,699.990 kHz. In other words, once you see the STORE mode is active, rotating the encoder control alters the frequency shown on line 1 of the LCD display. As you would expect, rotating clockwise raises the displayed frequency and rotating the control counter-clockwise lowers the displayed frequency. As you can see, the default frequency step is set to 10 Hz. (The default step is set with switches SW4 and SW5, discussed later. For now, we'll just use the default 10 Hz step.)

Storing a New Frequency
To store the new displayed frequency in record 20 of the EEPROM memory space, push the encoder control shaft. The new frequency is then stored for you. The program immediately resets itself to the BANDUP mode. We force the code to limit itself to a single write using the STORE mode just to make sure you don't overwrite some other frequency. Also recall that EEPROM has a finite number of write/erase cycles (about 100,000) before it can become unreliable, and we don't

want to waste those cycles. While that may sound like a lot of cycles, don't forget that if you are just cruising around the band and the frequency has changed in the past 10 seconds, that new frequency is written to EEPROM.

If you press any other switch before pressing the encoder switch, SW6, the STORE mode is canceled.

SW2, the Band-Up Switch (BUS)

The Band-Up Switch is used to increase the current frequency from the current band to the next higher frequency band. Therefore, if you are currently displaying a frequency of 3.501 MHz (80 meters) and you press the BUS, you advance frequency to 7.0 MHz (40 meters). Each press of BUS advances to the low edge of the next higher band. There is nothing to prevent you from advancing the frequency to the next band by rotating the encode control a bazillion times. It's just that the BUS gives you a faster way to move to the next highest band.

The supported bands are: 160, 80, 40, 30, 20, 17, 15, 12, and 10 meters. Once you reach the band you are interested in, you can rotate the encoder control to raise or lower the displayed frequency. Frequencies are always displayed on the first line of the LCD display, while "control information" appears on the second display line. As you rotate the encoder shaft, each "click" of the encoder control increases/decreases the frequency by an amount controlled by the STEP function (discussed in a few moments). If you advance the BUS past the 10 meter frequency, the displayed frequency wraps around to the lowest frequency (i.e., 1800 kHz or 160 meters).

Out-of-Band Condition

Suppose you have the BUS so that the VFO is displaying the lower edge of the 40 meter band (i.e., 7.000 MHz). If you have the STEP to (the default) 10 Hz and turn the encoder control counter-clockwise, you display a frequency of 6,999.990 kHz, which is outside of the range of legal operation for most hams. Because the frequency lies outside of the ham bands, and asterisk (*) appears in the first column of the first line on the LCD display. This indicates that the displayed frequency lies outside of the ham band shown on the second line.

If the (illegal) state of the VFO is that shown in Figure 16-15 and you press the BUS again, the display would show 10,100 kHz, which is the lower edge of the 30 meter band. Because this edge is within the band limits, the asterisk would no longer appear on the LCD display.

SW3, the Band-Down Switch (BDS)

The BDS is used to decrease the current frequency to the next lowest band edge frequency. That is, BDS is a mirror opposite of the BUS switch. You can interleave the BDS and BUS switches to move quickly throughout the VFO's amateur band spectrum. Again, you could rotate the encoder control counter clockwise and (eventually) get to the next lowest band edge. However, the BDS provides a much faster way to move to the next lower band edge.

FIGURE 16-15 Frequency display with out-of-band condition.

FIGURE 16-16   Using PSS.

SW4, Plus Step Switch (PSS)

We mentioned earlier that rotating the encoder control changes the displayed frequency plus or minus 10 Hz, depending upon which direction you rotate the control. The PSS can be used to alter the default frequency step amount. The allowable frequency steps are:

10 Hz   20 Hz   100 Hz   0.5 kHz   1 kHz   2.5 kHz   5 kHz   10 kHz   100 kHz   1 MHz

Because the default step is 10 Hz, if you press the PSS, the display changes to 20 Hz and the +STEP mode is displayed. This can be seen in Figure 16-16. Now each "click" of the encoder control changes the frequency by 20 Hz. This step can be used to increase or decrease the displayed frequency.

If the LCD display shows the PSS at 1 MHz and you press it one more time, the display changes to 10 Hz. That is, the PSS wraps around to the lowest step.

SW5, Minus Step Switch (MSS)

As you might expect, the MSS does the opposite of the PSS. If the LCD displays 5 kHz and the –STEP mode, pressing the MSS changes the step to 2.5 kHz. If you press the MSS while the 10 Hz step is displayed, the step wraps around to 1 MHz for the frequency increment. You can interleave PSS and MSS to move about in the step ranges very quickly.

Once you have set the STEP to your desired value, press the BUS or BDS to go back to the tuning mode. If you don't press one of these two switches, you'll just end up cycling through the various steps.

SW6, the Encoder Control

The encoder control is actually two parts: 1) a shaft that can be rotated in either direction to change the displayed frequency, or 2) to push the shaft to activate the switch component of the control. The switch component only has meaning when SW1 is in the STORE mode, as explained earlier. Therefore, the primary function of the encoder control is to alter the displayed frequency.

You now know how to use all of the controls on the DDS VFO. The software behind those controls is explained in the next sections.

The DDS VFO Software

One of the features of the DDS VFO is the ability for you to store your own favorite frequencies as part of the DDS VFO. You might, for example, have a frequency for a net that you check into each week, or some other frequency you use to keep in touch with another ham. All of us probably have favorite bands we like to use, and perhaps specific frequencies within those bands. As shown in Figure 16-13, you can store up to 10 such frequencies. Because you want these frequencies to persist even when you turn the DDS VFO off, we write those frequencies to EEPROM memory. Because EEPROM is non-volatile memory, those frequencies persist even after power is removed.

You might ask: Flash memory is also nonvolatile, so why not store your 10 favorite frequencies in Flash memory? True, that is possible, but what if you want to change one of those frequencies? Your only alternative is to load the DDS VFO source code, edit the frequency list, recompile, and upload the new program to the Arduino. So, yes, you could use Flash memory to hold the user-defined frequency list, but changing that list is not very convenient. It also assumes the user you lent your new DDS VFO to is also a programmer. Maybe ... maybe not. It just seems like a better design decision to change user-defined frequencies in EEPROM as the program is running rather than having to edit, recompile, and upload a new version of the program. However, when you have finished building your version of the DDS VFO, you can code your favorite frequencies into the source code so you don't have to edit them at runtime. This saves you from needing to use the REVIEW/STORE features the first time you run the DDS VFO.

EEPROM Initialization Program

Listing 16-1 presents a short program that initializes EEPROM memory to a set of frequencies. You should load this initialize EEPROM program into your Arduino and run it before you run the VFO program.

There are three EEPROM record blocks that the program initializes:

1. The 10 user-defined frequencies.
2. The nine frequencies that represent the lower band edges for the frequencies covered by the VFO.
3. The nine frequencies that represent the upper band edges for the frequencies covered by the VFO.

Unless the FCC changes things, you should not edit the last two EEPROM blocks. You should only edit the first, user-defined, EEPROM memory block. As the code is currently written in Listing 16-1, the 10 user-defined frequencies are the common North America QRP frequencies for each band. As we mentioned earlier, since there are only nine bands, we set the 10th address to the upper band edge for 10 meters. The reason for selecting this frequency for the 10th position is ... we don't have a reason. Make it anything you wish, as the user-defined frequencies are not integral to the needs of the rest of the program. If you have other frequencies you'd prefer to have in the list, you can edit the *qrp[]* array to reflect those new frequencies. Just make sure you end up with 10 frequencies in the *qrp[]* array ... no more, no less.

If you have less than 10 "favorite" frequencies, it's okay to leave them blank. However, if you do initialize less than 10 frequencies in this program and your EEPROM has never been set before, you may see 4,294,967,296 pop up as a displayed frequency. The reason is because unwritten EEPROM bytes are usually set to 0xFF, which, when viewed as a 4-byte unsigned value, is 4,294,967,296. If you make an error and attempt to write 11 user-defined frequencies, the code creates a small black hole in your EEPROM and the VFO disappears under its own weight. Well ... not really, but you will likely blow away one of the upper band edge values. If you do that, edit the extra frequency out of the user-defined list and rerun this initialization program.

You should *not* edit the other two array contents, *lowBandEdge[]* and *hiBandEdge[]* unless the FCC changes the band edges. The content of these arrays is used by the DDS VFO in different ways, and changing them would likely break some of the functionality of the DDS VFO. You should view these two arrays as etched in stone.

Wait a minute!

If the *lowBandEdge[]* and *hiBandEdge[]* arrays are fixed, why not simply hardcode them into the program as *static* or global arrays? Believe us, life would be much simpler if we could. However, because we wanted you to be able to run the VFO using a 328-based Arduino, we only have 2K of SRAM available for variables. Keep in mind that the stack and the heap use SRAM, which means the amount of unused SRAM ebbs and flows as the program runs. When we tried to run the program with the *lowBandEdge[]* and *hiBandEdge[]* arrays as part of the program space, the program did a lot of ebbing and *no* flowing. At one part of the program, we simply ran out of SRAM memory and the program died an inglorious death. Our only recourse was to place the arrays in EEPROM memory space.

```
#include <EEPROM.h>

#define READCURRENTFREQ      0           // The EEPROM record of the last
                                         // written frequency
#define READUSERCOUNTS       (READCURRENTFREQ + 1) // Reserved
#define READLASTBANDFREQ     (READUSERCOUNTS + 1)  // The starting EEPROM record
                                                   // address for the 9 bands
#define READUSERSTOREDFREQ   (READLASTBANDFREQ + 9)// The starting EEPROM record
                                         // address for 10 user-saved frequencies. Note
                                         // that READUSERSTOREDFREQ plus the number
                                         // stored at READUSERCOUNTS is the next empty
                                         // address.
#define READUPPERBANDEDGES   READUSERSTOREDFREQ + 10 // The upper band edges for
                                                     // the 9 bands

#define MAXUSERFREQS        10    // The max frequencies the user can store
#define MAXBANDS             9    // Number of ham bands covered

void setup()
{
  Serial.begin(115200);
  InitEEPROMAFrequencies();
}
void loop()
{
                         // Nothing to do here

}

void InitEEPROMAFrequencies()   // All the work is done here...
{
  int i;
  int offset;

  union {
    byte array[4];
    unsigned long val;
  } myUnion;
```

LISTING 16-1 Initialize EEPROM data.

```
//                    Code for writing default QRP frequencies

unsigned long qrp[] = {1818000,3560000,7030000,10116000,14060000,
                       18096000,21060000,24906000,28060000, 29700000};

for (i = 0; i < MAXUSERFREQS; i++) {
  myUnion.val = qrp[i];
  offset = (READUSERSTOREDFREQ + i) * sizeof(unsigned long);
  EEPROM.write(offset, myUnion.array[0]);
  EEPROM.write(offset + 1, myUnion.array[1]);
  EEPROM.write(offset + 2, myUnion.array[2]);
  EEPROM.write(offset + 3, myUnion.array[3]);
}
Serial.println("===================== QRP Done ================");

//               Code for writing lower edge band frequencies

unsigned long lowBandEdge[] = {28000000,24890000,21000000,18068000,
                               14000000,10100000,7000000,3500000,1800000};

for (i = 0; i < MAXBANDS; i++) {
  myUnion.val = lowBandEdge[i];
  offset = (READLASTBANDFREQ + i) * sizeof(unsigned long);
  EEPROM.write(offset, myUnion.array[0]);
  EEPROM.write(offset + 1, myUnion.array[1]);
  EEPROM.write(offset + 2, myUnion.array[2]);
  EEPROM.write(offset + 3, myUnion.array[3]);
}
Serial.println("===================== Low Edge Done ================");


//               Code for writing band frequencies upper edges

unsigned long hiBandEdge[] = {29700000, 24990000, 21450000, 18168000,
                              14350000,0150000,7300000,4000000,2000000};

for (i = 0; i < MAXBANDS; i++) {
  myUnion.val = hiBandEdge[i];
  offset = (READUPPERBANDEDGES + i) * sizeof(unsigned long);
  EEPROM.write(offset, myUnion.array[0]);
  EEPROM.write(offset + 1, myUnion.array[1]);
  EEPROM.write(offset + 2, myUnion.array[2]);
  EEPROM.write(offset + 3, myUnion.array[3]);
}
Serial.println("===================== High Edge Done ================");
}
```

LISTING 16-1   Initialize EEPROM data. *(continued)*

The program doesn't do all that much. Basically, the program takes the contents of the three arrays, writes them to EEPROM memory, and displays a few messages on the Serial object so you know it ran. Because there's nothing to repeat, the *loop()* function is empty. All of the work takes place in *setup()*.

There is one aspect of the program that is a little unusual. The code fragment:

```
union {
  byte array[4];
  unsigned long val;
} myUnion;
```

uses the C keyword *union*; a keyword you don't see that often in C programs. Perhaps the easiest way to think of a *union* is as though it is a shapeless chunk of memory that you've set aside to hold some piece of data. What makes a *union* unique is that you don't know exactly what it holds at any given moment. In our program, all we are saying is that we have defined the *union* so that it is large enough to hold *either* a 4-*byte* array *or* a single *unsigned long* data type. Functionally, a *union* is a buffer where we can place different types of data. It's up to the programmer to keep track of what's actually in the buffer.

Given that we have defined *myUnion* to hold either a 4-*byte* array or an *unsigned long*, both of which require four bytes of memory, *myUnion* ends up occupying 4 bytes of memory. If we had added an array of 10 integers (i.e., *myInts[10]*), *myUnion* would now require 20 bytes of memory because each *int* requires two bytes. The data definition would now look like:

```
union {
  byte array[4];
  int myInts[10];
  unsigned long val;
} myUnion;
```

In that case, *myUnion* now defines a memory space that would occupy 20 bytes of memory. The rule is: A *union* always requires an amount of memory equal to the largest data item defined in the *union*.

So, what does a *union* bring to the table? Well, first of all, we could have defined all three members of the union separately as:

```
byte array[4];
int myInts[10];
unsigned long val;
```

and used the data accordingly. However, doing so means we are now using 28 bytes of memory rather than just 20. A byte here ... a byte there, pretty soon it adds up. So, one advantage is that a *union* saves us a little memory. That's the good news.

The bad news is that when using a *union*, it can only hold a value for one of those three members at any given moment and you need to keep track of what's actually living inside the *union* when you go to use it. You use the *dot operator* when assigning or retrieving a value from a union:

```
// the definition of myUnion is in effect...
unsigned long lowFreq;

myUnion.val = 7000000;
lowFreq = myUnion.val;
```

We talked about the dot operator before in the integration chapter (Chapter 10). Note the dot operator appears between the *union* name (*myUnion*) and the union member we wish to use (*val*). You can verbalize the dot operator in the example above as: "Go to the *union* memory address space named *myUnion*, fetch *unsigned long* bytes of that memory space (4 bytes), and subsequently reference those four bytes as the *val* member of that *union*." The dot operator is like a key to get inside a black box. In Object Oriented Programming jargon, the dot operator is the key that opens a black box object named *myUnion* and, once inside the black box, allows you access to the object member named *val*.

Note how we use the *union* in Listing 16-1. We stuff each frequency element of the *qrp[]*, *lowBandEdge[]*, and *hiBandEdge[]* arrays into the *union* as an *unsigned long*, but extract that data from the *union* as though it is a 4-*byte* array of bytes. We do this because the EEPROM *read()* and *write()* functions in the EEPROM library can only read and write data one byte at a time. For example, suppose you knew that an *unsigned long* was stored at EEPROM memory address 10. That *unsigned long* occupies EEPROM memory addresses 10, 11, 12, and 13. Now, if we want to retrieve that *unsigned long* using the statement:

```
unsigned long frequency = EEPROM.read(10);
```

the program goes to the EEPROM memory address 10, reads the byte stored there, and assigns it into *frequency*. The problem is that *read()* only returned one byte—the byte stored at EEPROM address 10. The variable *frequency* now contains ¼ of an *unsigned long*! By using the *union*, we can stuff the frequencies into the *myUnion.val* member as an *unsigned long*, but then write them out to EEPROM memory as four separate bytes. We don't have to know if an *unsigned long* is stored with its bytes in ascending or descending order ... they are just bytes to us. Likewise, we can read the *unsigned long* from EEPROM memory as 4 separate *read()* function calls, and then extract the *unsigned long* from the *union* via the *val* member.

When you finish running the program in Listing 16-1, the EEPROM memory space is ready for use. We can now move on to the program that controls the KP VFO.

The KP VFO Software (VFOControlProgram.ino)

Unlike previous programs, we are not going to list the entire KP VFO software program. It's simply too long and you can download it at a McGraw-Hill web site (www.mhprofessional.com/arduinohamradio). Instead, we are going to discuss those sections that may be of interest to you should you want to change the way the VFO functions from a user interface point of view.

The code associated with the rotary encoder is largely taken from Richard Visokey's work with the AD9850 DDS chip. You can download his rotary library at

http://www.ad7c.com/projects/ad9850-dds-vfo/.

The program also uses the Adafruit port expander code, which can be downloaded from

https://github.com/adafruit/Adafruit-MCP23017-Arduino-Library/.

You should be pretty well versed in how to download and add these libraries to the Arduino IDE so we won't repeat those instructions here. After you have downloaded and installed the libraries called by the various *#include* directives, load the VFOControlProgram.ino sketch. The VFOControlProgram sketch also has two additional source code files (W6DQ_DDS_VFO.h and W6DQ_DDS_VFO.cpp). These files are necessary to create additional objects required by the program.

We should state at the outset that we are not totally happy with the structure of this program. Indeed, we desperately wanted to add a new class source code file to handle all of the user interface elements of the program. However, we also want the VFO to fit in the memory space allocated to the 328 type of Arduino boards, which is limited to 2K of SRAM. Our program runs near the edge of this memory limit. As a result, there's not much room for adding features to the extent they are going to use any SRAM space. If you're using an ATmega1280 or ATmega2560, you have 8K of SRAM and a lot of windows (no pun intended) open up for you. We assume, however, that is not the case so we stuck within the 2K limitation.

The program begins with a fairly long list of preprocessor directives and a number of globally defined variables, including the following objects:

```
Rotary r = Rotary(2,3); // Pins for rotary encoder.  Must be interrupt pins.
DDSLCDShield lcd = DDSLCDShield();
w6dqDDSVFO myVFO = w6dqDDSVFO();
```

The Rotary object, *r*, calls its constructor with two pin assignments, 2 and 3. These must be interrupt pins as they are used for calling the Interrupt Service Routine (ISR). The *lcd* object is different in that it uses the I2C interface. The display found in Chapter 3 uses six I/O pins, and combined with the control panel switches and the data pins for the DDS module, the Atmel ATmega328 Arduino runs out of IO pins. By switching to an I2C interface and using a port expander on the control panel board, we gain an additional 16 digital IO pins, more than enough for this project. Finally, the *myVFO* object is a new class we wrote to move some of the user interface elements into its own class. While we could have put other elements in the class, doing so eventually chewed up our SRAM space. As you know from the discussion of the Init program, we had to move some of the arrays into EEPROM memory space.

setup()

All of the code in the *setup()* function is Step 1, the initialization code, and sets the environment in which the VFO runs.

After the LCD display is initialized, the code establishes the interrupt service routines (ISR) and calls *sei()* to SEt Interrupts. As you probably know, an interrupt causes the normal program flow to stop and control is immediately transferred to the ISR. The function *ISR(PCINT2_vect)* is the ISR routine for our program and is called when the encoder shaft is turned. We changed this code quite a bit, so our apologies to the original authors. Basically, whenever the user turns the encoder shaft, the ISR is called and the *process()* method of the *rotary* class object (*r*) determines whether the shaft was turned clockwise (DIR_CW) or counter clockwise (DIR_CCW). Based upon what the user is doing (e.g., REVIEW or STORE mode), specific actions are taken.

Alas, our ISR is pretty ugly. ISR's are supposed to be tight little pieces of code that are done in an instant. The reason is because, while the processor is servicing the interrupt, nothing else can take place ... not even another interrupt. Our ISR is way too busy. Instead of being away for an hour lunch break, ours is like a two-week vacation. Still, because humans turning a shaft isn't a very fast event in the microcontroller world, the routine works. If someone with more ISR experience finds a more elegant way, we hope they would share it with the rest of us. Also note that the VFO limits are enforced in the ISR near the bottom of the routine.

The rest of the *setup()* function is used to set the pin modes. The *pulseHigh()* statements at the bottom of *setup()* look like function calls, but it's really preprocessor directive that results in what's called a macro substitution and it's worth learning. Look near the top of the program and you'll find the directive:

```
#define pulseHigh(pin) {digitalWrite(pin, HIGH); digitalWrite(pin, LOW); }
```

Recall that a *#define* causes a textual substitution to take place. While that's still true here, the difference is that the *#define* has a parameter associated with it, something named *pin*. (Sometimes at cocktail parties you will hear this kind of #define referred to as a parameterized macro.) What the preprocessor does is substitute whatever appears as *pin* in the macro definition of *pulseHigh()*. Consider the last statement in *setup()*:

```
pulseHigh(FQ_UD);
```

In this statement, *pin* is equal to FQ_UD. If we could slow down the sequence the preprocessor goes through in processing this statement, it would look like:

```
pulseHigh(FQ_UD);
#define pulseHigh(pin)  {digitalWrite(pin, HIGH); digitalWrite(pin, LOW); }
#define pulseHigh(FQ_UD) {digitalWrite(pin, HIGH); digitalWrite(pin, LOW); }
#define pulseHigh(FQ_UD) {digitalWrite(FQ_UD, HIGH); digitalWrite(FQ_UD, LOW); }
digitalWrite(FQ_UD, HIGH); digitalWrite(FQ_UD, LOW);
digitalWrite(FQ_UD, HIGH);
digitalWrite(FQ_UD, LOW);
```

Note the use of both braces and parentheses in the statement lines. In other words, the single *#define* macro expands to two program statements. You could do away with the parameterized macro and simply use the last two statements with exactly the same result.

loop()

The first thing we do in *loop()* is see if the user has changed frequency using the encoder shaft. If the frequency has not changed, we go on to the next task. If it has changed we call *NewShowFreq()* to display the new frequency on the first line of the LCD display. *DoRangeCheck()* is then called to see if the new frequency is within a ham band. If not, an asterisk is displayed in column 1 of the first line on the LCD display. If we didn't do this check-and-update sequence, you would likely be able to notice a little flicker in the display on each pass through *loop()*. If a new frequency has been set, the call to *sendFrequency(currentFrequency)* updates the AD9850 chip with the new frequency data.

Next, the code reads the current millisecond count via the call to *millis()*. If *DELTATIMEOFFSET* milliseconds (default is 10 seconds) have occurred since we last updated the current frequency stored in EEPROM, we call *writeEEPROMRecord()* to store that frequency in record 0 of the EEPROM memory space. However, because we don't want to do EEPROM writes unless we have to, we check to see if the current frequency is different from the last frequency we wrote. If they are the same, no write is done and control reverts to *loop()*. If you turn off the VFO, it is this EEPROM frequency from record 0 that is used to initialize the display the next time your power up the VFO.

Note that we could have performed the frequency check in *loop()* just as well as in *writeEEPROMRecord()*. However, because *loop()* is going to execute many, many times before the user has a chance to change the frequency, we thought it would be slightly more efficient to check the elapsed time first, since that is the more restrictive condition.

The last thing that *loop()* does is check to see if any buttons have been pressed via a call to *lcd.readButtons()*. If so, the button code is passed to *ButtonPush()* for processing. All switches, SW1 through SW6, have their associated events processed in *ButtonPush()*. Although the function is fairly long, there is nothing in it that you haven't seen before.

That's all there is to it. Now all you have to do is test the VFO and then hook up your new VFO to your rig!

Testing the DDS VFO

The most obvious thing that you should be thinking about right now is how do I know if this thing is actually working correctly? Maybe you have a lab full of test equipment and the question is moot. You KNOW whether or not the DDS VFO is working. But, not having the proper test equipment is not an obstacle. Rather, it is an opportunity to be creative. But the first step is to get the software compiled and loaded before you think about testing anything. Remember that you must compile and load the EEPROM initialization program first and only then should you compile and load the main program.

Thanks to the generosity of the folks at Agilent Technologies, Dennis was loaned several of their 4000-series oscilloscopes for evaluation. The instrument pictured in Figure 16-17 is the MSO-X 4054A, a 500 MHz, 5 GS/sec digital storage oscilloscope. Dennis was able to use this scope in a number of projects and is considering purchasing one for his lab. This scope made design and testing much simpler than using his older HP 1742A scope.

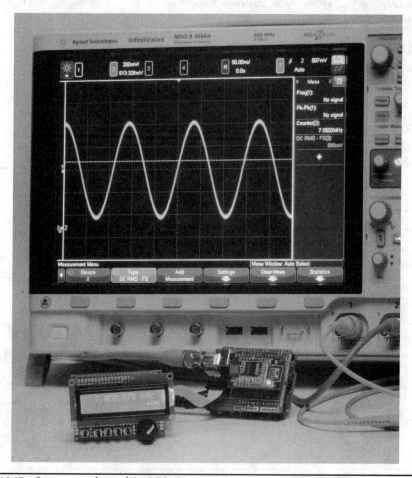

FIGURE 16-17 Output waveform of the DDS VFO.

Figure 16-18 shows a spectrum plot at 14.150 MHz from the DDS VFO. This display was captured using another Agilent instrument, an N9918A FieldFox. The plot shows that the second harmonic, at 28.325 MHz, is 49.6 dB below the fundamental. This is without any additional bandpass filtering on the output. This meets FCC Part 97 requirements for spectral purity in the amateur service, which is specified to be "at least 43 dB below the mean power of the fundamental emission."

The first step in testing, once the software is loaded, is to see that the user interface is operating correctly. This means that the Band+ and Band–, and the Step+ and Step– switches are all working correctly. Check that the shaft encoder is functioning by turning the encoder and checking that the frequency in the display increases and decreases. Now you can check that the memory functions are working.

If you are building this project, then you are also more than likely to own a receiver of some sort. Something that tunes the HF ham bands? There is your first means of testing the DDS VFO. The simple procedure is to connect a short piece of wire to the output connector of the DDS VFO and set the frequency to one that you can tune with your receiver. With the receiver set to receive CW or SSB, tune to the frequency you set. You may have to tune around a little bit depending on the accuracy of your receiver plus the frequency of the DDS may be off slightly. You should be able to detect a strong carrier on your receiver at the test frequency, meaning: Success!

Calibrating the DDS VFO

It is relatively simple to calibrate the frequency of the DDS VFO. Calibration does require an accurate receiver to note one frequency from the VFO. Lacking an accurate receiver, one that can

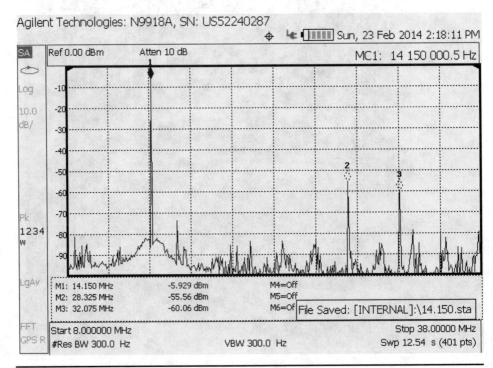

FIGURE 16-18   Spectrum display at 14.150 MHz.

receive WWV on one of their many frequencies can also be used. (WWV broadcasts on 2.5, 5, 10, 15, and 20 MHz.) For example, let's use WWV transmitting on 10 MHz. Tune the receiver to 10 MHz in the AM mode. Using a wire connected to the output pin of the DDS VFO, wrap the wire around the antenna lead to your receiver. As you tune the DDS VFO near 10 MHz, what you hear is a beat note or heterodyne between the WWV carrier and the DDS VFO. Adjust the DDS VFO so that the beat note is as close to zero as you can. (That is, the note is so low that you can no longer hear it.) Record the displayed frequency on your VFO. What you now know is the displayed frequency for a 10 MHz signal. Chances are good that the DDS VFO does NOT display 10.000.000 MHz. Note the difference between the displayed frequency and 10 MHz. The difference is the calibration offset for your VFO.

The calibration offset is set using a *#define* statement:

```
#define CALIBRATIONOFFSET    0      // this is used to adjust the output
                                    // frequency to match display frequency
```

A negative value is used when the display frequency is less than the output frequency. If the display frequency is greater than the output frequency, the value is positive. In the case of our DDS VFO, the offset is –100.

The software uses the following algorithm to calculate the data to load into the AD9850 to set the frequency. The AD9850 uses a 32-bit "tuning word" to set the frequency. The tuning word is a function of the desired frequency and the system clock. In our case the system clock is 125 MHz and the frequency is an *unsigned long* data type (32-bits). Analog Devices provides an algorithm in the AD9850 data sheet that we use:

$$f_{out} = (\Delta\ Phase * CLKIN) / 2^{32}$$

where: Δ Phase is the 32-bit tuning word, *CLKIN* is the input reference clock in MHz, and f_{out} is the output frequency in MHz.

We already know f_{out}. What we want is the tuning word, so for our program, we rearrange the algorithm to provide the tuning word as the result. A little algebra and we end up with:

```
freq = (frequency + CALIBRATIONOFFSET) * 4294967295/125000000;
```

In decimal, 2^{32} would normally be 4294967296. Since we are in a binary world using a 4-byte value that starts with "0," the value actually used is (2^{32} – 1) or 4294967295; all 1s. We express the clock frequency in Hertz (125000000) as well as the desired frequency: frequency.

We adjust the desired frequency adding CALIBRATIONOFFSET and the resulting tuning word, *freq*, is loaded into the AD9850.

Using the DDS VFO with Your Radio

Many QRP radios are crystal controlled and lack the ability to "tune around" even within a narrow frequency range. Some examples of this are the Pixie and the Blekok SSB radios. There are many others. This section describes how you can interface these and other radios to the DDS VFO project described in this chapter.

Any ham who started as Novice Class licensee in the days of "Incentive Licensing" was "rock bound" by regulation. Novice transmitters had to be crystal controlled. When the Novices moved up with their Technician or General Class licenses (or Advanced for that matter) they could now be "frequency agile" and use a VFO. The VFO plugged right into the crystal sockets on those old transmitters. The beauty of this was that the transmitter received new life! It did not have to be replaced right away. Novices could "stretch their legs" in the other portions of the bands now available due to their new frequency privileges.

The QRP community has embraced being rock bound for different reasons and not by regulation. A compact radio was the first order and crystal control lent itself well to building diminutive gear. Today, there is no reason that a crystal-controlled radio can't be "unleashed" with a VFO, just like in the days of moving up from the Novice Class license. The process is to remove the crystal from the radio and replace it with the DDS VFO. The former crystal oscillator now functions in part as a buffer between the DDS VFO and the radio.

In many cases, all that is required is to remove the crystal and plug in the DDS VFO, as shown in Figure 16-19. The DDS VFO project is a generic VFO in that it produces a fundamental frequency for each of the MF and HF ham bands. The typical QRP rig is single band and uses the fundamental frequency. However, there are QRP and other rigs that you may wish to use this VFO with that do not use the fundamental frequency, but rather, they use a range of frequencies that are mixed with other oscillators to produce the fundamental frequencies for a particular ham band. This process was discussed in some detail in Chapter 15 when attaching the Arduino Frequency Counter to a double conversion radio. We provide guidance on how to change the frequency range of the DDS VFO for such an application.

The Pixie QRP Radio

The Pixie is an ingenious transceiver that really takes the "minimalist" approach to radio design. From what we can gather, the design originated in England in 1982 when George Burt, GM3OXX, wrote about the "Foxx" in the summer 1983 edition of *SPRAT, The Journal of the G-QRP Club*.

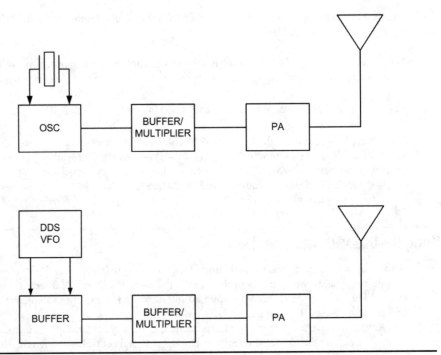

FIGURE 16-19 Connecting the DDS VFO to a typical radio.

The key feature of the design was the dual use of a single transistor as both PA and receiver mixer in a crystal controlled, direct conversion radio. The Foxx evolved into the Pixie and the Pixie has been evolving ever since. Today it is most commonly seen as the Pixie 2, consisting of one integrated circuit (an audio amplifier such as an LM386) and two transistors, one being the crystal oscillator and the other being the PA/mixer.

Today a Pixie 2 kit can be purchased on the Internet for as little as $9 and change. They are rockbound, often at a frequency of 7023 kHz. For our North American readers, this is in the Amateur Extra portion of the 40 meter band and not available to other license holders. In other parts of the world, however, this is a recognized 40 meter QRP frequency. Since most QRP activity takes place above 7040 kHz in North America, one must either replace the crystal with one more favorable to QRP operation in North America, or in either case (North America or elsewhere) add a VFO.

We secured two Pixie kits for 40 meters from eBay and assembled them. Assembly was quite straightforward and simple as this is a minimalist radio and has few parts. A partial schematic is reproduced in Figure 16-20 showing the crystal oscillator and PA, and how we connected the DDS VFO. The easiest way to add the DDS VFO is to disable the crystal oscillator by removing the oscillator transistor, Q1, and several surrounding components. The VFO is fed directly to the base of the PA/Mixer transistor, Q2, through capacitor C4.

We removed four components: Q1, C3, C7, and R4. When you build the kit, you can leave these four components out (along with the crystal Y1 and R1). If you have already built the kit, remove the four components indicated. They are shown in Figure 16-20 with dashed lines. Attach a piece of RG-174 to connect to the DDS VFO as shown in Figure 16-20.

In our testing, we discovered that the Pixie needs between 500 and 600 mV of drive at the base of Q2 for the receiver to function properly and for the transmitter to have the expected output. The DDS VFO may not have enough drive for this application so there are two alternatives. One alternative is to add the buffer stage described earlier in the chapter (Figure 16-10). Another option is to add a toroid transformer at T1 in Figure 16-20. T1 is designed to couple the output voltage of the DDS VFO. Wound on an Amidon FT-23-43 core, it has 5 turns for the primary and 20 turns for the secondary, both using 30AWG enameled copper wire. Follow the procedure described in Chapter 15 (Directional Wattmeter/SWR Indicator) for winding toroids and for tinning the ends.

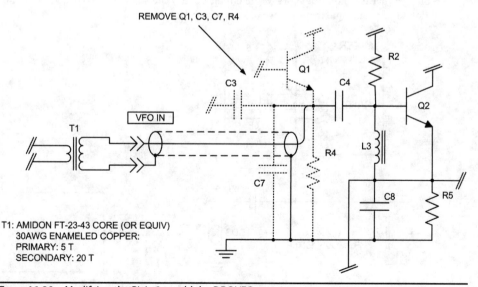

FIGURE 16-20 Modifying the Pixie 2 to add the DDS VFO.

Count one turn each time the wire passes through the center of the core and use a soldering iron to burn the insulation off of the end of each wire. The transformer is installed on the DDS shield between the output of the DDS breakout module and the output connector.

This method of attaching the DDS VFO to a radio in place of a crystal oscillator is usable for other radios for which you may wish to use the DDS VFO. Oftentimes, removing the crystal oscillator and installing the DDS VFO in its place is a good solution.

Blekok Micro 40SC

The Blekok Micro 40SC is a 40 meter Single Sideband (SSB) and CW QRP transceiver developed in Indonesia by Indra Ekoputro, YD1JJJ. Jack recently obtained his from eBay and we have found that it has good sensitivity and selectivity but the VFO can be a bit touchy. We have made minor modifications to the Micro 40SC, which allow the addition of the Arduino DDS VFO. The modifications include removing a 100 pF SMT capacitor and inserting a 2-pin header in a marked spot already silk screened on the printed circuit board.

The Micro 40SC uses an SA602 as mixer and VFO. The modifications, shown in Figure 16-21, and visible in the photographs in Figure 16-22, showing the chip cap removed, and Figure 16-23, showing the location of the new 2-pin header.

CRKits CRK 10A 40 meter QRP Transceiver

The CRK 10A is another inexpensive, highly portable QRP CW transceiver that operates on a fixed, crystal controlled frequency. The CRK 10 A is very easy to modify to use the DDS VFO for frequency control. The modifications remove the crystal, and install a capacitor and a connector for the VFO input.

Figure 16-24 shows a partial schematic of the CRK 10A and shows where the DDS VFO connects. We attached the coax as shown in Figure 16-25. We used a length of RG-174 with a BNC connector, drilled a ⅜₂-in. hole in the rear panel, and ran the coax to the pads that were for the crystal we removed, X3. The RG-174 is routed under the circuit board as shown in Figure 16-25. In the photograph, the center conductor is the lead on the left. We wrapped a piece of AWG22 solid wire around the shield braid, soldered the connection, and then covered it with heat-shrink tubing so that it would not short to the circuit board. The solid wire is covered with a piece of Teflon tubing, the same materials we use for wiring most of our shields.

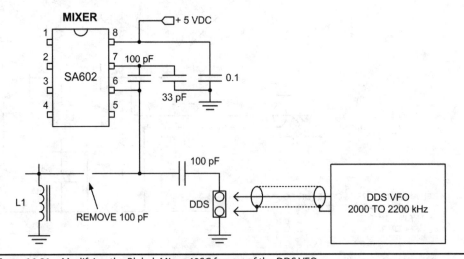

FIGURE 16-21   Modifying the Blekok Micro 40SC for use of the DDS VFO.

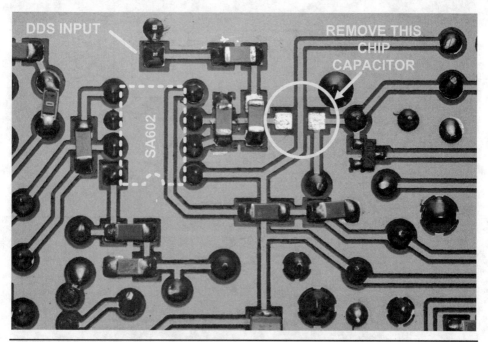

FIGURE 16-22   Remove the 100 pF SMT capacitor from this location.

FIGURE 16-23   Two-pin header added for the DDS VFO input.

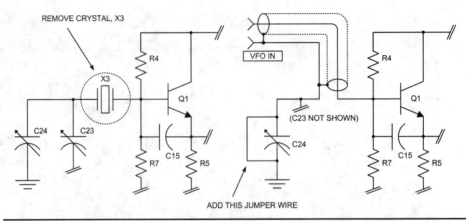

FIGURE 16-24   Partial schematic of CRK 10A showing connection for DDS VFO.

FIGURE 16-25   Modifications performed on the CRK 10A transceiver.

Other Applications of the DDS VFO and Additional Enhancements

The DDS VFO shown in this chapter can be used for a variety of applications. By adding an output attenuator, it becomes a low-cost RF signal generator with a range of 1.6 MHz to 30 MHz. The DDS VFO is not limited to use with QRP rigs either. It is also compatible with vacuum tube equipment, given one minor modification. The DDS VFO as it stands generates just under 1 V peak-to-peak output. A typical vacuum tube circuit requires a bit higher voltage, usually 6 to 8 V peak-to-peak. We showed one version that uses an output buffer stage, but to get the output to 8 Volts, the easiest solution is to add a toroid transformer on the output. Using a small toroid such as an Amidon FT-23-43, wind 5 turns for the primary and 20 turns for the secondary using 30 AWG enameled wire. This roughly doubles the output voltage and is more suitable for driving most tube type equipment.

There are some enhancements that can be added to the DDS VFO with minor changes to the software and hardware. One such modification would be to add a frequency offset while transmitting. By sensing when transmitting (using a digital input on the keying circuit) it is possible to slightly shift the transmitted frequency. Another possible enhancement would be to add a Receiver Incremental Tuning or RIT control in addition to the transmit sense. The RIT control can use a pot and an analog input to adjust the receive frequency over a range of 1 to 2 kHz.

Conclusion

As you can see, this is a versatile project with many uses with many pieces of equipment. Maybe you want to build a new transmitter or receiver. The DDS VFO provides a low-cost stable source of low-level RF in the HF ham bands.

CHAPTER **17**

A Portable Solar Power Source

A fairly consistent theme throughout this book has been the use of low power amateur radio equipment. One reason is because low power often means low cost. QRP operation, especially for CW, is possible with a relatively small capital expenditure. We also felt that, concentrating on less expensive equipment, which typically has fewer bells and whistles, would encourage the reader to try some of the projects presented in this book so they could add the bells and whistles they wanted at fairly low cost.

A solar power source is applicable to any ham radio equipment. Given enough solar panels and lots of sun, it would be possible to run a full kilowatt station on solar power. However, in this chapter we want to present a solar panel implementation that is capable of powering most QRP stations, providing the sun cooperates to some degree. The goal was to make the system reasonably portable, easy to set up for something like Field Day, and once set up, not require constant attention. Figure 17-1

FIGURE 17-1   A portable solar power unit.

379

shows the portable solar panel project, which can store the Arduino and most other components in the toolbox. When not in use, the solar panel quickly disconnects from the mount shown in Figure 17-1 and all components are stored within the toolbox except the panel itself for easy transport.

Our design uses a 10 W solar panel. What this means is that, in bright sunlight with the panel perfectly positioned at a 90 degree angle to the sun, the panel is capable of generating 10 W of power. Our system uses a homemade sensor and an Arduino to keep the panel facing the sun throughout the day. Power is stored in a small (less than 3 in. × 3 in. × 4 in.), relatively light (i.e., less than 5 lb) 12 V Sealed Lead Acid (SLA) battery with a 5 ampere/hour rating (see Figure 17-2). You could use the general design here with a somewhat larger battery and panels to generate more power, but those changes would make the system heavier and may change the system from portable to luggable. It would also require heftier components in the system, which would add to the cost.

We selected a lightweight, inexpensive, toolbox to house the battery, solar charger controller, and the Arduino board. These toolboxes are available at Lowes, Home Depot, and other retailers for about $15. Smaller, less expensive toolboxes may work, but you don't want to make it too small. The reason is because ours is 19-in. wide (see Figure 17-3), which means there is enough room left over in the toolbox for the stepper motor and solar sensor, a small QRP rig, simple antenna, and key or paddle. The PVC pipe support structure is designed to fit into the toolbox. Only the panel itself is external to the toolbox. Another consideration is the top of the toolbox. Because of the way we mount the solar panel (shown later in the chapter), you want to select a toolbox with a flat top. Finding such a toolbox, we think, is impossible. We settled for a toolbox with a fairly flat surface, but it still has some slope to the top. The sides are not parallel either, although we wish they were. Most inexpensive boxes are narrower at the bottom and slope outward toward the top. This makes mounting the stepper motor a little more ugly than it should be.

The solar panel we selected is a 12 V, 10 W panel with a retail price of less than $50 (see Figure 17-4). However, with some careful shopping, you can buy the panel for about $30. The

FIGURE 17-2 The 12 V, 5 Ah SLA battery.

Chapter 17: A Portable Solar Power Source

FIGURE 17-3 Toolbox used for solar system.

FIGURE 17-4 The 10 W, 12 V solar panel.

panel is polycrystalline and is mounted in a sturdy aluminum frame with mounting holes pre-drilled on the back side of the panel. The unit comes with a 10-ft. connecting cable. Solar panels do degrade over time, reducing their output capabilities. As a general rule, most vendors guarantee 90% output for 10 years and 80–85% for 25 years.

The Solar Sensor

The project uses a homemade solar sensor that is relatively cheap and simple to make. Even so, there are those who could argue the sensor isn't necessary or too expensive. Well, you do need something that keeps the panel facing toward the sun. Our experience is that, if it's around noon and the sun is fairly high in the sky and the panel is lying on the ground, flipping the panel over drops the output by about 95% ... it doesn't go to zero. We also noticed that the direction toward the sun has very little effect on the output *voltage* as you rotate the panel away from the sun. However, the *current* through the panel drops significantly as you rotate the panel away from the sun. This tells us that there is more of a power penalty than a voltage penalty if you don't orient the panel toward the sun. However, you can orient the panel to face the sun without a solar sensor.

If you know the date, longitude, and latitude where the solar panel is located, some fairly bright people have figured out the math to enable you to calculate the optimum angle the panel must be for a given time of day. So, the "you-don't-need-a-solar-sensor" critics are correct, you don't need a solar sensor. But where's the fun in that?

Other critics might argue that our sensor is too expensive to make. After all, you can take a Manual Graphite Display Generator (MGDG) and an Armstrong Position Mechanism (APM) and use it in place of our sensor at almost no cost. Most golf courses will give you a MGDG free ... just ask for a scoring pencil. Glue it to the solar panel at 90 degrees, then using your hands (APM), move the panel until the shadow of the MGDG is minimized. Again, almost zero cost but kind of a pain for the person who has to hold the panel all day.

You could also just prop the panel up facing toward the sun and, every 15 minutes or so, readjust it for maximum power. Of course, you know that Murphy's Law means that while you're out adjusting the panel, one of the four operators in Yemen will be 599 on your frequency due to a perfect skip and, since no one heard him, he moved on.

Figure 17-5 shows our solar sensor. It consists of two photoresistors, two resistors, a piece of balsa wood, two plastic bottles, an old 4-conductor answering machine cord, and some glue. Our cost was less than $5, and most of that was spent on the plastic bottles. (We picked the photoresistors up at the Dayton Hamfest in a bag of 20 for a dollar.) If you look closely at Figure 17-5, you might be able to see the piece of thin balsa wood that essentially divides the bottle into two halves. The balsa divider was "painted" with a black felt tipped pen to reduce glare and glued into place. The two photoresistors are placed such that one is more-or-less centered in its half of the bottle. Figure 17-6 shows how the photoresistors are positioned before they are glued in place.

Photoresistors behave such that in complete darkness they exhibit very high resistance, often upwards of 200 kΩ. However, in bright sunlight, their resistance drops close to zero. Often photoresistors are used in a voltage divider configuration and then mapped to the desired levels and read via one of the analog inputs. In our case, however, we just want to know if there is more or less sunlight on one photoresistor relative to the other photoresistor. The balsa wood divider helps to isolate the two photoresistors. If the sensor is pointed directly toward the sun, both photoresistors are getting the same amount of sunlight and their values will be similar (but not necessarily equal). If the sensor is horizontal with respect to the balsa wood divider, the bottom photoresistor will be

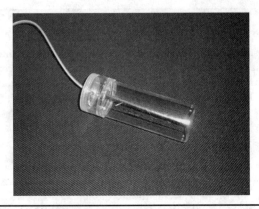

FIGURE 17-5   The homemade solar sensor.

Chapter 17: A Portable Solar Power Source 383

FIGURE 17-6 Placement of the photoresistors.

in shade and the top photoresistor in bright sunlight. By examining the two resistances from the photoresistors, we can rotate the solar panel until both sensors are in equal sunlight.

The photoresistors are connected to the system according to the schematic shown in Figure 17-7. The value of the resistors is not critical. We used 470 Ω resistors for R1 and R2 in the schematic.

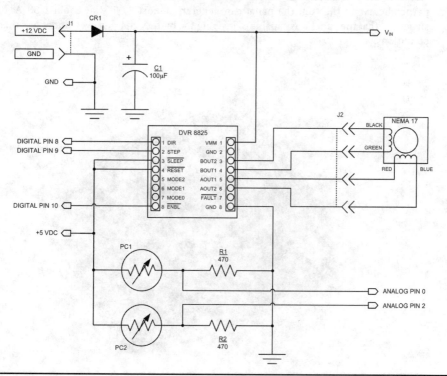

FIGURE 17-7 Schematic of solar sensor.

In hindsight, a better, and perhaps less expensive, method for building the solar sensor would be to use a short length of 1.5-in. PVC pipe rather than a plastic bottle. The end cap could be a standard PVC end cap built the same way as shown in Figure 17-5 and then cemented into place. The balsa wood divider would also be the same, only glue inside the PVC pipe. This would make a much more durable sensor in the long run.

Solar Charger Controller

You should also plan to have a solar charge controller attached to your system. The controller provides overload, short circuit, and lightning protection. It also safeguards against reverse polarity and discharging protection as well as over- and under-voltage protection. Figure 17-8 shows the unit we purchased on eBay for less than $10, including shipping.

The controller we purchased is capable of managing up to 180 W at 12 V, which is more than enough for our system. Small LEDs on the unit indicate whether the unit is charging or discharging and the state of the battery. The unit is small, measuring about 4 in. × 3.75 in. × 1.5 in. and easily fits into the toolbox. There are six terminals on the unit: two each for the panel, battery, and the load. Connecting it is a snap.

Figure 17-9 shows an early test of the solar panel, battery, and charge controller. As you may be able to tell from the shadows, the photo was taken late in the afternoon on a partly cloudy day with the panel lying flat. Still, the panel was putting out 13.48 V although the current is less than optimum. Although perhaps not discernible in the photo, the charger is powering the ATmega2560 board and is also spilling power into the battery. In bright sunlight with the panel perpendicular to the sun, the panel can generate almost 17.5 V at about 0.58 A. The tests suggested that the panel was capable of keeping the battery charged even under less than perfect conditions.

FIGURE 17-8   Solar charger controller.

FIGURE 17-9 Early test of solar charger.

Panel Positioning and Stepper Motor

To position the panel so it faces the sun at all times during the day would require two motors: one to control the azimuth position and another to determine the altitude. Given that the altitude doesn't change as much as the azimuth does for any given day, we simplified our positioning system to use a single stepper motor to rotate the panel. That is, when the panel is set up in the morning, the angle that determines the altitude is fixed for the day. The stepper motor controls azimuth, or east-west, movement. A threaded rod with a wing nut sets the altitude for the day (as explained later in this chapter).

The stepper motor is a NEMA 17 unipolar/bipolar motor (see Figure 17-10). The motor has six leads, and can be configured to run with either bipolar or unipolar stepper drivers. Some of the reasons we selected this stepper is because it is fairly small, yet provides 44 oz-in. of torque, and is relatively inexpensive (about $20). The stepper motor has a 5-mm "D shaft," which means it has a flat side machined into the otherwise circular shaft. This allows us to use the Pololu universal hub to attach the stepper motor to the solar panel. The motor is a 200 step motor, so each motor pulse allows the shaft to rotate 1.8 degrees, which is more than enough precision for our use.

Stepper Wiring

Figuring out which stepper wires to use is always a question that gets asked, particularly when the motor is purchased online and no documentation is provided. One of the best resources we've found that explains four, six, and eight wire leads can be found at

http://www.linengineering.com/resources/wiring_connections.aspx.

FIGURE 17-10   NEMA 17 stepper motor. *(Stepper motor courtesy of Pololu Robotics and Electronics)*

It appears that there is a *de facto* convention for stepper motor wire colors and the web site does a good job of explaining how to connect both unipolar and bipolar steppers. Don't always believe what you read (except in this book) and check the wiring yourself with an ohmmeter.

We did buy several stepper motors online and they all worked, but in one case, the wires emerging from the motors were all the same color. Because most of these motors had four leads, we could use an ohmmeter to figure out which leads were which. For those stepper motors with six leads, all followed the color coding for the wires mentioned in the linenginering.com link. Just make sure you buy a stepper that can be powered by a 12 V battery and has a decent amount of holding torque. Also, pay attention to the maximum current draw and make sure it doesn't exceed the capabilities of your stepper motor driver.

Stepper Motor Driver

The Arduino family of board has a limited amount of power that they can supply in any circuit, especially when you are using the USB connection as the only source of power. Most stepper motors, however, can draw a fairly substantial amount of power, far in excess of what the Arduino board can supply via the USB connection. For that reason, we use a stepper motor driver to control the stepper motor. (The driver has other advantages, too, such as micro-stepping.)

The motor driver we used in a bipolar configuration is the Pololu DR8825. The motor driver is about the size of a postage stamp (0.8 in. × 0.6 in.). However, don't let its small size fool you. The high current version of the driver can handle up to 1.5 A per phase without a heat sink or forced air cooling. In our tests, the driver ran cool to the touch at all times. We also tried two other drivers and one of them was hot enough to cook breakfast on. (Most of the drivers have thermal shutoff protection.) The cost of the driver is less than $15 and has additional features (e.g., six stepping resolutions) that we do not use. Much of the discussion that follows in this section can also be found in the documentation found on the Pololu web site (see http://www.pololu.com/catalog/product/2133).

If you need a motor driver that can supply more power, you might consider the Linksprite motor shield (LSMOTORSH_M35), which can handle up to 2 A of current. The cost is under $15 and it's built like a tank. See Appendix A for a photograph of the shield and contact details.

Chapter 17: A Portable Solar Power Source

FIGURE 17-11 The DRV8825 motor driver. *(Driver courtesy of Pololu Robotics and Electronics)*

The image shown in Figure 17-11 is actually "upside down" in that the reverse side has silk screened labels for each of the pin connections on the driver. So why show this side? The reason is that, when you have the driver in your circuit, you can use a small Phillips head screwdriver to adjust the potentiometer seen in the lower-left corner of the driver in Figure 17-11 to adjust the current flow to about 70% of that for the stepper motor coils (see the Current Limiting section of the Pololu documentation at the Web address mentioned above). If you mount the driver so you can read the silk screen legends, you can't get to the pot for adjustment while it's on the breadboard. (We mounted the driver "right-side up" in our circuit after a little trial-and-error adjustment of the current pot, as shown in Figure 17-17. In a second test, we did use the driver as it came from the factory and it worked fine with the NEMA 17 without adjustment. If your stepper has a heavy current draw, you probably need to adjust the pot control.)

Figure 17-12 shows the driver connections as presented in the Pololu documentation that's available for the DRV8825. The driver is shipped with header pins that you can connect to the board, which makes bread boarding pretty simple.

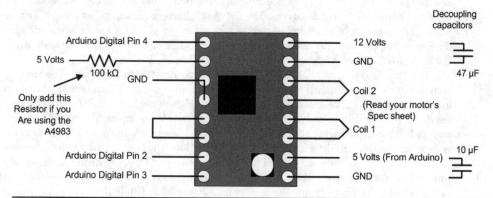

FIGURE 17-12 Connections for the DRV8825.

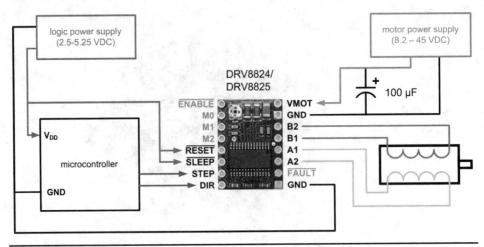

FIGURE 17-13 The DRV8825. *(DRV8825 courtesy of Pololu)*

In Figure 17-13 you can see that separate voltages are used to power the motor with the Arduino supplying the 5 V for the electronics. (Our NEMA 17 uses 12 V. Note that there are other flavors of NEMA 17s that use different voltages and have different torque ratings.) The driver board uses low-ESR ceramic capacitors, making it susceptible to voltage spikes, especially when the motor power leads are longer than a few inches. Since our power leads are fairly long, we did add a 100 µF capacitor across the motor power lead (VMOT) and ground. (Figure 17-12 suggests a 47 µF capacitor.) We did not add the 10 µF capacitor to the Arduino power lines, as suggested in Figure 17-12.

As mentioned earlier, stepper motors come in different configurations and may have four, six, or eight wires coming from the motor. Our NEMA 17 has six wires for the two motor coils. Our motor used the Red-Blue wires for coil 1 and the Black-Green wires for coil 2. (The Yellow-White wires are not used.) You should consult your motor's documentation and use your VTVM to check the leads to make sure you have the windings figured out correctly.

Control Inputs

Figure 17-13 shows the basic wiring for the DRV8825. Each pulse of the STEP input equals one step of the stepper motor in the direction specified by the DIR input. Because the NEMA 17 is a 200 step motor, the 200 steps produce one full revolution of the motor shaft. (It also follows that each step corresponds to 1.8 degrees. The DRV8825, however, is capable of "micro-stepping." If you implemented the quarter micro-step feature using the M0, M1, and M2 pins, for example, there would be 800 steps per revolution. Because such angular granularity in our application is an H-bomb-to-kill-an-ant, we leave the micro-stepping features unused.)

The RESET, SLEEP, and ENBL inputs control the power states of the board. The SLEEP pin is pulled low through an internal 1 M pull-down resistor while the RESET and ENBL pins are pulled low through internal 100K pull-down resistors. *These default states for RESET and SLEEP prevent the driver from operating and must be pulled high to enable the driver.* They can be pulled high directly by connecting them to the Arduino 5 V supply, as suggested in Figure 17-13, or you can use code to pull then high. By default, the ENBL pin is pulled low, which means it can be left unconnected.

In Figure 17-6 we show the ENABLE pin connected to pin 10 of the Arduino. However, if you want to have the ability to disable the stepper in software, you can change the state of the ENABLE pin to HIGH and effectively shut the stepper off. On the other hand, if you are going to program the software to reside on an ATtiny85 or a Digispark where pins are a tad scarce, simply leave the

ENABLE pin unconnected. The software necessary to control the panel easily fits into an ATtiny85 or Digispark. If you use the ATtiny85 chip, you need to provide a controlled 5 V source for the logic components in the system. The Digispark has an onboard 5 V regulator.

The DRV8825 sets the FAULT pin low whenever the H-bridge FETs are disabled because of thermal overheating. The driver connects the FAULT pin to the SLEEP pin through a 10K resistor, which acts as a FAULT pull-up resistor whenever SLEEP is externally held high. This means no external pull-up is necessary on the FAULT pin. (The FAULT line also has a 1.5K protection resistor in series to make it compatible with an earlier version of the driver.) Because our demands are such that we don't even make one full revolution during the day, we don't make use of the FAULT feature.

The DIR pin is used to set the direction of rotation for the stepper. The DIR pin uses HIGH and LOW for its two states. Therefore, if you want to reverse the present direction of the stepper, you can do so in software using the C statement:

```
dirPin = !dirPin;
```

In the software presented later in this chapter, you can see that we add small delays any time we change the state of the driver. We do this to insure that the board is in a stable state before attempting the next software instruction.

Solar Panel Support Structure

Clearly, there must be some form of support structure that holds the solar panel in place. We have constructed the support structure so everything except the panel itself fits inside the toolbox.

The first step is to secure the pivot rod to the back of the solar panel. This can be seen in Figure 17-14. The cable coming from the center of the panel is the panel's power feed. Eventually, the power cable is fed through the vertical PVC pipe and into the toolbox. Notice that the top of the vertical support pipe is notched so the threaded rod simply "drops" into place. You can also see why a flat surface on the top of the toolbox is useful. The threaded PVC collar you see at the base of the support pipe is glued to the top of the toolbox. The vertical PVC support pipe is NOT glued to the collar. Rather, the function of the collar is to serve as a bushing to align the pipe with the stepper motor fitting contained within the toolbox. (Details are given later in the chapter.) The collar is a threaded fitting to allow you to thread a cap onto the collar when the panel is removed to make it a little more weatherproof.

FIGURE 17-14   Fixing the pivot rod on the solar panel. *(Panel courtesy of UL-Solar)*

By the way, gluing the threaded PVC collar to the toolbox took some trial-and-error on our part. We had some "superglue," which we applied to the PVC collar and the toolbox. We let it sit overnight. The next day we opened the toolbox lid and the PVC collar "broke off." The glue would not bind the two plastic surfaces. So, not having any other type of glue handy, we used PVC cement to "weld" the collar to the toolbox. Same result. Finally, we bought some original superglue that specifically states it works on plastic and, voila, it stayed in place. Moral of the story: read the label on your glue before you use it!

If you look closely near the top of the panel in Figure 17-14, you can see something fuzzy near the center of the frame. That is a small strip of Velcro that holds the sensor shown in Figure 17-5. Gluing strips of Velcro on the sensor and the panel makes it simple to remove the sensor for storage in the toolbox. The four conductor wire from the sensor also is routed through the center of the PVC pipe and into the toolbox. The sensor wires ultimately connect to the Arduino. What is not shown in Figure 17-14 is a piece of threaded rod that is attached to the bottom edge of the panel and slides into the vertical PVC support pipe. A wing nut allows you to adjust the altitude of the panel relative to the sun.

The panel has six pre-drilled holes in the sturdy aluminum frame. We used the center holes on either side to mount two ½-in. L braces. The pivot rod itself is a threaded ¼-in. rod with matching nuts on either side of the brace locking it into place. We added a few drops of superglue to make sure the nuts don't shake loose.

Stepper Motor Details

Figure 17-15 shows how the Polulu universal hub is mounted to ½-in. PVC end cap. We drew diagonal lines across the top from corner-to-corner to locate the center of the cap. We then drilled a hole slightly larger than the diameter of the universal hub. Our hole is ¼-in. but may vary depending upon the universal hub you use.

FIGURE 17-15   Universal hub assembly. *(Hub courtesy of Polulu)*

FIGURE 17-16   Fixing the hub for drilling mounting screws.

Figure 17-16 shows how we placed the drill bit through the hub and into the cap. This allows us to center the hub on the cap while we drilled four holes for the mounting screws shown in Figure 17-15. After the holes were drilled, a few drops of superglue were placed in each hole and the screws attached through the hub to the end cap. Eventually, a piece of ½-in. PVC pipe is fitted into the end cap and is used to transfer angular motion from the stepper to the solar panel.

Mounting the Stepper Motor

Figure 17-17 shows how bad Jack is at fashioning a bracket to fasten the stepper motor to the toolbox. That's the bad news. The good news is that the bracket only has to keep the stepper motor from rotating in the toolbox. The bracket is fashioned out of aluminum strap that can be purchased at almost any hardware store.

The stepper motor bracket is attached to the side of the toolbox as shown in Figure 17-18. The bracket is attached to the side of the toolbox because this better aligns with a flat surface on the top of the toolbox lid. Because the sides are sloped, we wedged a wood shim between the toolbox and the stepper motor to make the shaft vertical. This is not a critical adjustment, but simply an attempt to make it easier for the stepper motor to rotate the panel.

You can also see three L brackets attached to the sides of the toolbox. These eventually support the wood platform that holds the upper part of the panel support plus the solar controller and the Arduino. Although the placement of the brackets is not critical, you do want them high enough to clear the stepper motor and the universal hub, but also low enough to house the panel controller and the Arduino.

FIGURE 17-17 Mounting bracket for stepper motor.

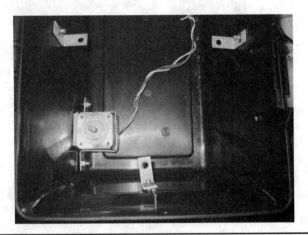

FIGURE 17-18 Mounting the stepper motor in the toolbox.

In Figure 17-19, you can see how a short piece of ½-in. PVC pipe is connected to the end cap that holds the universal hub. The PVC pipe with the universal hub piece rises through the wood platform. The actual length of the piece of ½-in. pipe depends on the depth of the toolbox. For our toolbox, the pipe extends about 3 ¼ in. above the outside edge of the end cap that holds the universal hub. (Figure 17-20 shows how the pieces fit together.) We suggest you read the rest of this chapter before cutting anything so you have a good idea of how everything is supposed to fit together. The "ring" that you see surrounding the ½-in. vertical pipe in Figure 17-18 is actually a coupling for a 1 ½ in. PVC pipe. We cut about ½ in. from the coupling and glued it to the wood platform. While this makes for a very loose fit for the vertical piece of 1-in. PVC pipe that holds the solar panel, it functions as a shim. The shim's purpose is simply to keep the vertical support pipe from sliding around on the board. The positioning of the various pieces of PVC pipe can be seen in Figure 17-20. We used small wood screws to attach the L brackets to the wood platform. After the wood screws are set, we then placed the platform in the toolbox using bolts secured to the sides of the toolbox.

In Figure 17-20 we have placed the 1-in. notched vertical support pipe for the solar panel over the ½-in. pipe attached to the stepper motor. This forms a sleeve using the two pipes. By placing a

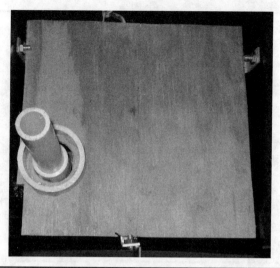

FIGURE 17-19   Wood support platform with stepper extension.

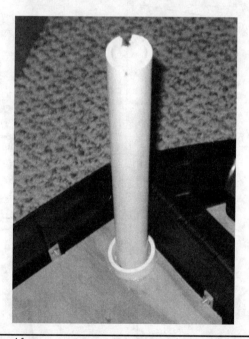

FIGURE 17-20   Sleeve formed from stepper motor ½-in. pipe and 1-in. panel support pipe.

cotter pin in the panel support pipe that rests on the thread collar, no weight is transferred to the stepper motor. Instead, the weight is born by the threaded coupling on the top of the toolbox.

A side view of the panel support is shown in Figure 17-21. The important thing to notice in the figure is that the weight of the panel is not transferred to the stepper motor. The sleeve allows the panel to rotate via the cotter pin within the toolbox, while the second cotter pin at the top of threaded coupling on top of the toolbox bears the weight of the panel.

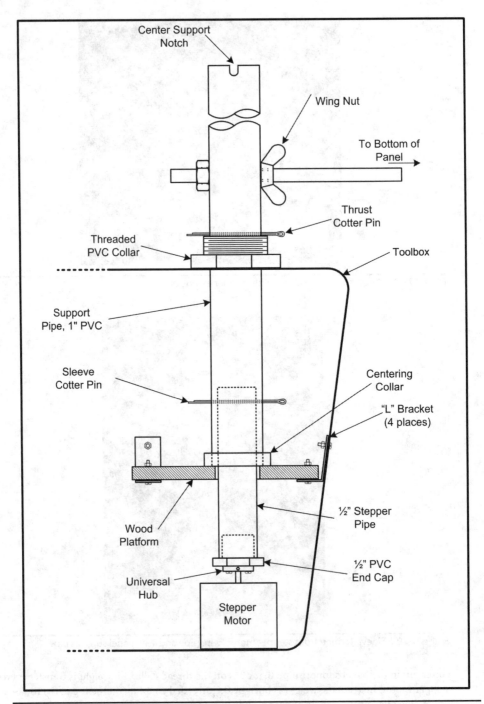

FIGURE 17-21   Solar panel support structure.

The support structure is fairly simple to set up and take down. The only tricky part is putting the interior sleeve cotter pin in place and removing it. The reason it is a bit tricky is because the lid has to be partially closed while changing the sleeve cotter pin. Attaching a small wire loop to the cotter pin makes is pretty easy to remove. At worst, it's a minute or two job to assemble and disassemble the system. As mentioned earlier, all of the parts except the solar panel itself fit within the toolbox, which makes it easy to carry in the field.

We did not "fan out" the inside cotter pin in order to make it easier to break the system down. While we never had either cotter pin work loose during testing, it could happen. Our solution was to use a thick rubber band, like you see holding a bunch of asparagus or broccoli together in a supermarket, to keep the interior cotter pin from moving. Simply take the rubber band and place it over the loop end of the cotter pin, stretch it around the side of the support pipe, and loop it around the open end of the cotter pin. The tension of the rubber band keeps the cotter pin from working loose.

Solar Panel Connections

The solar panel has an output cable that takes the power from the solar panel array and transfers it to the panel controller. A piece of Velcro is attached to the solar panel sensor (see Figure 17-4) and to the top of the panel. Both the power and sensor cables have quick connectors on them, making it easier to connect and disconnect the panel and the sensor. Both cables are fed through the support tube to the panel controller and the Arduino.

Note that the positioning of the Velcro on the sensor tube is important. In Figure 17-22, you can see that the balsa wood divider is more-or-less parallel to the vertical edge of the panel. (It looks like half of the tube is painted black, but that is just the balsa wood divider being distorted by the tube's shape.) This means that one photoresistor is to the left of the balsa divider, while the other photoresistor is to the right of the balsa divider. Suppose the panel position is fixed at 90 degrees to the rising sun. In this position, both photoresistors are receiving equal amounts of sunlight. As the sun moves from East to West, a shadow is cast on the left photoresistor by the balsa wood divider while the right photoresistor starts to receive more and more sunlight. The difference in resistance between the two photoresistors causes the Arduino to activate the stepper motor and rotate it to a point where the shadow disappears on the left photoresistor. If you mounted the Velcro strip such that the balsa divider was horizontal to the edge of the panel, both photoresistors would always be in sunlight and the stepper motor would not be activated to orient the panel toward the sun.

FIGURE 17-22 Solar panel and sensor position.

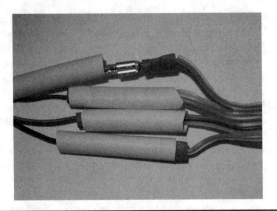

FIGURE 17-23   Stepper motor quick disconnect connectors.

The cable that is attached to the solar panel is 10 ft long, which leaves you plenty of length to work with. The telephone cable we used for the sensor started out to be 25 ft long, so we can trim that to just about any length we wish. While you could permanently attach the sensor to the panel with some form of bracket, the Velcro allows you to just "rip" the sensor loose when you are finished using the panel, coil the cable, and place it in the toolbox. Because the sensor is very light, the Velcro keeps the sensor in place.

Placing the Quick Connectors

The question, however, as to where to place the "quick disconnect" connectors remained. If we placed the quick connectors outside the toolbox and close to the panel, they are readily accessible. Assembly and disassembly is pretty easy. The downside is that the connectors are exposed to unexpected weather conditions that often accompany Field Day and similar excursions. Placing the connectors inside the toolbox improves the exposure aspect, but means you can't free the cables until after the solar panel assembly is taken down. It also means that the cables need to be a bit longer to give you some slack while you open the toolbox. In the final analysis, we opted to place the connectors outside the toolbox near the panel. Most of us can see bad weather approaching long before the time it takes to take the panel down. Indeed, you can just grab the toolbox by the handle and move it indoors with the panel still in place if need be. It's your system ... place the connectors wherever they make sense to you.

Figure 17-23 shows the quick disconnect connectors that are used for the stepper motor. Sadly, Jack wasn't smart enough to stagger the connectors before he trimmed them, so he slipped some Teflon shrink tubing over the quick connectors. The tubing is not heated and shrunk because we want to be able to disconnect the cable, coil it up, and store it in the toolbox when we're finished using the panel. However, the tubing keeps the connectors from contacting each other. The tubing is small enough it doesn't slide easily on the wires. (A second panel used Molex connectors ... a much better choice.)

The Motor Controller Shield

Figure 17-24 shows the motor controller shield prototype that holds the stepper motor controller. The barrel connector on the right provides the power source to the motor driver. This connector provides enough voltage and current to drive the stepper motor and usually exceeds the 5 V

Chapter 17: A Portable Solar Power Source

FIGURE 17-24   Stepper motor shield.

needed to power the Arduino. However, in the field, power for both the Arduino and the motor driver comes from the 12 V Load terminals of the panel charge controller. If the panel is not generating enough power, the charge controller extracts the power from the battery.

Routing Power Cables

While not shown here, the power source for any equipment attached to the system should also be taken from the Load terminals of the charge controller, mainly because of the protection features of the controller. Some of you may wish to route the power output cables to drive your rig with the toolbox lid securely fastened. Of course, this means holes need to be drilled into the walls of the toolbox to pass the cables through. Others may simply opt to have the lid "loosely closed," routing the cables over the lip of the toolbox to the rig. The advantage of this approach is that the toolbox remains a little better sealed against the weather. However, the toolbox isn't weatherproof anyway, so how you route the cable is a personal choice.

Motor Controller Shield Wiring

On the left side of Figure 17-24, you can see the 4 conductor cable that goes to the stepper motor. Also note that there are two 4-pin headers on the left edge of the board. These were used during testing and made it easy to test the stepper motor and the solar sensor values. Because the sensor cable is made from a lightweight 4-conductor phone cable, we glued it in place to the board once the tests were complete as a form of *strain insulator* for the cable. (The motor cable is heavy enough we didn't think it was necessary to glue it.) We could have removed the headers at that point, but we left them in place in case we need to read either the sensor or motor connections at some future time. Because of the ease of future troubleshooting, we decided to use the proto board instead of building a "pretty" motor controller shield. If you do build a shield, you might consider adding the two headers for the cables.

The proto shield is a true shield, even though it's "homemade." The motor shield piggybacks onto your Arduino board. Many of the header pins you used in most projects are 2.54-mm long, but if you use 17-mm pins, they are long enough to mate with the Arduino board. The "sandwich stack" ends with this shield, however, because we mounted the motor driver in its own socket that we built up from header strips like those made for the cables. As a result, the motor driver sits about ¾ in. off the deck, making it impossible to stack another shield on top. (But then, why would you need another shield?) The shield is wired as shown in the schematic in Figure 17-7.

Altitude Positioning

Figure 17-25 shows how the solar panel is held in place relative to the altitude of the panel. A hole was drilled in the center of the bottom edge of the aluminum frame. A small L bracket is attached to the bottom of the panel frame with a ¼-in. bolt, lockwasher, and nut, then tightened. A threaded ¼-in. rod attached to the other opening of the L bracket using two nuts and lock washers. A wing nut is then threaded on the open end of the rod. The end of the rod is fed through a ½-in. hole in the vertical support pipe, just above the support cotter pin. You may need to bend the L bracket a little to make things line up correctly. When the panel is first set up, the wing nut is moved in or out to make the panel at 90 degrees to the sun. It shouldn't need to be adjusted again for the rest of the day.

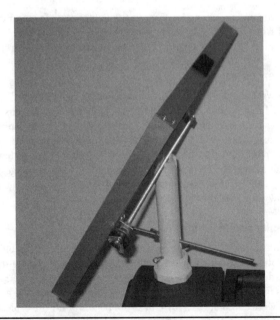

FIGURE 17-25   Altitude positioning mechanism.

The Software

The software that controls the system is actually pretty simple. The source code is presented in Listing 17-1. As usual, the code begins with a preprocessor directive for the *DEBUG* flag. You should comment this out (or *#undef DEBUG*) when you get your system and its software stable. The rest of the preprocessor directives are used to minimize the number of magic numbers in the code. Several global variables are also defined after the preprocessor directives. The *setup()* method uses *pinMode()* to place the various pins in the proper state. We do a *digitalWrite()* to the *SLEEPPIN* and *RESETPIN* to set them HIGH. This is actually done more for purposes of documentation more than anything else since those pins are wired to the 5 V supply (see Figures 17-6 and 17-12). Forgetting to tie these pins on the motor driver HIGH when using the DRV8825 renders the motor driver inoperable ... not good.

```
/*****
Code based upon:
  http://forum.pololu.com/viewtopic.php?f=15&t=6424

*****/

#define DEBUG 1                    // Used to toggle debug code

#define SETTLESTEPPER 10           // Time for stepper to change state
#define READSENSORDELAY 5          // Time for sensors to stabilize

#define TOPSENSORPIN A0            // Pin for top sensor
#define BOTTOMSENSORPIN   A2       // Pin for bottom sensor

#define SLEEPPIN 11                // Sleep pin on driver
#define RESETPIN 12                // Reset pin      "

#define STEPPIN    9               // Step pin       "
#define DIRECTIONPIN 8             // Direction pin  "
#define ENABLEPIN 10               // Enable pin     "

#define SENSORSENSITIVITY 60       // Sensor difference that causes a rotation
#define STEPPERINCREMENT 10        //  This many steps (18 degrees) to move panel

int directionVal = HIGH;

int topPhotoValue = 0;             // reading from the top photoresistor
int bottomPhotoValue = 0;          //           "           bottom      "
int previousValue = 0;             // Previous reading from sensor

// ============================ setup() ==================
void setup()
{
```

LISTING 17-1 Solar panel system code.

```
#ifdef DEBUG
  Serial.begin(9600);
#endif
  pinMode(SLEEPPIN,OUTPUT);              // Initialize stepper pins on uC
  pinMode(RESETPIN,OUTPUT);
  pinMode(DIRECTIONPIN,OUTPUT);
  pinMode(STEPPIN,OUTPUT);
  pinMode(ENABLEPIN, OUTPUT);

  pinMode(TOPSENSORPIN, INPUT);          // Set sensor pins
  pinMode(BOTTOMSENSORPIN, INPUT);

  digitalWrite(SLEEPPIN, HIGH);          // Make sure defaults are set
  digitalWrite(RESETPIN, HIGH);
}


// ============================ loop() ===================
void loop()
{
  ReadSensors();                         // Sensor read to get top and bottom
                                         // sensor values
#ifdef DEBUG
  Serial.print("bottom sensor = ");
  Serial.print(bottomPhotoValue);
  Serial.print("   top sensor = ");
  Serial.print(topPhotoValue);
  Serial.print("   dir pin = ");
  Serial.print(directionVal);
  Serial.print("   previous value = ");
  Serial.println(previousValue);
#endif

  if (abs(bottomPhotoValue - topPhotoValue) > SENSORSENSITIVITY)
  {
    RotateStepper(STEPPERINCREMENT);
    delay(1000);
  }
  previousValue = topPhotoValue - bottomPhotoValue;
  if (previousValue < 0)                                 // Wrong direction??
  {
    directionVal = !directionVal;
    while (previousValue < 0) {
      RotateStepper(STEPPERINCREMENT);
      ReadSensors();
      previousValue = topPhotoValue - bottomPhotoValue;
#ifdef DEBUG
  Serial.print("**************** bottom sensor = ");
  Serial.print(bottomPhotoValue);
  Serial.print("   top sensor = ");
```

LISTING 17-1 Solar panel system code. (*continued*)

```
      Serial.print(topPhotoValue);
      Serial.print("   dir pin = ");
      Serial.print(directionVal);
      Serial.print("   previous value = ");
      Serial.println(previousValue);
#endif
    }                                      // End while (previousVal < 0)
    directionVal = !directionVal;
  }                                        // End if (previousVal < 0)
}

/*****
  This method rotates the stepper motor steps units, where a unit is
  defined by the type of stepper motor used. For a NEMA 17, each step
  corresponds to 1.8 degree, or 200 steps to make a single revolution.
  The Pololu motor driver is capable of microsteps less that 1.8 degrees,
  but this code does not make use of microstepping.

  Parameters:
    int steps    the number of steps we wish to rotate the motor

  Return value:
    void
*****/
void RotateStepper(int steps)
{
  int i;

  digitalWrite(ENABLEPIN, LOW);            // LOW turns stepper driver on
  delay(SETTLESTEPPER);
  digitalWrite(DIRECTIONPIN, directionVal);
  delay(SETTLESTEPPER);

  for (i = 0; i < steps; i++)
  {
    digitalWrite(STEPPIN, LOW);
    delay(SETTLESTEPPER);
    digitalWrite(STEPPIN, HIGH);
    delay(SETTLESTEPPER);
  }
}

/*****
  This method reads the values from the two photoresistors that make up
  the sensor. It is assumed that the sensor is positioned correctly to
  the solar panel for reading.

  Parameters:
    void
```

LISTING 17-1   Solar panel system code. *(continued)*

```
   Return value:
      void
*****/
void ReadSensors()
{
  topPhotoValue = analogRead(TOPSENSORPIN);   // Get values for top - bottom
                                              // sensors
  delayMicroseconds(READSENSORDELAY);         // Prevent "digital jitters"
  bottomPhotoValue = analogRead(BOTTOMSENSORPIN);
  delayMicroseconds(READSENSORDELAY);
}
```

LISTING 17-1   Solar panel system code. *(continued)*

The *loop()* method first calls the *ReadSensors()* to get the values returned from the two photoresistors that comprise the solar sensor. You'll have to play around with the *SENSORSENSITIVITY* value because you can expect some variance in each photoresistor's value even when they are both under the same sunlight conditions. For our sensor, the value was about 60 Ω. Your mileage will vary, so adjust *SENSORSENSITIVITY* accordingly.

The code then compares the two photoresistor values, adjusts for the sensitivity and, if the variance is too large, the *RotateStepper()* method is called, passing it *STEPPERINCREMENT* as an argument. Because each step is 1.8 degrees, we opted for 10 steps to adjust the panel position, or a panel rotation of about 18 degrees. The *if* test on *previousValue* checks to make sure we are rotating the panel in the correct direction, since the *DIRECTIONPIN* can reverse it if needed. If we're rotating in the wrong direction, we reverse *directionVal* and rotate the panel in the new direction. If the direction was changed, we set it back to its original value so we don't start rotating "backwards" when the next adjustment is needed.

The DEBUG statements can be used to help you see how the panel moves with changes in sunlight. It's easiest to play with the code before actually attaching the sensor to the panel assembly. You can just hold the solar sensor in your hand, point it toward a light source, and rotate it toward and away from the light. You can use the serial monitor in the IDE to observe the changes in the sensor values. Pay attention to the values as they determine where to place the strip of Velcro on the sensor. (If you place the Velcro on backwards, you can always reverse the logic tests for the sensor values.)

The *RotateStepper()* method is used to pass control information to the motor driver and, hence, to the stepper motor. As you can see in Figure 17-13, the ENABLE pin defaults to logic LOW, so the call to *digitalWrite()* really is more of a documentation element than a necessary step. However, if you use a different motor driver, you may need to set some other pin before the motor can be moved. The call to *digitalWrite()* may help you to remember to check this for other motor drivers.

The *delay(SETTLESTEPPER)* is a small time slice to insure that the state of the pin is stable. The *DIRECTIONPIN* is also updated at the start of the method. The *for* loop simply moves the motor *step* steps. The Pololu driver requires the *LOW* to *HIGH* state transitions to function properly. Different motor drivers may have different requirements. Check your driver's documentation for details.

Once the stepper is moved to its new position, the *loop()* method code simply continues to execute this sequence throughout the day. Eventually, the output from the panel decreases as the sun sets and the panel can no longer supply a working voltage to the charge controller, which effectively shuts the system down. (However, power can still be drawn from the Load terminals.) While you could write code that monitors the voltage and, upon sensing a shutdown, rotates the panel to a "sunrise" position to prepare for the next day, that's left as "an exercise for the reader." If you add this feature, be sure to let the rest of us know about it.

Our method is to manually shut the system down and have the person with the least seniority in the Field Day party get up at sunrise and manually restart the system with the panel facing the sunrise. Much simpler, albeit totally unfair.

Final Assembly

Figure 17-9 shows the electronics of the system producing about 13.48V in the afternoon. Earlier that morning, the voltage was 13.1V. A half hour later, the panel was putting out 16.3V, while the regulated output to the battery remained at about 13.5V. The panel current was about .55A. The output load voltage remained unchanged.

There are two power plugs coming from the load pins on the charge controller. One plugs into the shield power connector and is used to power the stepper motor. (The third connection to the Load terminals used to power the rig is not shown.) The second connector plugs into the Arduino board.

The motor driver can handle voltage far in excess of what the panel can generate, so you don't need to worry too much about the voltage to the DRV8825. However, putting the load voltage to the Arduino board power connector or Vin are probably the only two ways you should do it. For example, if you have a spare USB connector, but no barrel power connector, you might toy with the idea of using the USB connector to supply the needed power through the USB port. Not a good idea. There is no voltage regulator on the USB port, but there is on the power connector and Vin. Since we are supplying around 13 V to a board that only needs 5 V, use the power connector or Vin so the voltage is regulated.

If you want to build the system around an ATtiny85, you will need to fashion a voltage regulator for it to reduce the 13 V from the panel to the 5 V expected by the ATtiny85. An LM7805 is an inexpensive regulator that is robust, but easy to implement. You can find hundreds of sample circuits on the Internet. You could also use a Digispark in lieu of the Arduino. You wouldn't have to worry about a voltage regulator in that case because the Digispark has one on board.

Assembly and Disassembly

When the project is complete, the small ½-in. piece of plywood forms the support platform that holds the charge controller and Arduino board and motor controller shield to the toolbox using small L brackets attached to the sides of the toolbox (see Figure 17-18). We placed the SLA battery at the other end of the toolbox to serve as a counterweight to the panel. The battery is tied to the toolbox using a strap similar to the one constructed for the stepper motor and is bolted to the toolbox sides. The small Allen wrench used to remove the universal hub from the stepper motor is tightly sandwiched between the battery case and the side of the toolbox. Although we have never needed to remove the stepper motor, the wrench is small enough to get lost easily. At least we know where the wrench *should* be when we need it.

Wing nuts are used to attach the adjusting arm of the solar panel. When you remove a wing nut from its bolt, it's a good idea to place them in a plastic storage box and place them in the toolbox. Otherwise, they behave like socks and you end up with one missing.

Because we have the quick connectors outside of the toolbox, it's a simple matter to pull them apart. If you also forgot to stagger the leads, store the Teflon pieces with the wing nuts. (Better still, use Molex connectors from the git-go.) Now open the toolbox and remove the internal cotter pin from the sleeve between the stepper motor and panel. Store the cotter pins in the plastic box, too. Now pull the panel support pipe from outside the toolbox slowly so you don't rip off the quick connectors on the two cables. Once the panel support pipe is removed, simply pull the two cables through the hole and store in the toolbox. The panel support pipe also fits in the toolbox. Finally, screw on the support pipe cap and you're done.

At this point, everything associated with the solar panel is stored in the toolbox, except for the panel itself. Even with the components stored in the toolbox, there is still enough room for a QRP rig, a key, and a wire antenna. Now you can go home and take a hot shower.

While we missed this year's Field Day exercise, we are pretty confident that even with the relatively small battery we are using, a sunny day will supply enough power to keep almost any QRP rig running throughout the day and the night. If you find that is not the case with your equipment, you could either get a larger solar panel with a higher wattage output, or you could buy a larger battery or connect two of the small ones in parallel. Of the two choices, we'd opt for a larger battery. The next size up for a solar panel seems to be a 20 W model. That model is 22 in. × 13 in. and weighs 6 lb. The 10 W model we are using is 13 in. × 12 in. and weighs 3.1 lb. The increased size and weight would likely mean that the stepper motor would also have to be larger and, hence, more expensive. That also likely means a heavier current draw, which may force a different choice in the stepper motor driver. The ripple effect of a larger solar panel just doesn't seem worth the effort and expense.

Conclusion

This chapter presented a reasonably priced solar panel system that can be used for emergency power or for remote operation, such as Field Day. Given Jack's limited construction talents, it's pretty safe to say almost anyone can build (and likely improve) the system. There are, of course, lots of improvements that could be made to the system. An automatic shutdown of the system when the voltage drops below a given level (e.g., darkness). It would also be nice if the panel rotated back to its sunup position after sitting idle all night. While the system is pretty easy to deploy and take down, there are likely hundreds of ways to improve the mechanical aspects of the system. Still, as ugly as it is, it works and getting free power from the sun and an extra multiplier during contests is kind of nice.

This brings us to the end of the last project. However, we hope this is not your last project. We're sure many of you have other ideas on projects that would be useful and provide greater enjoyment from the hobby we share. To that end, if you do complete a project that you think would be of interest to other hams, please submit it to our web site at www.arduinoforhamradio.com. We hope to hear from a lot of you!

APPENDIX A
Suppliers and Sources

This appendix presents information about where you can go for further information of some of the many Arduino compatible boards, sensors, components, and other peripheral devices as well as QRP transceivers and other equipment of interest to hams.

Parts and Component Suppliers

Cooking Hacks

This company supplies several different types of shields, including a GPS module pictured here.

We're including it here because we know you readers will think of some clever use for it that we didn't have the time or space to devote to it. The web site also provides a tutorial on using the module as well as downloadable software for testing purposes (http://www.cooking-hacks.com).

DFRobot

As you might guess, this company has a lot of products that tie into the robotics field plus a boatload of sensors and shields. However, the shield pictured below is what brought this company to our attention. This is a really clever piece of engineering and is perfect for integration projects where you want multiple shields driven by a single µC. The Arduino Mega-multi Expansion shield means that you can plug four different shields at the same time. All I/O pins of the Mega board are expanded so that you can plug up to four shields and not have to worry about the I/O line conflicts. Although it's hard to see in the figure below, all of the I/O and other lines are clearly silk screened onto the board, which makes integrating this board a snap. Well worth a look if your project requires multiple shields or you think it might grow to that state.

The company can be found at http://www.dfrobot.com.

Diligent Inc.

The Uno32 board takes advantage of the powerful PIC32MX320F128 µC. This µC features a 32-bit MIPS processor core running at 80 MHz, 128K of Flash program memory, and 16K of SRAM data memory. There is a modified IDE that is an Arduino lookalike and is available for Windows, Mac, and Linux. (The board supports all three.) The modified IDE can be downloaded free. For additional platform-specific support for your chipKIT, visit:http://www.chipkit.org/forum/. We have tried numerous sketches and all ran without modification on the chipKIT.

However, the compiler has some differences ... all of them good! For example, an *int* data type for this board uses 4 bytes of storage and a *double* is 8 bytes, versus 2 and 4 for most Atmel boards. Depending upon your app, this could be a real plus in terms of range and precision. If you need a bunch of I/O lines and a very fast processor, this is a great choice and clearly worth investigating. We also found the placement of the reset button very convenient. Also, this is the board upon which the TEN-TEC Rebel is based. For information, check http://www.digilentinc.com.

Linksprite Technologies Inc.

Until we started this project, we were not aware of this company ... our bad. This company supplies some very sophisticated sensors, proto and screw shields, camera, lighting, and other types of electronic devices and components. They have some nice project enclosures, too, and are the manufacturer of the case used for the Dummy Load project. The drive controller for a stepper motor is shown below. Compared to other motor drivers we've used, this one's built like a tank.

However, as impressed as we were with the motor shield, it was the pcDuino that blew us away. We were expecting another μC. Wrong. The pcDuino is a complete computer with USB inputs for mouse and keyboard, 1 Gb of DRAM, 2 Gb of Flash memory, HDMI video interface, onboard SD card slot, network interface with RJ45 and USB WiFi Dongle, and an ARM CPU clocked at 1 GHz ... 16 times faster than an Arduino, with support for Linus or Android OS. We include it here because you can interface Arduino compatible shields to the pcDuino.

We've seen a number of posts on Arduino forums that needed a faster processing speed because sensor data was coming in faster than they could process them. Our guess is that this board may be their answer. The price is around $60. More information can be obtained at http://www.pcduino.com. To view the Linksprite product line go to http://www.linksprite.com.

OSEPP

The company offers a number of Arduino compatible boards and sensors. The board received was the OSEPP Mega 2560 R3 Plus, as seen in the photo below. The board is well constructed and is designed to easily attach external devices and sensors. (Notice the mini-USB connector on the left edge of the board plus the Molex connector in the top left corner for connecting to Osepp sensors and I2C devices.) This series features 256K of Flash memory, 8K of SRAM, 4K EEPROM, 54 digital I/O pins, and 16 analog pins.

You can get further information at http://osepp.com.

Pololu Robotic and Electronics

As their name suggests, this company is a great source for electronics in general, but especially motors and sensors. The stepper motor used to rotate the solar panel shown in Chapter 17 is a NEMA 17 supplied by Pololu. The company also manufactures numerous types of sensors, relay, and other types of boards that would be of interest to hams. Their quality is excellent and good delivery times and service (http://www.pololu.com).

Seeed Studio

Suppliers of many reasonably priced µC boards and shields. They submitted their Seeed Mega 2560 and SD shield for evaluation. Their 2560 Mega board has one of the smallest footprints we've seen for this board. We have also used several of their other shields, including their SD shield (with micro adapter), and everything has been of very high quality and performed as advertised. They also offer a proto shield kit designed specifically for their Mega board. The kit comes with just about all headers and pins you'll ever need to use the shield.

Web: http://www.seeedstudio.com

Tinyos Electronics

This company has a 2560Mega compatible board, but also supplied us with an ATmega328 board. As you can see relative to the pen cap in the photo below, this is one of the smallest boards we received, but it ran all sketches perfectly. The board is well constructed and reasonably priced. Also note that the microcontroller chip is removable. This means you could load software onto the board, remove the chip, and place it in a bare-bones board with only a chip and a few other components if you wanted to do so.

They also sell a wide variety of shields, sensors, and other products for the Arduino boards. I used this board a lot while writing this book, mainly because of its size. For further information, visit http://tinyosshop.com.

Transceiver and Device Suppliers

Blekokqrp 40 meters SSB/CW Transceiver

This transceiver is amazingly small (approximately 2.75 in. × 5.25 in.) yet has a 2–4 W output signal on both SSB and CW. It can be adjusted between 7.0 and 7.3 MHz with an internal ferrite adjustment or 170 kHz with the supplied pot. The rig is a semi-kit in that the onboard parts are already mounted and only the external controls need to be added. Jumpers and connectors are supplied. This is a very nice kit at a very affordable price of around $75 (http://blekokqrp.blogspot.com).

CRK-10A CW and KN-Q7A SSB Transceivers

The CRK-10A transceiver kit can be ordered for 40 meters or 20 meters. We received an assembled version ($70), but the manufacturer states that the kit ($55) can be completed in 4 hours. The instruction manual is extremely well written, so we do think the kit can be built in 4 hours. It includes CQ memory and has an RF output of about 3 W at 12 V. Power supply range is 9 to 15 V. TX current is about 500 mA, and RX current is about 15 mA (measured at 12 V). The built-in MCU (12F629) can generate side tone of about 700 Hz, switch RX/TX, and act as a keyer for not only paddle (normal paddle or bug key simulation mode), but also straight key.

The receiver is a direct conversion receiver, but the sensitivity is very good because of a two-pole crystal filter in the receiver front-end and an audio filter, which block interference and filter out background noise. The MCU automatically shifts TX frequency, generates side tone and acts as the keyer, which makes the whole radio quite practical. The power supply polarity protection and the high SWR protection make the radio durable.

While most of the narrative above is taken from their advertising literature, we were stunned at the receiver's sensitivity. If you're looking for a rig that's about the size of a deck of cards and relatively inexpensive, this is a great choice, especially with the addition of the VFO and LCD frequency display projects added. We really had fun with this rig!

You can read the construction manual at http://crkits.com/crk10amanual.pdf.

The KN-Q7A is an 8-10 W SSB transceiver that can be ordered for either the 40 meter or 20 meter bands.

The transceiver is a little larger than the CRK-10A and requires a heftier power supply. This receiver is also a hot performer and the power output is more than adequate to punch through the QRM on either band. The 40 meter band can be ordered with frequency segment slices that fit your favorite frequencies, which span about 20 kHz. The 20 meters version spectrum slice is fixed at 14.200 to 14.230 MHz. The kit is available for $115 while the assembled version is $165. We would suggest ordering their mic, as it works well with the rig. We also found Larry, the rep for the US distributor, very helpful and quick to respond to questions. The US distributor can be contacted at http://www.larvell.net.

Iambino and Peaberry

This is a clever combination of several useful ham devices. Not only is it an Arduino compatible LCD shield, it also features a DAC, speaker, radio output, pot, and a professional-quality keyer kit.

All the code controlling the device is General Public License (GPL) so you can access and read the code at https://github.com/AE9RB/iambino.

The Peaberry SDR V2 kit is a Software Defined Radio using a 96 kHz digital I/Q interface common to most SDRs. There are several things that make the Peaberry different from other SDR kits. First, you select the band combinations (e.g., 160/80/75, 60/40/30/20) you want at construction time. All parts are included with the kit. Second, all ADC and DAC circuits are included, so you don't need to add sound cards to your computer. Finally, the firmware is Open Source, allowing you to modify it if you wish. (The picture is shown with the optional acrylic case.) For current prices and order information, see http://AE9RB.com.

MFJ Enterprises

MFJ has a number of Cub kits available for 80, 40, 30, 20, 17, or 15 meters. Some of the specs are: 0.2 µV receiver sensitivity, crystal filter and shaped audio, differential-mode AGC, good AF output for phones or speaker, adjustable transmitter with RF output is variable from zero to 2 W out through 20 meters (1 W on 17/15 meters), full QSK, sidetone, shaped keying, 36 mA receive, 380 mA transmit using any regulated 12-15 VDC power source.

Prices are $100 for the kit versions and $150 completely assembled (http://www.mfjenterprises.com).

ozQRP MST2 SSB Transceiver Kits

The ozQRP MST2 series are 5W SSB transceiver kits configurable for operation on 80 meters, 40 meters, or 20 meters. The transceivers feature a sensitive superhet receiver with a 5-pole crystal filter, very effective AGC, and built-in tone generator for antenna tuner adjustment. The IC's are pre-mounted SMDs on the 165 mm × 110 mm PCB. The board with all onboard components sells for $85. A DDS VFO kit specifically designed for the MST2 features a rotary encoder for frequency selection that is displayed on a 2 × 16 LCD. The display also shows current voltage and adjustable frequency step value and is available for $65. Finally, an innovative LED S meter kit accurately displays receiver S units and output power on a 7-segment bar graph for $30. Kits are complete except for enclosure (requires metal rear panel as heat sink for PA) and external controls and connectors. All three kits can be purchased for $170.

I (Jack) received this late in the book writing, but I'm really glad I had a chance to use it. I received the 20 meters version, virtually identical to the one pictured below. This is a very nice rig. The receiver works well and the LCD display is easy to read. The signal strength/output indicator adds a touch not found on most QRP rigs and is surprisingly useful. I was able to make SSB contacts on both coasts with a Hamstick vertical dipole! This is a (non-opal) little gem from down under. Comprehensive construction manuals plus more helpful information is available at www.ozqrp.com.

TEN-TEC

TEN-TEC manufactures a complete line of QRP CW transceivers for the 20, 30, 40, or 80 meter bands. Each features about 3 W power output, QSK, and can cover a 50 kHz segment determined by you at the time of construction. Kit includes all required components and professional silk screened and painted enclosure. Single conversion superhet receiver performance is superlative and QSK is just what you expect from the hams at TEN-TEC. Each kit is priced at less than $125.00. We were impressed with the performance of the 1300 series.

Appendix A: Suppliers and Sources 415

TEN-TEC is not sitting on its laurels, however, as evidenced by their new Rebel transceiver. The most interesting part of the Rebel is that it is controlled by the chipKIT Uno32 microcontroller chip from Digilent. Designed for 40 and 20 meters the Rebel defaults to 7.030 MHz or 14.060 MHz on start-up. However, since the Uno32 is running the show, you can reprogram it to alter the way frequencies are managed. Software changes can be made using the free IDE provided by Digilent. The Rebel has a USB port for uploading software changes. Inside, TEN-TEC has made it very easy to fiddle with the Uno32 by providing two rows of headers, which align with an Arduino-type shield. There are also convenient tie points for reading battery voltage, RF power, CW, and CW speed. We started playing around with the Rebel source code and in less than 25 minutes, we interfaced the LCD display from Chapter 3 into the Rebel, as can be seen in the figure below.

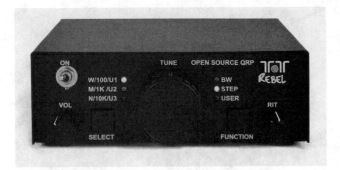

We think hams will have a lot of fun with this one! We also think this design will be more and more popular in the future as hams discover what hardware/software marriages bring to the table. The price for the Rebel is $200.

TEN-TEC announced the Patriot at the 2014 Dayton Hamfest. The rig is so new that we do not have the full specs on it yet. However, it has coverage of 40 to 20 meters on both SSB and CW without bandswitch jumpers, is Open Source using the chipKIT Uno32 like the Rebel, and had a DDS VFO. We believe the market price will be around $400. For complete info, contact http://www.tentec.com.

Small Wonder Labs

This company produces the Rock-Mite CW transceiver kit. As you can see in the figure below, the transceiver is quite small, yet has a surprising number of features, many of which exist because of the 8-pin PIC microcontroller onboard. It has a built-in Iambic keyer, automatic T/R switching, 700 Hz sidetone, and a half-watt output with a 12 V source. It's available for 80, 40, 30, and 20 meters with calling frequency choices. The price is $40! The Rock-Mite series is now distributed by http://www.qrpme.com/.

UL-Solar

This is the 10 W, 12 V solar panel that is used in Chapter 17. The panel is roughly 14 in. × 12 in. × 1 in. and weighs 3.11 lb, which is lighter than some other 10 W panels we've seen. The cells are polycrystalline and have a 10 year limited warranty of 90% output and 25 years at 80% output. The suggested retail price is $49.95, although we've seen them on sale for a little less than that. They also carry larger panels up to 210 W at 24 V, plus various types of mounting hardware. For more information, see http://www.ul-solar.com/.

Amateur Radio Equipment Retailers

These are the major retailers of amateur radio gear in the United States. They are a great source for the radio-related odds and ends that you can't obtain from other sources. All three provide international service.

Amateur Electronics Supply. Internet sales and 4 retail stores. http://www.aesham.com

Ham Radio Outlet. Internet sales and 12 retail stores. http://www.hamradio.com

R&L Electronics. Internet and retail sales located in Hamilton, OH. http://www.randl.com

Texas Towers. Internet and retail sales in Plano, TX. http://www.texastowers.com

Universal Radio. Internet and retail store located in Rynoldsbur, OH. http://www.universal-radio.com

Test and Measurement Equipment

These days there are many companies producing test equipment and if you are on a tight budget, look for older, used test equipment to suit your needs at a local ham flea market or on eBay. Names from the past like Heathkit, Eico, and Knight Kit are available for modest expenditure through eBay. Look for Hewlett-Packard (Agilent), Tektronix, and other high end equipment there as well. These manufacturers provide the tools that we use today:

Agilent Technologies. Formerly branded as Hewlett-Packard and "spun off" in 1999, this company has been a premier provider of test and measurement equipment since 1939. http://www.agilent.com

B&K Precision. Affordable, broad range gear such as multimeters, generators, power supplies, frequency counters, and oscilloscopes. http://www.bkprecision.com

Fluke. Digital multimeters and portable oscilloscopes. http://en-us.fluke.com

Rigol. Good quality, low cost lab equipment such as oscilloscopes, spectrum analyzers, power supplies, and multimeters. http://www.rigol.com

Tektronix. Another major US test equipment manufacturer. Founded in 1946. http://www.tek.com

Other Suppliers and Sources

There are a number of places where you can go to purchase electronic components for your projects. Some of the ones we have used are listed below. You should also use eBay as a source and a reference for parts. With almost 100 purchases there, including many foreign suppliers, we have never had a problem.

Adafruit Industries. Suppliers of both hardware (shields) and software, plus a number of good tutorials on how to use their products. http://www.adafruit.com.

All Electronics. Component Supplier. Worth getting on their email list for special deals. http://www.allelectronics.com

Allied Electronics. Component supplier. No minimum order. http://www.alliedelec.com

Antique Electronics Supply. Good source for carbon composition resistors and other hard-to-find parts. http://www.tubesandmore.com

Debco Electronics. Component supplier. Small distributor run by husband-wife team, both hams. Strong on ICs, custom made cables, RF connectors. Like an old hardware store with hundreds of parts bins ... fun place to visit. Fast mail order and reasonable prices on small orders. http://www.debcoelectronics.com

Digi-Key Electronics. Component supplier. No minimum order. http://www.digikey.com

Intertex Electronics. Source for the Philmore microphone connectors used in the Directional Coupler / SWR Indicator project. http://intertexelectronics.com

Jameco Electronics. Component supplier. http://www.jameco.com

Martin P Jones & Associates. Component supplier. Monthly email specials are interesting, and they are a good source for all components including power supplies. They are one of the best sources for project cases that we have found. http://www.mpja.com

Mouser Electronics. Component supplier. No minimum order. http://www.mouser.com

Newark. Component supplier. No minimum order. http://www.newark.com

Parts Express. Components and specials. Good source for project cases. http://www.parts-express.com

Radio Shack. Component Supplier. Great for when you forgot to order that one part that makes it all work. http://www.radioshack.com

APPENDIX B
Substituting Parts

If you have been involved in hobbyist electronics and amateur radio for a long time and have done any home-built projects from published plans, you know that sometimes we are not always able to find all of the parts as specified by the original designer. The projects in our book are no different in that many times we try to use parts that are on hand, substituting what we have for those parts specified to be used. If you have experience and a well-stocked "junk box" or a large inventory of discrete parts in your workshop, you may try to use parts you have rather than ordering the exact parts from a supplier. This is a perfectly fine practice and one that hams have done for a very long time. But when is it OK to substitute parts? When is it not OK? Here we provide some guidelines to help the newcomers to the art of "homebrew" substitutions.

Hams are always frugal. It's the nature of the hobby. We scrounge for parts for our projects at the swap meets and ham fests. We purchase surplus equipment to be "parted out" for the valuable bits and bobs they hold within. But just how do we know what will work, or not?

The most common parts to substitute would be resistors and capacitors. But there are many different types and styles of resistors and capacitors. We have almost entirely, with one exception being the Dummy Load/Wattmeter, specified the use of ¼ W metal film resistors. The parts we obtain are generally of 5% tolerance, meaning that the actual value of the component, were we to measure it, would be within 5% of the marked value. And when we design a circuit, one of the criteria we look at is how well the circuit behaves if the tolerances are off on components. This process is called sensitivity analysis and can become quite rigorous in critical circuits. In the case of the projects in our book, we want you to be successful in building the projects so we do consider what parts may be used. Let's examine a few examples.

As you examine the designs, notice that common values are used. For instance, we tend to use 220 Ω ¼ W series limiting resistors for backlighting the LCD panels. What if we use a different value? The next higher and lower standard 5% resistor values from 220 Ω are 200 Ω and 240 Ω respectively. What if we use one of those instead? Intuitively, we know that if we use the lower value the backlight is brighter (lower resistance, higher current) and with the higher value, dimmer. But what of the current in the resistor? Do we stay within the ¼ W rating?

Using Ohm's law ($E = I \times R$) and the Power law ($P = I \times E$) we estimate that the current through the resistor is 5 VDC minus the drop across the LED (approximately 0.6 V) divided by 220 (the value of the resistor). The resulting normal current is 20 mA and the power dissipated is 0.088 W, much less than ¼ W. With 200 Ω the current is 0.022 mA and the power is 0.0968 W, well within the ¼ W rating. Without making any calculation, we know that the higher value produces less current, therefore less heat so we are also within the ratings of that value. The conclusion is that any of the three values is acceptable. Series resistors for LEDs for the most part are very forgiving in this manner. If we are driving the LED with a digital output pin on the Arduino, we are also safe because 22 mA is well within the maximum acceptable ratings for per-pin current for the ATmega328.

Consider other resistor values such as those in the panel meter that set the gain for the op amp. The gain is determined by two resistor values: The first (called the *feedback* resistor) connects the output to the inverting input, and the second (the *gain* resistor) connects the inverting input to ground. The feedback resistor includes a potentiometer so that the gain is adjustable, making it easier to set the precise gain, given the variation in component values due to tolerance. Doing the math, the potentiometer compensates for a wide variation in the feedback resistor values. Hence it is again possible that these values are not cast in concrete. Another way to look at the resistors setting the gain is that the gain is merely a ratio and that the individual values can be varied so long as the ratio is maintained.

Capacitors are another part that is subject to substitution. There are many types of capacitors: ceramic disk, molded ceramic, monolithic, electrolytic, Tantalum, and so on. They also come in many different values of working voltage. Capacitor tolerances tend to be wider than those of resistors and we see typical values of 20%.

Many times, the application determines the type of capacitor that we use. For instance, we wouldn't use an electrolytic as a coupling capacitor in an RF circuit, but we might do so in an audio circuit. We tend to use monolithic caps of 0.1 µF at 50 VDC for bypassing the power in our circuits. What about using a 0.22 µF or a 0.047 µF instead? With a working voltage of 16 V, or for that matter a ceramic disk with a working voltage of 100 V? Our operating voltage in most circuits is either 5 VDC or 12 VDC so these are perfectly fine substitutions.

Another example would be coupling capacitors in several projects. When we are dealing with RF, such as in the Frequency Counter or the DDS VFO, we generally use 0.001 µF coupling caps. A 0.0022 would be perfectly acceptable as would a 470 pF capacitor. The net result of those substitutions would be that there could be a low frequency "roll off," meaning that the signal level passing through the capacitor drops as the frequency goes lower, but in general this would occur at much lower frequencies than we are using.

Places where we tend to stick to the design values are where the capacitor is used for timing or frequency determination. In a practical sense, we tend to design using commonly available values, but sometimes you run into special situations where you must use a specified part. We have not included any projects with these criteria in our book.

We've discussed the passive components but what about active devices like transistors and integrated circuits? There are many transistor substitutions that are possible. For example, the universal relay shield uses a component that was nearing the end of production and as we completed the book, we were finding the part (a 75492 hex display driver) was becoming hard to find. What do we do? When we started looking for a substitute part for the rotator controller, we had two options: First, we could find a replacement IC, or second, we could use discrete transistors. We did find a quad driver that was a perfect part substitution albeit a quad device, but this was fine since we were only driving four relays. But what about transistors? Could we have used them instead? Of course, and there are any number of transistors that could have been used as a driver. One of Dennis's favorites is a part that has been around for decades and is perfect for digital circuits and this is the MPS A12. You can find these on eBay for literally pennies. There is a new version that is recommended for new designs but it essentially the same device. By the way, when you hear that phrase "not recommended for new designs" that is an indicator that the manufacturer has determined that this part is nearing what we call "end of life." They may have a newer and better part to replace it so it is always worthwhile to investigate that.

So, why does Dennis like the MPS A12? It is an NPN Darlington device in a plastic TO-18 package that is capable of handling upwards of 1 Amp of collector current. Because of the

Darlington configuration, the gain (Beta or HFE) is in the 10,000 range. It is a great switch or driver. Four of these could have been used instead of the parts we did choose, the STI ULN2068B. The ULN2068B itself uses Darlington-configured transistors and has much greater current handling capacity than the 75492 we originally used.

In the Frequency Counter design, we discussed the prescaling divider circuit using a Dual D-Type Flip-Flop and mentioned that there are two parts, the SN7474 or the CD4013, that are functionally equivalent, albeit different technologies (TTL versus CMOS, respectively) with different pinouts. However, either part is acceptable. Consider also that TTL comes in many "flavors," such as Low-power (L), Low-power Shottky (LS), High-speed (H), and so on. The LS version of the 7474 would be an SN74LS74. Again, because of the lower frequencies at which these projects function, any of these different types of TTL parts would be usable.

Another part that is open to substitution are the op amps we use. LM358 and LM324 are dual and quad general purpose op amps, respectively. The reason we like to use these parts is that they are both internally biased for a single supply and operate easily at 5 V. Many op amps are used with split or "bipolar" power supplies, meaning that there are both a positive and a negative power supply with respect to a common ground. Since the Arduino provides us with a 5 VDC source, a single voltage op amp is a good choice. Most op amps are usable with a single supply, but they would require an external bias circuit to provide the correct mid-voltage reference or half the supply voltage as applied to the non-inverting input. In high gain circuits, setting the bias point is tricky in that op amps are DC amplifiers and the slightest amount that the bias is off shows up greatly amplified at the output. Biasing an op amp used in this fashion requires the use of multi-turn potentiometers and is prone to drift from temperature change. So, the 358 and 324 are much more attractive! But, there is no reason why others could not be used.

One substitution to be aware of is that involving the LCD. The drivers that we use are specific to LCDs using the Hitachi HD44780 controller. As we mentioned in Chapter 3, the LCD Shield, there are other display controllers out there, but it is highly probable that they fail to work with these drivers. To avoid problems, it is a good idea to verify that the LCD you use does use the HD44780 controller. There are some LCD displays on the surplus market that have a very attractive cost and use the HD44780 controller, but be aware, some have odd pin spacing and will not fit properly with the header pins and prototyping shields we have used. If you are trying to keep a project's size very small, you could also investigate using a small OLED display instead of an LCD. Some OLEDs are less than an inch square, yet are quite readable and do not require backlighting.

Hopefully, we have provided you with some guidance regarding part substitution. One should not hesitate to try different parts within reason if you understand the probable outcome as we have discussed. A well-equipped junk box goes a long way to construct many of the projects in our book. Above all, don't be afraid to experiment. After all, that is what hobbyist electronics and amateur radio are about. Be sure and let us know about your adventures on our web site www.arduinoforhamradio.com.

APPENDIX C
Arduino Pin Mapping

When designing shields for Arduino, it is common to find conflicts between pins when combining different projects. The use of a spreadsheet to keep track of pin assignments greatly reduces the risk of conflict. We created Table C-1 for our projects in this book. It allowed us to keep track of where specific functions are assigned and enables us to see where potential conflicts will happen, for example, building a frequency counter using the hardware timer and adding the LCD shield. Pin 5 is the timer input but is also used for the LCD (DB4). The table allows us to move the conflicting display pins, as depicted in the row named "MAPPED."

Project	Introduction of New Features	TX 0	RX 1	2	3	4	5	6	7	8	9	10	11	12	13
				PORTD						PORTB					
LCD Shield				DB7	DB6	DB5	DB4						E	RS	
RTC Shield	I2C									S1					
Meter Shield															
Relay Shield								K1	K2	K3	K4				
DDS Shield										CLK	FUD	DTA	RST		
Front Panel Board	Port Expander			CW	CCW										
Panel Meter															
Meter Shield	AREF														
LCD Shield				DB7	DB6	DB5	DB4						E	RS	
RTC/Timer															
RTC Shield	I2C									S1					
LCD Shield				DB7	DB6	DB5	DB4						E	RS	
Frequency Counter	Hardware Timer						F_IN								
LCD Shield				DB7	DB6	DB5	DB4						E	RS	
MAPPED							F_IN	DB7	DB6	DB5	DB4		E	RS	
Keyer															
Morse Decoder															
LCD Shield				DB7	DB6	DB5	DB4						E	RS	
PS/2 Keyboard Keyer				CK	DT										
System Integration															
RTC/Timer										S1					
LCD Shield				DB7	DB6	DB5	DB4						E	RS	
Morse Decoder															
LCD Shield				DB7	DB6	DB5	DB4						E	RS	
PS2 Keyboard Keyer				CK	DT										
MAPPED		Pins are mapped using the Dfrobot Mega Expansion Shield													

TABLE C-1 Table Used to Manage Arduino IO Pins

Appendix C: Arduino Pin Mapping

	Analog			I2C		MCP23017 Port Expander															Notes
0	1	2	3	4	5	GPA0	GPA1	GPA2	GPA3	GPA4	GPA5	GPA6	GPA7	GPB0	GPB1	GPB2	GPB3	GPB4	GPB5	GPB6	GPB7	
		PORTC																				
4	15	16	17	18	19																	
				SDA	SCL																	
	IN																					
				SDA	SCL	FNC	S1	S2	S3	S4	SEL				DB7	DB6	DB5	DB4	E		RS	
	IN																					
				SDA	SCL																	
																						a
																						b
					IN																	
				SDA	SCL																	
					IN																	c

[a] Only pin 5 usable as hardware timer input.
[b] ATtiny85/Digispark.
[c] Morse input needs a new analog input pin. Conflicts with the I2C bus. Use of the Mega Expansion shield allows pin conflicts to be resolved because the four R3 positions are mapped to unique pins where needed.

Project	Introduction of New Features	Digital													
		TX	RX											
		0	1	2	3	4	5	6	7	8	9	10	11	12	13
		PORTD								PORTB					
SWR / Wattmeter								S1	S2						
LCD Shield				DB7	DB6	DB5	DB4						E	RS	
LCD Shield (Alt)				DB7	DB6	DB5	DB4			B	G	R	E	RS	
Sequencer															
Relay Shield									K1	K2	K3	K4			
Rotator Controller															
Panel Meter Shield															
Relay Shield	HW Interrupt			CW	CCW				K1	K2	K3	K4			
Front Panel Board	Port Expander														
DDS VFO															
DDS Shield											CLK	FUD	DTA	RST	
Front Panel Board				CW	CCW										
Solar Tracker															
Sensors															
DVR 8825											DIR	STP	EN		

TABLE C-1   Table Used to Manage Arduino IO Pins (*continued*)

Appendix C: Arduino Pin Mapping

Analog						MCP23017 Port Expander																Notes
				I2C																		
0	1	2	3	4	5	GPA0	GPA1	GPA2	GPA3	GPA4	GPA5	GPA6	GPA7	GPB0	GPB1	GPB2	GPB3	GPB4	GPB5	GPB6	GPB7	
		PORTC																				
	IN1	IN2																				
IN																						
				SDA	SCL																	
						FNC	S1	S2	S3	S4	SEL				DB7	DB6	DB5	DB4	E		RS	
				SDA	SCL	FNC	S1	S2	S3	S4	SEL				DB7	DB6	DB5	DB4	E		RS	
		S																				

Index

Note: Page numbers for figures and tables are shown in italics.

"", 209
<, 24
}, 209
., 215–216
:, 209
::, 211
{, 209
/, 20
/<<, 192
<>, 209
//, 21
   HelloWorld, 56
/&, 192
/*, 20
   HelloWorld, 56
#, 207
1.0.5 IDE, 138
1N4001, 104
1N4148, 122, 130
3-to-1 timing ratio, 161
4N26 optoisolator, 144, 151
   isolating Arduino from transmitter, 194, *194*
6-pin Mini-DIN, 174
7-to-1 ratio, 155
10-bit device, 113
10K pot:
   adjusting, 52
   installing, 39
16-pin header, soldering, 44–45, *44*, *45*
24AWG, 84, 106
30 dB, 285
200 step motor, 385
328. *See* ATmega328
9600 baud, 27
75492, 248, 420, 421

A

Accessibility, 74
Accuracy, 328
AD8307:
   advantages, 286
   detector design, 288
   sensor board, 292
AD8307A, 298
AD9850, 352
Adafruit RTClib library, 85, 87–89
ADC, 102, 104
   code walkthrough, 113
   dummy load, 130
   element of DDS, 350
AGC, 159
Ah-Ha moment, 191
Alarm functions, 324–325
Algorithm:
   defined, 19
   processing CW signal, 160
Allen wrench, 403
Altitude, 385
Altitude positioning, 398

Index

Amateur radio equipment retailers, 417
American Standard Characters for Information Interchange (ASCII), 191
Amidon FT-82-67, 329
Ammeter, *116*, 117
Analog bar graph, 104
Analog Reference, 103
Analog to Digital converter, 102
   code walkthrough, 113
   key element of DDS, 350
AnalogIn, 103
   panel meter, 114
AnalogReference(EXTERNAL),
   dummy load, 126, 129
AnalogReference(INTERNAL), 103
Angle brackets, 57, 207
Angular granularity, 388
Anhydrous Isopropyl Alcohol, 50–51
Antenna rotator controller, 246–279
Antennas, 227
API (Application Programming Interface), 279–280
APM, 382
Arduino, 2–3
   reasons to choose, 3
Arduino antenna rotator controller, *245*, 246–256
   connecting, 256–260
   control panel, 253–256
   panel meter shield, 248–253
   relay shield, 247–248
      adding I2C interface, 256
   software, 260–279
   supported rotators, 246–247
   world beam headings, 279–282
Arduino Beam Heading Software, 260–261
Arduino Board, connecting to your PC, 11–15
Arduino Expansion shield, 199
Arduino Libraries, 27
   additional, 36
   structure of, 36
Arduino Reference, 28
Arduino Sandwich, 199
ArduinoISP, 137, 138
AREF, 102–103
   jumper pin to disconnect, 325
   warning, 110
AREF pin, 126
   default behavior, 126
Armstrong Position Mechanism, 382
Aronsen, Ragnar O., 160–161

ARRL code practice, 170
Arrow icon, 54
ASCII, 191
ASCII table, 191
Asterisk, 360
Asterisk-slash (*/), 20
ATmega328, 3, *4*
   Arduino Pro Mini, 4
   input/output lines, 5
   rolling your own board, 5
   Tinyos Electronics, 410, *410*
ATmega1280, *4*
   memory and I/O pins, 5
ATmega2560, *4*
   memory and I/O pins, 5
   for project integration, 199
Atmel, 2
Atom, 160
Atom space, 150
AttachInterrupt(), 323
Attenuator, 288–289
   characteristic impedance, 329
ATtiny directory, 133–134
ATtiny85, 130, *133*
   adding new features, 153
   advantages, 131, 153
   connecting to your Arduino, 134–136
   cost, 132
   feature set, *132*
   pinouts, *139*
   programming sequence, 137–138
   required software, 133–134
   sequencer, 228
ATtiny85 keyer, 131, *132*
   construction, 151, *151*
ATtiny85 labels, 135
Autoscroll, 28, *29*
   unchecking, 30
AUX, 240
Avrdude, 137
AWG24, 290
Azimuth, 246
   solar panel positioning, 385

B

Backlighting, 37
Balsa wood, 382
Banana jack, 125–126
   binding posts, 122, *123*
Band-Down Switch, 360

Band-Up Switch, 360
Barrel power connector, 396
Baud rate, serial monitor, 27–28
BAV21, 121–122
BC547B, 333
BCD, 80–81
BcdToDec(), 99
BDS, 360–361
Beat Frequency Oscillator, 342
Begin() class method, 58
Begin() function, 27
   LiquidCrystal class, 58
Begin Transmission(), 98
Beginning C for Arduino, 13, 17
"Belongs to" operator, 211
BFO, 342
Bidirectional drive motor, 256
Binary, 161
Binary bit pattern, 169
Binary Coded Decimal, 80–81
Binary search, 160
Binary sketch size, 140
Binary tree, 160–161, *161*
Bipolar stepper, 385
Bit mask, 193
Bit shift left operator, 192–193
Bit shifting, 161
   some bit-fiddling, 192–193
Bitwise AND operator, 192
Bitwise operators, 33
Black wire, 47
Blekok Micro 40SC, 374, *374*, *375*
Blekokqr 40 meters SSB/CW transceiver, 410
Blink program, 12, *13*
   comments in C, 20–21
   loop function, 25–26
   setup(), 24–25
   when to comment, 21–22
Blink sketch:
   modifying, 26–31, *26–27*
      output, *29*, 30
   source code for, 20, *20–21*
BNC connector:
   attaching lid to resistor pack, 123
   DDS VFO shield, 353, 356
   fabricating lid connections, 122
Board menu, 134, 140
Boatanchor, 219
Body capacitance, 144
Boolean data, 32–33

Boolean data type, 198
Bootloader:
   described, 14
   software, 6
Box of rocks, 72
Bracket, 391
Brake, 246
Brass sheet, 121
Breadboard, acquiring, 6–7, *6*
Break, 192–193
Breakaway header pins, 40–41, *41*
Brick, 103
BS2, 137
Bug, 75
   Directional Wattmeter, 329
Burt, George, GM3OXX, 372
Bus, 40
   I2C interface, 79
BUS (Band-Up Switch), 360
Buses:
   soldering, 40, *40*
   solderless prototyping boards, 37–38
Bushing, 389
Byte data type, 32–33
Byte value, 191

C

.c, 23
µC, 1
   board origins, 2
C progamming language, 3, 12–13
   Arduino programming essentials, 20
C Standard Library, 20
C++ compiler, 20
   project integration, 206
C++ functions, class constructors, 57
C++ header files, 207
C++ LcdBarGraph class constructor, 114
C#, 3, 279
Cache, 150
Calibrate, 304–307
Calibrating the DDS VFO, 370–371
Calibration offset, 371
Call Sign Prefix, 280
Call sign prefixes, 260
Capacitance Sensors, 144–145
Capacitive reactance, 131
Capacitor substitutions, 420
Capital city, 280
Case statement, 198

CD4013, 421
CDE, 257–258
Change the scale, 117
Check mark icon:
  compiling source code, 13
  HelloWorld program, 54
  right-pointing arrow versus, 14
ChipKIT Uno32, 2, *2*, *4*
  Arduino-compatible, 4
  decoder modifications, 159–160
  Diligent as supplier, 406–408
  memory and I/O pins, 5
  Rebel, 60, 416
Chirpy warble, 349
Churchward, Budd, 161
Circular buffers, 326
Clamp, 104
Clamp-on ferrite beads, 244
Class, 57
  defined, 214
  as description of an object, 215
Class constructor, 57
  panel meter, 114
Class constructor method, 211
Class declaration, 209–210
Claxon, 100
Clock Mode, 97
Clock pin, 174–175
Clock speed, 5
  I2C bus, 79
Closing brace, 209
Code Read, 159
Code Speed, 144
  changing, 197–198
Code Walk-Through, 56–59
  frequency counter, 338–341
  on Listing 9-1, 189–190
  Panel Meter, 113–115
  sequencer, 238–239
Colon, 209
Comment character pair, 56
Comparator, 332
Compile:
  Blink program, 12, *13*
  defined, 13
Configuration jumpers, 239–240
Const, 73
Contesting, 245–246
Continuous Wave, 173
Cookie cutters, 57
Cookie dough, 57

Cooking Hacks, 405–406
Cornell-Dublier Electronic, 257–258
Coupling capacitors, 420
Coupling factor:
  design of directional coupler/remote sensor, 288
  SWR, 284–286
*.cpp, 23, 200, 206, 211, 214, 215
.cpp, 23, 210–211
cpp files, 210–211
  suitability, 200
CR2032, 83
CR2477, 143
CRK-10A transceiver kits, 411–412
CRKits CRK 10A, 374, *376*
Crystal controlled, 371–372
Ctrl-F, 61
Ctrl+U, 14
CUR, 278
CURRENT field, 278
Current limiting resistor, 190, 198
Current line number, 190
Current working directory, 209
CW:
  capabilities, 173
  cost of transceiver, 173
  radio signal, 155
CW Decoder Shield:
  expansion board, 203
  project integration, 202, *203*
CW keyer, 143–144
  adjusting code speed, 144
  capacitance sensors, 144–150
  volatile keyword, 150–151

D

D shaft, 385
Dah(), 193
DAHTRIGGER, 144, 151
D'Arsonval, 102
Data, 19
Data declaration, 22
Data definitions, 22
  "static", 76
Data Input Step, 129–130
Data pin, 175
Data points, 324
DBm, 327
DC, 346–347
DDS, 350

Index

DDS VFO, 349–377
  circuit description, 352–357
  DDS, 350
  EEPROM initialization program, 362–366
  functional description, 357–361
  kit for MST2, 414
  module used, 350–351
  other applications and enhancements, 376–377
  schematic, *350*
  software, 361–362
  testing, 369–371
  using with your radio, 371–376
  VFOControlProgram.ino, 366–368
Debounce function, *92–93*
  sequencer, 239
Decibels milliwatt, 327
Declare, 22
Decoder software, 160–170
DecoderPoll(), 217–218
DecToBcd(), 89
Default frequency step, 361
Default include directory, 109
Default keyer speed, 150
Default startup frequency, 64
Define, 22
#Define, 71–72
#DefineDEBUG, 144
DEG field, 278
Delay():
  function call, 76
  panel meter, 115
  station timer, 71
Delay(1000), 26
Demilanove, *4*
  using decoder with, 159
DFRobot, 199, *200*
  as supplier, 406
Digilent, 60, 159, 415
Digispark, 131, *132*
  advantages, 132
  chip, *133*
  compiling and uploading programs, 140–143
  construction, 152–153, *152*
  memory and I/O pins, 5
  parts placement, *153*
  pinouts, *139*
  sequencer, 228, *229*
  sequencer program for, 235–237
  using, 138–140
  wire connectors, 131

Digispark keyer, 138–143
Digispark Wiki, 234
Digistump, 131, 152, *152*
Digital pin 5, 332
DigitalWrite() function, 26
Dilemma, 76
Diligent:
  IDE, 60
  as supplier, 406–407
  using decoder with, 159
Diode, 122, 130
Diode detector, 286
DIP, 37
DIR pin, 389
Direct Conversion, 346–347
Direct Digital Synthesis, 350
  DDS VFO project, 350–357
Directional Watt and SWR Meter, 283–329
  constructing, 286–303
  further enhancements, 329
  obtaining antenna system SWR, 284–286
  parts list, *287–288*
  software walkthrough, 307–329
  SWR and measuring it, 284
  testing, 304–307
Directivity:
  coupler, 284–286
  defined, 284
Disk drive cable, 61
Dit(), 193
Dit spacing, 170
DITADJUST, 150
DITTRIGGER, 144, 151
Dot operator, 59
  assigning/retrieving value from union, 365–366
  fetch the memory address of, 215–216
  in OOP, 366
Double Conversion, 342–343
Double pole double throw, 220, 221
Double quotation marks, 31
  LiquidCrystal class, 57, 58
  project integration, 209
DPDT, 220, 221
DS13-7 Registers, 80–81
DS1307, 67, *68*
  choice of, 79–80, *80*
  device address, 80
  initializing registers, 89–96
  resetting internal register pointer, 99
  slave address, 79

DS75492 hex driver chip, 221
    rotator controller, 248
    sequencer, 228
dtostrf(), 114
Dual In-line Package, 37
    station ID timer, 80
Dual-winding motor, 246
Due board, 4
Duemilanove, 3, 4, 137, 201
Dummy load (DL), 119–130
    attaching lid to resistor pack, 123–124
    electronic construction, 124
    fabricating lid connections, 122–123
    math, 124–126
    mechanical construction, 120–121
    resistor pack spacing, 121–122, *121*
    software, 126–130
Dupont jumpers, 7, *8*

E

EAS-100, 230
EEPROM:
    defined, 5
    setting a new heading, 279
EEPROM Initialization Program, 362–366
EEPROM memory, 32
Electrical Erasable Programmable Read-Only
    Memory. *See* EEPROM
Electronic components suppliers, 418
EMI suppressors, 244
Enable (EN), 202, 388–389
ENBL pin, 388–389
Encapsulation, 74–75
Encoder Circuit, 195–196
Encoder Control, 361
Encoder control shaft, 361
Encoder shaft switch, 279
Encoder switch, 253
#endif, 207
EndTransmission(), 98
Expansion Shield, DFRobot, 406, *406*
Expansion Shield Pin Mapping, 203–204
EXTERNAL, 103
External power source, 203
External reference voltage, 102
    dummy load, 126

F

F() Macro, 33
    DDS VFO project, 350

Fan out, 395
Farnsworth timing, 169–170
FAULT pin, 389
FC-6P connectors, 297
FCC, 173
Federal Communications Commission (FCC), 173
Feedback resistor, 420
Ferrite cores, 289–290, 329
Field Day, 379
Filter, 156
Filter network, 157
First mixer, 342
Fist, 155, 171
Five Program Steps, 18–20
Flash, defined, 5
Flash memory:
    bootloader use of, 14
    nonvolatile, 5
    Saving Memory, 32
    storing frequencies in, 362
    USB cable, 11
Flat forehead mistake, 73–74
    ATtiny85, 138
    yet another, 100
Flickering, dummy load, 130
For loop, 23
    bit shifting, 192–193
    three expressions, 23–24
Forums, 3
FORWARD, 284, *285*
FreeRam(), 34
Frequency Control Register, 350
Frequency Counter, 331–347
    circuit description, 333–334
    code walkthrough, 338–341
    constructing the shield, 334–337
        alternate design, 337–338
    displaying tuned frequency on display-less
        QRP rig, 342–347
    other radio applications, 347
    parts list, *334*
    substitutions, 421
    tools needed, 332
Frequency coverage of HF ham bands, 351
Frequency readout to 10 Hz, 351
Funafuti, 280
Function prototypes, 209–210
Function scope, 74
Function signature, 209
Functions, 22
    in C, 23
    function type specifier, 25

Index

G

G-800SDX/DXA, *245*, 259–260
G-1000SDX/DXA, 259–260
G-2800DXA, 259–260
Gain:
    decoder, 169
    formula, 102
Gain resistor, 420
Gauge, 46
Ghost constructor, 211
    primary responsibility of, 211
Gibberish (serial monitor), 28
Glawson, Dave, WA6CGR, 293
Global scope, 74
    public keyword, 209
Global search, 74
GO TO field, 278
Google Maps, 279–280
Ground braid, 123

H

.h, 23
H-bridge, 389
Ham 2, 258
Ham IV, 258
Ham-M, 248
HardwareSerial.ccp, 34
HD44780, 36
    substitutions, 421
Header files, 23
    PS2Keyer.h, 207
Header pins, 37
    breakaway, 40–41
    motor shield, 398
    soldering headers, 41
Header socket, 40, *42*
    soldering, 49–50, *51*
Headers:
    breakaway header assembly, 40–41
    parts list, 37
    placement, *44*
    stackable and standards, *42*
Headings[] array, 278
HelloWord program, 54–55, *55–56*
    code walk-through, 56–59
Hertz, 332
Heterodyne, 371
Hex driver, 248
Hiccup, 99–100
HIGH, 25–26

High nibble, 81
High power alarms, 326
Highlighter pen, 46
Hitachi, 36
HyGain, 258

I

I2C, 67
    charge for registering device address, 80
    interface, 78–79
I2C bus connector, 256
I2C interface:
    adding to relay shield, 256
    for control panel board, 253
    for RTC, 201–202
    VFO, 367
Iambino combination device, 412–413
Ibg objects, 113–114
Icd objects, 113–114
IDE. *See* Integrated Development Environment (IDE)
IDE download, 8–9
IF, 342
#Ifdef, 100
#ifdef DEBUG, 213
#ifndef, 207
#include, 56–57
    project integration, 208
Indra Ekoputro, YD1JJJ, 374
Infinite loop, 24, 25
Information booth, 59
Init(), 23
    function call to, 24
Initialization, 18–19
    RTC at startup, *90–96*
    setup(), 25
Initialization Step, 18–19
    dummy load, 129
INITIALIZEEEPROM, 261
*.ino, 23, 60, 205–206
    IntegrationCode, 209, 211–214
Input, 19
Input Step, 19
Insertion loss, 284, 285
Instantiation, 57
    Icd and Ibg objects, 113–114
    object, 214, 215
    project integration, 214
Int data type, 32–33
INT0, 213

INT1, 216
Integers, 22
  as whole numbers, 25
Integrated Development Environment (IDE), 3
  Arduino 156 directory, 10–11, *10*
  Blink program, 12, *12*
  downloading, 8–9
  installing, 9–11
    avoiding default folder, 9–10, *9*
    USB device driver, *10*
  running your first program, 11–15
    connecting Arduino to your PC, 11–15
  selecting the Arduino board, 11, *11*
Integration issues, 199, 200–202
IntegrationCode, 211–214
Inter-Integrated Circuit interface (I2C), 67, 78–79
  bus, 79, *79*
Intercept, 326
  calculating, 326
Intermediate frequency, 342
INTERNAL, 1–3
Internal reference voltage, 102
International Dateline, 282
Interrupt service routines, 54
Interrupts, 54
Isalpha(ch), sendcode() method, 190–191
Isolation, coupler, 284, 285–286
Isopropanol, 50
ISR, 54
  VFO, 367

J

Jacobi, Moritz von, 283
Jacobi's Law, 283
Jameco Electronics, 6–7, *6*
Jumpers, acquiring, 7, *8*

K

K4EAA, 119
K6HX, 161
Kemski, Ken, 119–120
Keyer, 131
Keyer speed, 144
KeyIsDown(), 169
KeyIsUp(), 169
Keywords, 20
KN-Q7A SSB transceivers, 412

KP VFO Software, 366–369
Kraft paper, 121

L

L brackets, 391, 398
Latching relays, 226
Latitude, 280
Layout drawing, 39–40, *40*
LCD Bar Graph library, 110
LCD display, 35, *35*
  caution with substitutions, 421
LCD library, 36
LCD Shield Project, 35–65
  adding components using a schematic, 46–51
  alternative design, 51–52
  assembling, 39–45
  checks before installing LCD, 49
  Code Walk-Through, 56–59
  loading and testing, 53–56
  parts list, 37–39
  parts placement, *44*
  using with TEN-TEC Rebel, 59–65
LcdBarGraph.h, 113
LCDFrequencyDisplay() function, 63, *63–64*
Lcd.print() function, 58
Lethal voltages, 246
Libraries directory, 189
Linksprite motor shield, 386
Linksprite project case, 125, *125*
Linksprite Technologies Inc., 407–408
LINUX, 3
LiOn battery, 82–83
  replacing, 96
LiquidCrystal, 36
  testing the display, 54, *55–56*
LiquidCrystal library:
  code walkthrough, 113
  other examples, 59
LIR2032, 83
LM324, 102–104
LM328, 157
LM358, 251
LM567, 156, *157*, 158, 159
LM7805, 403
LMB, 230
LNA, 227
Logarithmic scale, 324
Longitude, 280

Index

Loop(), 23
   forward power sensor, 326
   modifying, 30–31, *31*
   panel meters, 114–115
   project integration, 217–218
   Rebel, 62–63
   RTC timer program, 98–99
   solar sensor, 402–403
   station ID timer, 75–76
Loop end, 395
LOW, 26
Low nibble, 81
Low noise amplifier (LNA), 227
Lvalue, 190

M

mA, 102, 104
   scale definitions, 113
Machine code instructions, 13
Macro definitions, 325
Macro substitution, 367–368
Magic numbers, 70–71
   fixing, 71–73
Main() function, 22–23
   primary purpose, 24
Main.cpp file, 23, *23*
Manual Graphite Display Generator (MGDG), 382
Marked band edges, 351
Master, 201
Master Device, 79
Maximum range, 103–104
McEwen, Charlie, 17
Medieval kings, 75
Mega shield, 106, 108–109, *108*, *109*
Members:
   defined, 209, 214–215
   defining specific, 210
Memcpy(), 170
Memset(), 325
   initializing buffers, 325
Messages, 142
Methods, 27
   class, 58, 210
   defined, 215
MFJ, 258
   Cub kits available, 413
MFJ 9340 Transceiver, 343
Micro-stepping:
   DRV8825, 388
   stepper motor driver, 386

Microcontrollers:
   Arduino-compatible, *2*
   choosing, 1–3
   table of Arduino, *4*
   See also µC
Micronucleus, 142
Microsoft Visio, 39
Microwave, 245
Milliampere, dummy load, 130
Milliamps, 117
Millis(), 97
   loop() function, 98
Millivolts, 130
Mineral oil, 123
Mini-DIN connector, 259
Minus Step Switch, 361
Misleading comment, 71
Molex, 403
Monk, Simon, 13, 17
Morse Code, FCC requirements dropped, 173
Morse Code decoder, 155–171
   Farnsworth timing, 169–170
   hardware design, 155–160
   Morse decode program, *162–169*
   parts list, *157*
   parts placement, *158*
   software, 160–170
Motor controller shield, 396–398
Moving the beam, 277–279
MPIDE, 60
MPS A12, 420–421
MSS, 361
Multi-band radio, 347
Multi-line program comments, 20–22
   LiquidCrystal, 56
Multiple source code files, 212
Multiple source files, 205

N

Nano-acres, 81
   Expansion shield, 199
NCO, 350
Negative SWR, 328
NEMA 17, 385, *386*, 388
Netduino, 3
Newline character, 28
Nibble, 80–81
"Non-IDE" library, 36

Index

Non-inverting amplifier, 158
Nonvolatile memory, 5
Norcal 40A, 345–346, *346*
Nouns, class members as, 215
NPN Darlington, 420–421
Numerically Controlled Oscillator, 350
NXP, 78–79

O

Object, 214–215
Object constructor, 114
Object of the class, 57
Object Oriented Programming (OOP), 20
    jargon, 210
    LiquidCrystal library, 57
    project integration, 206–207
Off board, 125
Ohm's Law, 115
    dummy load, 124
    resistor substitutions, 419–420
Omega MCU Systems ProtoPro-B protoyping shield, 104–105
OMS ProtoPro shield, 298
    assembling DDS VFO, 353
Onboard voltage regulator, 139
One memory per amateur band, 351
OOP. *See* Object Oriented Programming (OOP)
Op amp, 102, *103*, 104
    signal preprocessing circuit, 157
    substitutions, 421
Open-collector, 221
Open end, 395
Open Source, 1, 3
    C++ compiler, 20
    new libraries to, 36
    TEN-TEC Rebel, 59
Opening brace, 209
Operational amplifier, 102, *103*, 104
& operator, 114
Optoisolator, 151
    isolating Arduino from transmitter, 194–196
OSEPP Mega 2560 R3 Plus, 408, *408*
Out-of-Band Condition, 360
OUTPUT, 24
Output, 19
Output Step, 19
    dummy load, 130
Overloaded Methods, 190
Overrange values, 115
OzQRP MST2 SSB transreceiver kits, 414

P

Paddle sensors, 131
PAGEL, 137
Paint can, 120–121
Palm Paddle, 152
Panel Meter, 101–118
    changing meter range and scale, 116–117
    circuit description, 102–105
    code walkthrough, 113–115
    construction, 104–109
    loading and testing software, 110–112
    parts list, *104*, *105*
    testing and calibration, 115–116
Panel meter shield, 248–253
Parameterized macro, 368
PARIS, 189
Parts and components suppliers, 405–410
    electronic components, 418
Parts, substituting, 419–421
PCB (print circuit board), parts list, 37
PcDuino, 3
    faster processing speed, 408, *408*
Peaberry SDR V2 kit, 413
Pegged, 104
Phase-Locked-Loop, 158
    synthesizers, 349, *349*
Phillips Semiconductor, 78
Photoresistors, 382–383, *383*
Pic QRP, 2
Pin assignments, 54
    analog, 113, 425
    ATmega2560, 201–202
    ATtiny85, 134–135, *136*
    checking, 138
    KP VFO software, 367
    pin mapping, 423–427
    Rebel-to-LCD, 61, *62*
Pin mapping, 423–427
PinMode(), 24
Pivot rod, 389–390, *389*
Pixie QRP Radio, 372–374
Plated-thru holes, 37
PLL, 349
Plus Step Switch, 361
Pointer variable, 190
Polar plot, 282
Polling, 150
Pololu DR8825, 386
Pololu Robotics and Electronics, 409
Polycrystalline, 381
Port expander, 253

Index

Portable Solar Power Source, 379–404
    altitude positioning, 398
    assembly and disassembly, 403–404
    final assembly, 403
    motor controller shield, 396–398
    panel positioning and stepper motor, 385–389
    software, 399–403
    solar charger controller, 384
    solar panel connections, 395–396
    solar panel support structure, 389–390
    solar sensor, 381–384
    stepper motor details, 390–391
    stepper motor mounting, 391–396
Potentiometer, 37
    adjusting, 53
    placement, *44*
    soldering, 45
Pound sign, 207
Pow(), 328
Power, 121–122
Power law, 419
"Pre-operating" tasks", 19
Preprocessor directives, 56–57
    #define, 71–73
    const, 73
    signs used, 207
    station ID timer, 71–73
    toggling the debug code, 99–100
Prescaler, 337–338, 346
Print Bearing List, 282
Print() method, 58–59
Printed circuit board (PCB), 37
Private, 209
Private members, 209
Processing, 19
Processing Step, 19
    dummy load, 130
Progamming shields, ATtiny85, 136, 138
Program an ATtiny85 chip, 137–138
Program bug, 75–76
Program comments, defined, 20
Project integration, 199–218
    CW Decoder Shield, 202
    expansion board, 203, *204*
    integration issues, 200–201
    Real Time Clock Shield, 201–202
    software project integration, 205–206
Project Stacking, 201
Proto shield, 37–39, *38*
    wiring, 47–48

PS2 boxes, 174, *175*
PS2 connector:
    pin out, 173–175, *174*
    testing, 175
PS2 keyboard, 173–175
    encoder circuit schematic, *195*
    encoder software, 176, 189
        program, *176–188*
    testing, 196
PS2 keyboard keyer, project integration, 202
PS2 Library Code, 176
PS2 pin out, 173–175, *174*
PS2 sockets, 174, *174*
PS2keyboard keyer, class source code, 210
PS2KeyboardLibrary, 176–189
PSS, 366
Public, 209
Public members, 209
PVC pipe, 384, 392

Q

QTH, 279–280
Quad op amp, 102, 157
QUEUEMASK, 198
Quick connectors:
    disassembly, 403–404
    placing, 396, *396*
Quiescent voltage, 328

R

RaspberryPi, 3
RCA jacks, 240
ReadCapacitivePin(), 144
Real Time Clock, 57, *58*
    instead of software clock, 78–79
    integration issues, 201–202
Rebel, 2
    chipKIT Uno32, 4, 416
    display, 64, *65*
    Open Source, 4
    software modifications, 61–65
    source code, 59–60
    splash screen, 62
    under the hood, 60–61, *60*
    using decoder with, 159–160
    using LCD display with, 59–65
    which wires to connect, 61–62
Rebel_506_Alpha_Rev02.pde, 60

Receiver Incremental Tuning, 377
Record, 357
Recursive function call, 32
Red wire, 47
REFLECTED, 284, *285*
Register pointer, 81
    resetting, 99
Register Select (RS), 202
Resistor:
    placement, 44, *44, 45*
    "pull-up", 79
    quad op amp, 158
    requisite value of (math), 124–126
    substitutions, 419–420
Resistor pack, 120–121, *120*
    figuring series resistor, 124–126
    spacing, 121–122
Resistors:
    common values, 419
    feedback and gain, 420
Responsibility of the ghost constructor, 211
REVIEW Mode, 358–359
RF interference, 244
RG-8/U, 329
RG-8X, 289, 304
RG58 cable, 121, 123
RGB backlit display, 301, *302*
RGB LCD, 301, *302*
Right-pointing arrow, 14
    HelloWorld program, 54
RIT, 377
RMS voltage, *129*, 130
Rock bound, 371–372
Rotator controller, 245–282
    Arduino antenna rotator controller, *245*, 246–256
    connecting, 256–260
    control panel, 253–256
    software, 260–279
    substitutions, 420
    world beam headings, 279–282
Routing Power Cables, 397
RTC, 81–83, 205
    initializing, 89–96, *90–96*
    integration phase, 201
    Timer Program, 98–99
RTC/TIMER Shield, 81–85, *82, 86*
RTCPoll(), 217
Rubber band, 395
Rvalue, 150

S

SA602, 343
Sampling ports, 284
Saturated, 104
Save Coordinates, 280
Saving Memory, 32–34
Scale definitions, 113
Scale factor, *111*, 113
    dummy load, 130
Schematic, 46
    adding components using, 46–51
    LCD shield, 39, *39*
SCL, 79, 201
Scope resolution operator, 211
Screw terminals, 106, 224
SDA, 79, 201
Sealed Lead Acid battery, 380
Search-and-replace, 72
Section 97.119(a), 67
Seeed Studio, 409
Sendcode(), 190–191
Sensitivity analysis, 419
Sensor Position, 395, *395*
Sequence order, 239
Sequencer, 227–244
    assembled board, 232, *232*
    code walk through, 238–239
    constructing, 229–234
    defined, 228
    design, 228–229
    loading sequencer program and testing, 235–240
    modifying relay shield, 240–243
    parts placement, *231*
    programming and testing, 234
    timing, 228, *229*
    wiring diagram, *231*
Serial baud rate, 11–12
Serial library, 27
Serial monitor, *29*
    baud rate, 27–28
    loading, 27, *28*
Serial Port, 12
Serial.begin(9600), 25
    debugging, 100
SerialEventRun(), 23
Serial.print(), 25
    errors with, 28
    LiquidCrystal, 59
    toggling the debug program, 100

Serial.println(), 28
SetCursor(), 58–59
Setting a new heading, 279
Setup(), 23
   function body, 24–25
   function scope, 74–75
   LiquidCrystal class, 58
   special, 25
   as Step 1, 25
SetWordsPerMinute(), 150–151
Sharp sign, 207
Shazam, 191
Shielding, 244
Shields, 2
   "add-on" modules, 36
   assembled decoder, *159*
   assembly, 39–45
   prototyping, 37–39, *38*
   See also Progamming shields
Shunt resistor, panel meters, 116, 117
Sidetone, 153
   generating, 198
   to monitor sending, 342
Signal preprocessing circuit, 156–157, *156*
   description, 157–159
Signature, 190
Silicone caulk, 122–123
Sine wave, 332, *332–333*
Single-line comments, 21
Single Sideband, 374
Sizeof() operator, 238
Sketches:
   header files, 23
   transferring programs, 4
SLA, 380, *380*
Slash/asterisk pair, 20
Slash characters, 20
Slave, I2C interface, 201
Slave address, 79
Slave Devices, 79
SLEEP pin, 388–389
Slope, 326
   calculating, 326
Small Wonder Labs, 416
SMD, station ID timer, 80
SMD (Surface Mounted Devices), 37
Smoothing, 325
SMT, 293–296
SN7474, 337–338, *337*
   substitutions, 421

SO-239, 297
   connector, 122
Socket header, 41, *42*
Software, general types of, 7–8
Software Defined Radio (SDR)
   kits, 413
Software library, 36
Solar charger controller, 384, *384*
   early test, *385*
Solar panel, 380–381, *381*
   connections, 395–396
   UL-Solar, 416–417
Solar panel support structure, 389–390, *394*, 395
Solar power, 379, *379*
Solar sensor, 381–384
Solder bridge, 43
Solder flux removal, 50–51
Solder sucker, 43
Soldering iron, acquiring, 7, *7*
Source code, defined, 12
Speed adjustment, 197–198
Speed setting, 197–198
Splash message, 62
Splash screen:
   DL wattmeter, 129, *129*
   panel meters, 114
Splitter jack, 155–156, *156*
SRAM:
   defined, 5
   fitting VFO into, 350, 367
   monitoring, 32
   string literals cluttering, 34
SRAM memory, 32
SSB, 374
   transceiver kits, 414
Stackable header, 40–41
Standing Wave Ratio (SWR), 284
Start program, 140
State, 115
State machine, 169
Static, 76
   data type specifier, 76–78
Static random access memory (SRAM), defined, 5
Station, 201
Station ID timer, 67
   LCD timer program 1.0, *68–70*
   LCD timer program 1.1, *77–78*
   magic numbers, 70–71
   software version, 68–70

Station Timer, 67–100
Step, 388
Step 1, 18–19
    setup(), 25
Step 2, 19
Step 3, 19
Step 4, 19
Step 5, 19–20
Stepper motor:
    details, 390–391
    panel positioning, 385–389
Stepper motor driver, 385–388
Stepper wiring, 385–386
Stockton Bridge, 284
Stockton, David, G4NZQ, 284
STORE Mode, 359
Storing a new heading, 279
Strain insulator, 397
Stray RF, 244
    frequency counter, 334–335
String class, avoiding use of, 33
String literal, defined, 33
Stubby pins, 42–43
    "top" of shield, 44
Substituting parts, 419–421
Suppliers and sources, 405–418
Supported bands, 360
Supported rotators, 246–247
Surface Mount Devices, 37
    station ID timer, 80
Surface mount parts, 293–296
SWR:
    formula for, 285
    how it is measured, 284
SWR meter, 283
Symbol table, 73
Symbolic constants, 73
    DEBUG, 196

T

T2B, 280
Table of contents, dot operator as, 216
Tailtwister T2X, 258
Teflon tubing, 46–48
    panel meters, 106
    relay shield, 222
    RTC/timer circuitry, 84
Telex, 258
Template, 122
Ten-pin header, 256

TEN-TEC, 2, 159
    Patriot, 416
    transceivers, 414–416
TEN-TEC Rebel:
    latest transceivers, 415–416
    using decoder with, 159–160
    using LCD display with, 59–65
Terminal block, 124, 126
Terminal connector block, 106
Terminal posts, 122
Termination, 19–20
Termination Step, 19–20
    dummy load, 130
    in most μC programs, 26
Test and measurement equipment suppliers, 417–418
TEST symbolic constant, 261
Textbox, 27
Textual data, 58
Thermal overheating, 397
Thermal shutoff, 386
Threaded PVC collar, 389–390
Tie points, 6–7
Timeout message, 142
Timer Mode, 97
Timer reset switch, 84
Timing:
    design values, 420
    sequencer, 228, *229*
Tinyos Electronics, 410
Tolerance rating, 117
Tone decoder:
    filtering function, 157
    phase locked loop as, 158
Toolbox, 380, *381*
Tools → Serial Monitor menu, 27
Torque, 385–386
Toupper(ch), 191
TR-44, 248
Transistor substitutions, 420
Transceiver kits, 414
    transceiver and device suppliers, 410–417
Transverter, 228
TTL, 421
Tuvalu, 280–281
TWI, 79
Two colons, 211
Two slash (//), 21
Two Wire Interface (TWI), 79
Type ahead buffer, 189
Type checking, 210
Type specifier, 211

Index

U

UFSS, 358–360
UHF, 245–246
UIP, 357
UL-Solar panel, 416–417
ULN6028B, 248
Union:
   data definition, 365
   dot operator, 365–366
   using dot operator, 365–366
Unipolar stepper, 385
Universal hub:
   assembly, 390–391, *390*
   stepper motor, 385, 403
Universal Relay Shield, 219–226, *221*
   construction, 221–224
   parts list, 222
   rotator controller, 247–248
   substitutions, 420
   testing, 224–225
Uno, 159
   input/output lines, 5
Upload icon, 142, *142*
   described, 14
USB:
   hub, 234
   omission of port, *2*, 4
   PS2 keyboards versus, 173–174
USB Device Driver, installing, 10, *10*
User Frequency Selection Switch, 358–360
User Interface Program, 357

V

Vacuum tube voltmeter, 120, 123
VandeWettering, Mark, 161, 176
Variable Frequency Oscillator (VFO), 349
VCO, 349–350
Velcro, 390, 395–396
Verbs, class methods as, 215
VFO, 349
VFO Circuit Description, 353
Visibility, 74
Visual Studio Express (VSE), 3
Void, loop() function, 25
Void setup(), 24–25
   LiquidCrystal class, 58
   panel meters, 114
Volatile keyword, 150–151
Voltage Controlled Oscillator (VCO), 349–350
Voltage divider, 116

Voltage drop, 130
Voltage regulator, solar panel, 403
Voltage spikes, 388
Voltage SWR, 284
Voltmeter, 116–117, *116*
VSWR, 284
VTVM, 120, 123

W

Wall wart, 99
   integration stage, 203
   for panel meters, 104
   PS2 keyboard encoder circuit, 196
WarnID(), 76
WB7FHC, 161
When to Comment, 21–22
Wikipedia, on SWR, 284
Windmilling, 246
Wing nuts, 403
Wire colors, 386
Wire library, 85, 89
   I2C interface, 201
Wire.beginTransmission
   (RTCI2CADDRESS), 99
Wire.read(), 99
Wire.requestFrom(), 98
Wires:
   connecting, 46–49
   gauge, 46
Wong, Leonard, 121
Wood shim:
   mounting stepper motor, 391
   PVC coupling as, 392
Words per minute:
   changing WPM sent by PS2 keyer, 216
   CW decoder, 171
   default code speed, 197
   timing for a dit, 189
Working directory, 209
World beam headings, 279–282
WWV, 371

Y

Yaesu, *245*, 259–260
   varying speed, 252

Z

Zero-based values, 58

Check out these TAB books for building, tweaking, and boosting radios!

Arduino Projects for Amateur Radio
Jack Purdum, Dennis Kidder
Expand your ham radio's capabilities using low-cost Arduino boards.
0-07-183405-2

Ham and Shortwave Radio for the Electronics Hobbyist
Stan Gibilisco
Get up and running as a ham radio operator—or just listen in on the shortwave bands.
0-07-183291-2

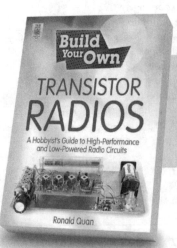

Build Your Own Transistor Radios
Ronald Quan
Create sophisticated transistor radios that are inexpensive yet highly efficient.
0-07-179970-2

 Learn more. Do more: MHPROFESSIONAL.COM   TAB_DIY TABBooks TAB.books